NATURAL HISTORY
UNIVERSAL LIBRARY

西方博物学大系

主编：江晓原

THE COMMON OBJECTS OF THE SEA SHORE
COMMON OBJECTS OF THE MICROSCOPE
THE COMMON OBJECTS OF THE COUNTRY

小生灵三部曲

[英] 约翰·乔治·伍德 著

华东师范大学出版社

图书在版编目（CIP）数据

小生灵三部曲：英文 /（英）约翰·乔治·伍德著.
— 上海：华东师范大学出版社，2018
（寰宇文献）
ISBN 978-7-5675-8152-4

Ⅰ.①小… Ⅱ.①约… Ⅲ.①动物-英文 Ⅳ.①Q95

中国版本图书馆CIP数据核字(2018)第179988号

小生灵三部曲
The Common Objects of the Sea Shore · Common Objects of the Microscope
The Common Objects of the Country
（英）约翰·乔治·伍德

特约策划	黄曙辉　徐　辰
责任编辑	庞　坚
特约编辑	许　倩
装帧设计	刘怡霖

出版发行	华东师范大学出版社
社　　址	上海市中山北路3663号　邮编 200062
网　　址	www.ecnupress.com.cn
电　　话	021-60821666　行政传真 021-62572105
客服电话	021-62865537
门市（邮购）电话	021-62869887
地　　址	上海市中山北路3663号华东师范大学校内先锋路口
网　　店	http://hdsdcbs.tmall.com/

印 刷 者	虎彩印艺股份有限公司
开　　本	787×1092　16开
印　　张	43.25
版　　次	2018年8月第1版
印　　次	2018年8月第1次
书　　号	ISBN 978-7-5675-8152-4
定　　价	798.00元（精装全一册）

出版人　王　焰

（如发现本版图书有印订质量问题，请寄回本社客服中心调换或电话021-62865537联系）

总　目

《西方博物学大系总序》（江晓原）	1
出版说明	1
The Common Objects of the Sea Shore	1
Common Objects of the Microscope	249
The Common Objects of the Country	459

《西方博物学大系》总序

江晓原

《西方博物学大系》收录博物学著作超过一百种，时间跨度为15世纪至1919年，作者分布于16个国家，写作语种有英语、法语、拉丁语、德语、弗莱芒语等，涉及对象包括植物、昆虫、软体动物、两栖动物、爬行动物、哺乳动物、鸟类和人类等，西方博物学史上的经典著作大备于此编。

中西方"博物"传统及观念之异同

今天中文里的"博物学"一词，学者们认为对应的英语词汇是Natural History，考其本义，在中国传统文化中并无现成对应词汇。在中国传统文化中原有"博物"一词，与"自然史"当然并不精确相同，甚至还有着相当大的区别，但是在"搜集自然界的物品"这种最原始的意义上，两者确实也大有相通之处，故以"博物学"对译Natural History一词，大体仍属可取，而且已被广泛接受。

已故科学史前辈刘祖慰教授尝言：古代中国人处理知识，如开中药铺，有数十上百小抽屉，将百药分门别类放入其中，即心安矣。刘教授言此，其辞若有憾焉——认为中国人不致力于寻求世界"所以然之理"，故不如西方之分析传统优越。然而古代中国人这种处理知识的风格，正与西方的博物学相通。

与此相对，西方的分析传统致力于探求各种现象和物体之间的相互关系，试图以此解释宇宙运行的原因。自古希腊开始，西方哲人即孜孜不倦建构各种几何模型，欲用以说明宇宙如何运行，其中最典型的代表，即为托勒密（Ptolemy）的宇宙体系。

比较两者，差别即在于：古代中国人主要关心外部世界"如何"运行，而以希腊为源头的西方知识传统（西方并非没有别的知识传统，只是未能光大而已）更关心世界"为何"如此运行。在线性发展无限进步的科学主义观念体系中，我们习惯于认为"为何"是在解决了"如何"之后的更高境界，故西方的分析传统比中国的传统更高明。

然而考之古代实际情形，如此简单的优劣结论未必能够成立。例如以天文学言之，古代东西方世界天文学的终极问题是共同的：给定任意地点和时刻，计算出太阳、月亮和五大行星（七政）的位置。古代中国人虽不致力于建立几何模型去解释七政"为何"如此运行，但他们用抽象的周期叠加（古代巴比伦也使用类似方法），同样能在足够高的精度上计算并预报任意给定地点和时刻的七政位置。而通过持续观察天象变化以统计、收集各种天象周期，同样可视之为富有博物学色彩的活动。

还有一点需要注意：虽然我们已经接受了用"博物学"来对译 Natural History，但中国的博物传统，确实和西方的博物学有一个重大差别——即中国的博物传统是可以容纳怪力乱神的，而西方的博物学基本上没有怪力乱神的位置。

古代中国人的博物传统不限于"多识于鸟兽草木之名"。体现此种传统的典型著作，首推晋代张华《博物志》一书。书名"博物"，其义尽显。此书从内容到分类，无不充分体现它作为中国博物传统的代表资格。

《博物志》中内容，大致可分为五类：一、山川地理知识；二、奇禽异兽描述；三、古代神话材料；四、历史人物传说；五、神仙方伎故事。这五大类，完全符合中国文化中的博物传统，深合中国古代博物传统之旨。第一类，其中涉及宇宙学说，甚至还有"地动"思想，故为科学史家所重视。第二类，其中甚至出现了中国古代长期流传的"守宫砂"传说的早期文献：相传守宫砂点在处女胳膊上，永不褪色，只有性交之后才会自动消失。第三类，古代神话传说，其中甚至包括可猜想为现代"连体人"的记载。第四类，各种著名历史人物，比如三位著名刺客的传说，此三名刺客及所刺对象，历史上皆实有其人。第五类，包括各种古代方术传说，比如中国古代房中养生学说，房中术史上的传说人物之一"青牛道士封君达"等等。前两类与西方的博物学较为接近，但每一类都会带怪力乱神色彩。

"所有的科学不是物理学就是集邮"

在许多人心目中，画画花草图案，做做昆虫标本，拍拍植物照片，这类博物学活动，和精密的数理科学，比如天文学、物理学等等，那是无法同日而语的。博物学显得那么的初级、简单，甚至幼稚。这种观念，实际上是将"数理程度"作为唯一的标尺，用来衡量一切知识。但凡能够使用数学工具来描述的，或能够进行物理实验的，那就是"硬"科学。使用的数学工具越高深越复杂，似乎就越"硬"；物理实验设备越庞大，花费的金钱越多，似乎就越"高端"、越"先进"……

这样的观念，当然带着浓厚的"物理学沙文主义"色彩，在很多情况下是不正确的。而实际上，即使我们暂且同意上述"物理学沙文主义"的观念，博物学的"科学地位"也仍然可以保住。作为一个学天体物理专业出身，因而经常徜徉在"物理学沙文主义"幻影之下的人，我很乐意指出这样一个事实：现代天文学家们的研究工作中，仍然有绘制星图，编制星表，以及为此进行的巡天观测等等活动，这些活动和博物学家"寻花问柳"，绘制植物或昆虫图谱，本质上是完全一致的。

这里我们不妨重温物理学家卢瑟福（Ernest Rutherford）的金句："所有的科学不是物理学就是集邮（All science is either physics or stamp collecting）。"卢瑟福的这个金句堪称"物理学沙文主义"的极致，连天文学也没被他放在眼里。不过，按照中国传统的"博物"理念，集邮毫无疑问应该是博物学的一部分——尽管古代并没有邮票。卢瑟福的金句也可以从另一个角度来解读：既然在卢瑟福眼里天文学和博物学都只是"集邮"，那岂不就可以将博物学和天文学相提并论了？

如果我们摆脱了科学主义的语境，则西方模式的优越性将进一步被消解。例如，按照霍金（Stephen Hawking）在《大设计》（The Grand Design）中的意见，他所认同的是一种"依赖模型的实在论（model-dependent realism）"，即"不存在与图像或理论无关的实在性概念（There is no picture- or theory-independent concept of reality）"。在这样的认识中，我们以前所坚信的外部世界的客观性，已经不复存在。既然几何模型只不过是对外部世界图像的人为建构，则古代中国人干脆放弃这种建构直奔应用（毕竟在实际应用中我们只需要知道七政"如何"运行），又有何不可？

传说中的"神农尝百草"故事，也可以在类似意义下得到新的解读："尝百草"当然是富有博物学色彩的活动，神农通过这一活动，得知哪些草能够治病，哪些不能，然而在这个传说中，神农显然没有致力于解释"为何"某些草能够治病而另一些则不能，更不会去建立"模型"以说明之。

"帝国科学"的原罪

今日学者有倡言"博物学复兴"者，用意可有多种，诸如缓解压力、亲近自然、保护环境、绿色生活、可持续发展、科学主义解毒剂等等，皆属美善。编印《西方博物学大系》也是意欲为"博物学复兴"添一助力。

然而，对于这些博物学著作，有一点似乎从未见学者指出过，而鄙意以为，当我们披阅把玩欣赏这些著作时，意识到这一点是必须的。

这百余种著作的时间跨度为15世纪至1919年，注意这个时间跨度，正是西方列强"帝国科学"大行其道的时代。遥想当年，帝国的科学家们乘上帝国的军舰——达尔文在皇家海军"小猎犬号"上就是这样的场景之一，前往那些已经成为帝国的殖民地或还未成为殖民地的"未开化"的遥远地方，通常都是踌躇满志、充满优越感的。

作为一个典型的例子，英国学者法拉在（Patricia Fara）《性、植物学与帝国：林奈与班克斯》（*Sex, Botany and Empire, The Story of Carl Linnaeus and Joseph Banks*）一书中讲述了英国植物学家班克斯（Joseph Banks）的故事。1768年8月15日，班克斯告别未婚妻，登上了澳大利亚军舰"奋进号"。此次"奋进号"的远航是受英国海军部和皇家学会资助，目的是前往南太平洋的塔希提岛(Tahiti，法属海外自治领，另一个常见的译名是"大溪地"）观测一次比较罕见的金星凌日。舰长库克（James Cook）是西方殖民史上最著名的舰长之一，多次远航探险，开拓海外殖民地。他还被认为是澳大利亚和夏威夷群岛的"发现"者，如今以他命名的群岛、海峡、山峰等不胜枚举。

当"奋进号"停靠塔希提岛时，班克斯一下就被当地美丽的土著女性迷昏了，他在她们的温柔乡里纵情狂欢，连库克舰长都看不下去了，"道德愤怒情绪偷偷溜进了他的日志当中，他发现自己根本不可能不去批评所见到的滥交行为"，而班克斯纵欲到了"连嫖妓都毫无激情"的地步——这是别人讽刺班克斯的说法，因为对于那时常年航行于茫茫大海上的男性来说，上岸嫖妓通常是一项能够唤起"激情"的活动。

而在"帝国科学"的宏大叙事中，科学家的私德是无关紧要的，人们关注的是科学家做出的科学发现。所以，尽管一面是班克斯在塔希提岛纵欲滥交，一面是他留在故乡的未婚妻正泪眼婆娑地"为远去的心上人绣织背心"，这样典型的"渣男"行径要是放在今天，非被互联网上的口水淹死不可，但是"班克斯很快从他们的分离之苦中走了出来，在外近三年，他活得倒十分滋润"。

法拉不无讽刺地指出了"帝国科学"的实质："班克斯接管了当地的女性和植物，而库克则保护了大英帝国在太平洋上的殖民地。"甚至对班克斯的植物学本身也调侃了一番："即使是植物学方面的科学术语也充满了性指涉。……这个体系主要依靠花朵之中雌雄生殖器官的数量来进行分类。"据说"要保护年轻妇女不受植物学教育的浸染，他们严令禁止各种各样的植物采集探险活动。"这简直就是将植物学看成一种"涉黄"的淫秽色情活动了。

在意识形态强烈影响着我们学术话语的时代，上面的故事通常是这样被描述的：库克舰长的"奋进号"军舰对殖民地和尚未成为殖民地的那些地方的所谓"访问"，其实是殖民者耀武扬威的侵略，搭载着达尔文的"小猎犬号"军舰也是同样行径；班克斯和当地女性的纵欲狂欢，当然是殖民者对土著妇女令人发指的践踏；即使是他采集当地植物标本的"科学考察"，也可以视为殖民者"窃取当地经济情报"的罪恶行为。

后来改革开放，上面那种意识形态话语被抛弃了，但似乎又走向了另一个极端，完全忘记或有意回避殖民者和帝国主义这个层面，只歌颂这些军舰上的科学家的伟大发现和成就，例如达尔文随着"小猎犬号"的航行，早已成为一曲祥和优美的科学颂歌。

其实达尔文也未能免俗，他在远航中也乐意与土著女性打打交道，当然他没有像班克斯那样滥情纵欲。在达尔文为"小猎犬号"远航写的《环球游记》中，我们读到："回程途中我们遇到一群黑人姑娘在聚会，……我们笑着看了很久，还给了她们一些钱，这着实令她们欣喜一番，拿着钱尖声大笑起来，很远还能听到那愉悦的笑声。"

有趣的是，在班克斯在塔希提岛纵欲六十多年后，达尔文随着"小猎犬号"也来到了塔希提岛，岛上的土著女性同样引起了达尔文的注意，在《环球游记》中他写道："我对这里妇女的外貌感到有些失望，然而她们却很爱美，把一朵白花或者红花戴在脑后的髪髻上……"接着他以居高临下的笔调描述了当地女性的几种发饰。

用今天的眼光来看，这些在别的民族土地上采集植物动物标本、测量地质水文数据等等的"科学考察"行为，有没有合法性问题？有没有侵犯主权的问题？这些行为得到当地人的同意了吗？当地人知道这些行为的性质和意义吗？他们有知情权吗？……这些问题，在今天的国际交往中，确实都是存在的。

也许有人会为这些帝国科学家辩解说：那时当地土著尚在未开化或半开化状态中，他们哪有"国家主权"的意识啊？他们也没有制止帝国科学家的考察活动啊？但是，这样的辩解是无法成立的。

姑不论当地土著当时究竟有没有试图制止帝国科学家的"科学考察"行为，现在早已不得而知，只要殖民者没有记录下来，我们通常就无法知道。况且殖民者有军舰有枪炮，土著就是想制止也无能为力。正如法拉所描述的："在几个塔希提人被杀之后，一套行之有效的易货贸易体制建立了起来。"

即使土著因为无知而没有制止帝国科学家的"科学考察"行为，这事也很像一个成年人闯进别人的家，难道因为那家只有不懂事的小孩子，闯入者就可以随便打探那家的隐私、拿走那家的东西、甚至将那家的房屋土地据为己有吗？事实上，很多情况下殖民者就是这样干的。所以，所谓的"帝国科学"，其实是有着原罪的。

如果沿用上述比喻，现在的局面是，家家户户都不会只有不懂事的孩子了，所以任何外来者要想进行"科学探索"，他也得和这家主人达成共识，得到这家主人的允许才能够进行。即使这种共识的达成依赖于利益的交换，至少也不能单方面强加于人。

博物学在今日中国

博物学在今日中国之复兴,北京大学刘华杰教授提倡之功殊不可没。自刘教授大力提倡之后,各界人士纷纷跟进,仿佛昔日蔡锷在云南起兵反袁之"滇黔首义,薄海同钦,一檄遥传,景从恐后"光景,这当然是和博物学本身特点密切相关的。

无论在西方还是在中国,无论在过去还是在当下,为何博物学在它繁荣时尚的阶段,就会应者云集?深究起来,恐怕和博物学本身的特点有关。博物学没有复杂的理论结构,它的专业训练也相对容易,至少没有天文学、物理学那样的数理"门槛",所以和一些数理学科相比,博物学可以有更多的自学成才者。这次编印的《西方博物学大系》,卷帙浩繁,蔚为大观,同样说明了这一点。

最后,还有一点明显的差别必须在此处强调指出:用刘华杰教授喜欢的术语来说,《西方博物学大系》所收入的百余种著作,绝大部分属于"一阶"性质的工作,即直接对博物学作出了贡献的著作。事实上,这也是它们被收入《西方博物学大系》的主要理由之一。而在中国国内目前已经相当热的博物学时尚潮流中,绝大部分已经出版的书籍,不是属于"二阶"性质(比如介绍西方的博物学成就),就是文学性的吟风咏月野草闲花。

要寻找中国当代学者在博物学方面的"一阶"著作,如果有之,以笔者之孤陋寡闻,唯有刘华杰教授的《檀岛花事——夏威夷植物日记》三卷,可以当之。这是刘教授在夏威夷群岛实地考察当地植物的成果,不仅属于直接对博物学作出贡献之作,而且至少在形式上将昔日"帝国科学"的逻辑反其道而用之,岂不快哉!

2018年6月5日
于上海交通大学
科学史与科学文化研究院

小生灵三部曲

出版说明

《小生灵三部曲》是《海滨小生灵》《微观小生灵》和《乡间小生灵》三书的合刊本，作者为英国著名博物学家、作家约翰·乔治·伍德（John George Wood，1827-1889）。伍德生于伦敦，由于体质羸弱，少年时不得不在家中接受教育。1830年，他随父母迁居牛津。

十七岁时考入牛津大学默顿学院。1851年，他出版了自己的第一本著作《自然图志》，全书厚达2400页，配有大量精美铜版画，涉及的动物既有乡间常见的品类，也有当时对英国人而言远在天边的珍奇物种，至今仍是博物学的传世名著。他在三年后成为牧师，但对博物学的热爱使他辞去神职，开始专门从事博物学研究与写作。伍德一生笔耕不辍，出版过十多部博物学专著以及相关的文学作品。有别于那些艰涩的科研论文，他的著作文字秀美，以细腻的观察与富于情感的笔触，抒发了对故土与自然生灵的热爱，成就了一部又一部博物学史上的杰作。

《小生灵三部曲》中的《海滨小生灵》是伍德于1856年出版的通俗博物学书籍，以自己在英国海滨的生活与科研经验为基础，细致介绍日常可见的海鸟、水生动物及植物。他对故土乡野的热爱，使他的叙述有了一种超越性的气质，行文间淡淡的乡愁，引人入胜。《乡间小生灵》是伍德于1857年出版的通俗博物学书籍，以自己在英国，特别是牛津田野乡间的生活与科研经验为基础，细致描述了人们每天都可看到，却又不受重视的花鸟鱼虫。此书风格与前一种相似，是历史上最畅销的博物学著作之一，初版推出一周，便售出十万册。《微观小生灵》是伍德于1861年出版的通俗博物学书籍，带领读者一窥显微镜下的微观世界，同样经久不衰。三书均配有多幅精美彩色珂罗版插画及黑白铜版插画。

今据原版影印。

THE
COMMON OBJECTS
OF THE
SEA SHORE

A Cheap Edition of this Book, with the Plates Plain, price 1s., *is also published.*

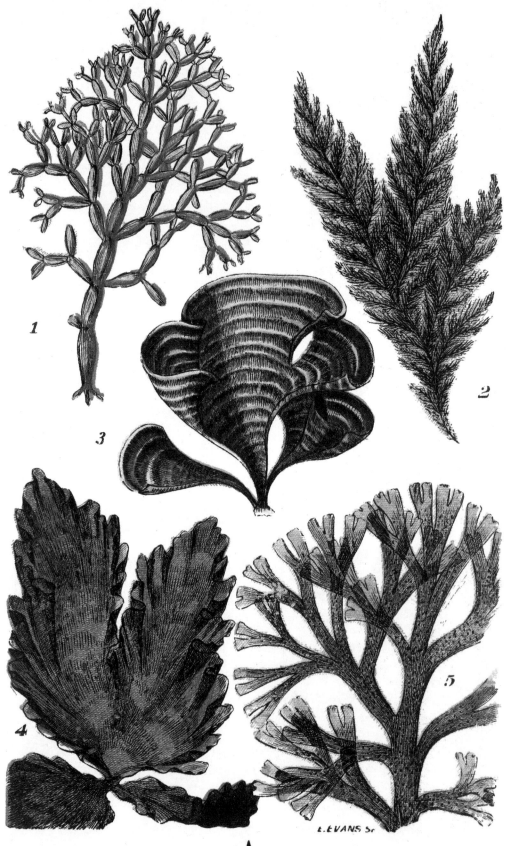

A

THE
COMMON OBJECTS

OF THE

SEA SHORE

INCLUDING HINTS FOR AN AQUARIUM

BY THE

REV. J. G. WOOD, M.A., F.L.S., ETC.

AUTHOR OF THE "ILLUSTRATED NATURAL HISTORY"
ETC.

With Coloured Illustrations from Designs by Sowerby

LONDON
GEORGE ROUTLEDGE AND SONS
THE BROADWAY, LUDGATE
NEW YORK: 416, BROOME STREET

BY THE SAME AUTHOR,

UNIFORM WITH THIS VOLUME.

COMMON OBJECTS OF THE COUNTRY.
COMMON OBJECTS OF THE MICROSCOPE.
COMMON BRITISH MOTHS.
COMMON BRITISH BEETLES.
THE FRESH AND SALT WATER AQUARIUM.
CALENDAR OF THE MONTHS.

PREFACE.

THIS little work is simply a popular account of the "Common Objects of the Sea-Shore," and is restricted to those objects which every visitor to the sea-side is sure to find on every coast. For descriptions of those creatures which only inhabit certain localities, and those whose lives are passed in the deep water, requiring the dredge, the net, or the drag to bring them to the light of day, the reader is referred to those magnificent and comprehensive works that have been written for the purpose of illustrating particular branches of science.

During my visits to the sea-coasts for the last six or seven years, I have taken note of the questions put to me by persons who were anxious to know something of the curious objects that everywhere met their eyes; and the following pages are, as nearly as possible, the condensed conversations that then took place.

The whole of the illustrative plates were drawn expressly for this work by Mr. Sowerby, whose name is a sufficient guarantee for their truth.

LONDON, *May*, 1857.

CONTENTS.

CHAPTER I.

MARINE BIRDS—PORPESSE 1

CHAPTER II.

WHELK — COWRY — COCKLE — PHOLAS — LIMPET —SEA-WEEDS ON SHELL—BALANUS—PURPURA— MUSSEL— PERIWINKLE, YELLOW — PERIWINKLE, COMMON—TROCHUS 16

CHAPTER III.

MARINE ALGÆ, OR SEA-WEEDS 41

CHAPTER IV.

RED-SPORED AND GREEN-SPORED ALGÆ 53

CHAPTER V.

Eggs of Marine Animals — Cuttles and their Habits 74

CHAPTER VI.

Sea Anemones and other Zoophytes 98

CHAPTER VII.

Star-fishes and Sea-Urchins 125

CHAPTER VIII.

Annelids — Barnacles, and Jelly-Fish . . . 144

CHAPTER IX.

Crabs — Lobsters — Shrimps — Prawns, and Fish . 173

Common Objects of the Sea-Shore.

CHAPTER I.

MARINE BIRDS—PORPESSE.

WHETHER the sea is approached by land or by water, the first indications of its existence are generally to be found in the air. On some days, the electric clouds that skirt the cliffs map out, as it were, the sea-coast; and when such signs fail, the marine birds give evident tokens that the sea, their great store-house, is close at hand. With the birds, then, we will commence our observations of the sea and its shores.

The bird that usually presents itself as the ocean's herald is the Common Gull (*Larus canus*). There are some twelve or thirteen species of British Gulls, including the Kittiwake and the Iceland Gull. The bird represented in the accompanying figure is the Great Black-backed Gull (*Larus marinus*), which is tolerably frequent on our coasts, but not so often seen as the Common Gull, nor does it form such large societies as those in which its more sociable relations love to congregate.

Why the word "gull" should be employed to express stupidity I cannot at all comprehend, for the gulls are very knowing birds indeed, and difficult to be deceived. If a piece of bread or biscuit be thrown from a boat, it remains but a very short time on the surface of the water before it is carried off by a gull, although previously not a bird was visible. But if a number of gulls are flying about, and a piece of paper or white wood be thrown into the water, there is not a gull who will even stoop towards it, although to the human eye the bread and the paper appear identical. The cry of the gull is very curious, being a kind of mixture of a wail, or scream, and a laugh, and on a dark stormy day adds wonderfully to the spirit of the scene. Its flight is peculiarly quiet, combining great power of wing with easy elegance of motion. It is a very bold bird, and for many miles will follow boats so closely that the very sparkle of its eyes is plainly visible, as it twists its wise-looking head from side to side while watching the voyagers.

BLACK-BACKED GULL.

The gull is an exclusively marine bird, being found only on the sea-shore or at the mouths of large rivers, although more than usually violent storms occasionally drive it inland, where it wanders about for some time very miserable, and quite out of its element, until it gets shot by some rustic sportsman.

The next bird to which our attention will be directed is the Common Tern (*Sterna hirundo*), or Sea Swallow, as it is very appropriately called. It belongs to the gull family, and has many of the gull habits, but is readily distinguishable from the gulls, even at a considerable distance, on account of its rapid, darting flight. It is not at all unlike a swallow in general shape,

TERN.

for the wings are long and pointed, the body is rather large in front, and tapers to a point, and its tail is forked like that of a swallow.

It is extremely dexterous in its capacity as a fisher, for in its swift flight over the waves it darts upon any small fish that may be unfortunate and curious enough to come near the surface, and scoops it up, as it were, from the water, without seemingly interrupting the speed of its course. The nest, if it can be dignified with the title, is merely a hollow scooped in the sand, well above high-water mark; and in this hollow two or three

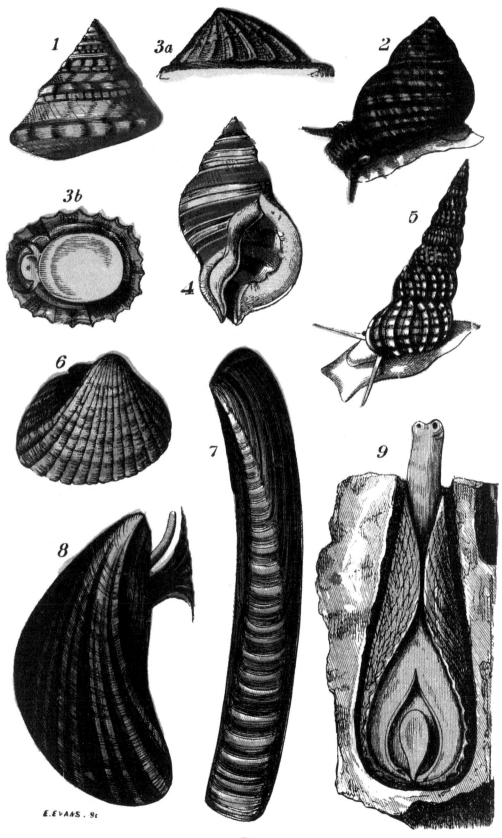

B

eggs are deposited. On the Scottish and northern coasts the Common Tern is not often found, but the Arctic Tern (*Sterna macroura*) comes to supply its place. There are ten or eleven species of British Terns.

On many of the English coasts, especially those that look towards the Channel, may be seen a tolerably large black-feathered bird, having a yellowish countenance, a decidedly long and rather hooked bill, and a pair of green eyes.

This is the Cormorant (*Graculus carbo*), one of our three British representatives of the Pelican family. The enormous pouch which decorates the lower bill of the white pelican is only rudimentary in our British pelicans, probably because there would be no use for it, as the birds live on or close to the coast.

The other English pelicans are the Gannet, a figure of which will be given shortly, and the Common Shag, a bird of a monosyllabic English cognomen, but who ought to consider himself recompensed by the scientific name given to him by certain naturalists, namely, *Phalacrocorax cristatus;* the epithet *cristatus*, or crested, being due to a tuft of reverted green feathers that decorates the head. This tuft, however, is only worn during the breeding season, when most animals put on their gayest apparel, and is lost as soon as the young *Phalacrocoraces cristati* take their places as independent members of society.

The cormorant is a persevering fisher, insatiable in appetite, and almost unparalleled in digestion. The pike and the shark among fish appear to possess much the same proportionate digestive power as the cormorant among birds. The cormorant is not content with sitting, like the heron, on the edge of the water, and snapping up the fish that may enter the shallows; or even, like the gulls, with seizing them from the surface of the waves; but he boldly defies his prey in its own element, plunges into the water, dives below the surface, and actually proves himself a more expert swimmer than the very fish themselves. In former days the cormorant was employed in England for the purpose of catching fish; and such is still the case in China. The Chinese cormorant, however, is not the same species as that which is found on our coasts. It is rather a curious circumstance, that one of the mammalia, namely, the otter, and some of the birds, should be enabled to carry on a successful subaqueous chase, and that both beast and bird have been pressed into the service of man.

The cormorant is sometimes found inland, especially in the winter season, and exhibits its powers among the fresh-water fish.

Although the pouch is comparatively small in the cormorant, it still exists, and is useful in giving elasticity to the throat and neck; a property which is much required, for the cormorant is a very greedy bird, and often swallows fish of so large a size that a throat of twice its dimensions seems incapable of permitting the passage of so bulky a body. In order to swallow a fish, the cormorant generally seizes it crosswise, tosses it in the air, and then catches it as it descends with its head downwards. One of these birds, however, has been seen to miss its aim, and to catch the fish with its head upwards; in this position the cormorant endeavoured to swallow its prey, but when the fish had passed about half-way down its captor's throat, the sharp fins prevented its further progress, and both bird and fish were soon dead. The poor cormorant seemed terribly distressed, and made violent struggles, but all to no purpose; for the fish was immovably fixed, and could neither be swallowed nor rejected.

The feathers of the cormorant, although they appear to be of a dusky black, are really of a very deep green, so deep, indeed, as to appear black at a little distance, something like the plumage of the magpie. The nest is composed of dried sea-weed, and is usually placed on lofty rocks, but is sometimes built among the branches of trees. The eggs are remarkable for a thick coating of chalk, which seems to envelope the shell quite independently, and can be easily removed with a knife. There are from three to five eggs in each nest.

Our remaining English pelican is the Gannet (*Sula bassanea*), also called the Solan Goose. It is to be found on many of our shores, but especially on the well-known Bass Rock, at the entrance of the Frith of Forth. There is no difficulty in identifying this bird, even at some distance, as it has very much the appearance of wearing spectacles; a circumstance that has earned it the title of the Spectacled Goose, although it is not a goose at all.

THE GANNET.

It is in the search after these birds, their eggs and young, that the St. Kilda cragsmen imperil their lives year by year.

The Auk family find representatives in the Guillemot, and the comical little Puffin. The Guillemot (*Uria troile*) is a common bird on many of our coasts, and may be seen, in the breeding season, sitting with extraordinary gravity and importance over a solitary egg. The egg is often laid, and the young hatched, on such a narrow ledge of rock, that it is quite a wonder

GUILLEMOT.

how the egg can escape a fall, or how the young bird can even open its big beak without toppling over the precipice.

The guillemot has earned the epithet of "foolish," because, when sitting on this solitary egg, it will suffer itself to be taken by hand, rather than forsake its duty. I would suggest that the word "faithful" be substituted for "foolish." The egg is a very handsome one, very large, and variable both in colour and shape. It is generally covered with large irregular blotches, of a brown colour, on a pale-green ground.

As to the Puffin (*Fratercula arctica*), it is generally to be found in company with the guillemots, and indeed lives in much the same

PUFFIN.

manner. It is a lively little bird, easily distinguished by its large beak; from which feature it has derived the popular name of Sea Parrot. This beak is very useful to the bird for three

DUNLIN.

especial purposes: the first being, to catch fish; the second, to dig the burrows in which its egg is laid; and the third, to fight the ravens and other foes who try to get at the egg.

The favourite food of the puffin appears to be the common sprat, which it chases under water, and of which it generally secures six or seven, all arranged in a neat row along the puffin's beak, and hanging by their heads.

DUNLIN.

There are many more marine birds that are often seen, but those already mentioned are the most common. Yet there is one other bird that I must notice, because it has not so much the marine aspect as the gulls and cormorants: this is the Dunlin Sandpiper (*Tringa alpina*), a very interesting little bird, that frequents the sandy shores in great numbers, for the purpose of feeding on the insects and small crustaceans that are found in such profusion, either buried in the sand, or hidden under stones and drifted sea-weeds.

It is quite aware that on the edge of every wave may be found the various substances which constitute its food, and so skirts the very margin of the sea, running hither and thither; and occasionally venturing a few paces into the retiring waters in chase of some detached limpet, some houseless worm, or tiny crab, as restlessly, and almost as untiringly, as the many-voiced waves themselves.

There is no difficulty in watching the habits of this, or indeed of any other bird. All that is required is perfect stillness and silence, and the birds will come and pick up their food almost within arm's reach.

In the hot summer months the observer may watch the sands without seeing a single Dunlin, for they then desert the sea-shores in favour of inland moors, where they lay their eggs, and hatch the young; returning with their offspring towards the end of August. This bird is sometimes called the Purre.

If we now leave the sands for a time, and ascend the cliffs, we shall probably be indulged with a transient glimpse of a very singular animal. Some little distance from the shore a number of black objects may be seen partly emerging from the water, executing a summersault, and disappearing below the surface. These

PORPESSE.

are Porpesses (*Phocæna communis*), and very curious creatures they are, belonging to the mammalia; forced, therefore, to breathe atmospheric air, and yet permanently inhabiting the sea, with something of the form and many of the habits of the fish. It is a curious connexion of the two most distant links of the chain of the

Vertebrates, the mammalia being the highest, and the fish the lowest.

Some people say, that as it looks like a fish, and lives like a fish, to all intents and purposes it *is* a fish. So it is, if the diver at the Polytechnic Institution is a fish; for it holds its lease of life on precisely the same tenure. Both diver and porpesse must breathe atmospheric air, or they would die; and therefore each finds means to supply himself with that indispensable material. The diver surrounds himself with a supply of fresh air, with which to renovate his blood; but the porpesse is able to renovate a surplus amount of blood, that lasts him for some time: so the chief difference is, that the diver takes down with him oxygen externally, and the porpesse internally. The man goes down inside the diving-bell, but the diving-bell goes down inside the porpesse.

Yet the porpesse has no reservoir in which atmospheric air is retained, for such a formation would make it too buoyant. There is, however, in the *cetaceæ*, to which family the porpesse belongs, a reservoir of blood, which is renovated by the atmospheric air, and is passed into the system as required. Even man has the same power, although in a limited degree. In general, a man cannot hold his breath more than one minute, and it is not every man who can do even that. But if he thoroughly renovates his blood by expelling all the impure air that remains in the minuter tubes of his lungs, and takes a succession of deep inspirations, he will be able to abstain from breathing for a much longer period. I have just made the experiment myself, and held my breath without difficulty for a minute and a-half; and had there been any necessity, could have done so for another half minute.

The porpesse is rather a sociable animal, being generally seen in shoals, or schools, as the sailors call them. I should hardly have said so much about so common a creature, were it not for the purpose of pointing out these remarkable facts in its structure and habits: and even though it be common, it is not so well known as might be imagined. Not long ago, as I was on board a steamer, a worthy old lady began to exhibit symptoms of nervousness and alarm. I thought she was fearing a storm, and told her that there was not the least danger of any commotion of the elements, for the barometer had been steadily rising for the last two or three days. However, it was a different subject that caused her uneasiness. She had heard that there were porpesses in those parts, and wished to know whether we were likely to meet one. I told her that we probably should meet several, but not so many as if a gale were impending. At this reply her fright evidently increased, and she asked, in much trepidation, whether, if we did meet one, it would upset the ship!

CHAPTER II.

WHELK—COWRY—COCKLE—PHOLAS—LIMPET—SEA-WEEDS ON SHELL—BALANUS—PURPURA—MUSSEL—PERIWINKLE, YELLOW—PERIWINKLE, COMMON—TROCHUS.

DESCENDING to the shore, we shall probably see at our feet many shells, or fragments of shells, which have been washed upon the beach by the advancing tide, and which having lodged behind a stone, or being sunk into the wet sand, remain behind when the waves retreat. These shells are almost invariably empty, their inhabitants having either died a natural death, or having fallen victims to some ravenous inhabitant of the sea.

The strong house with which most of these creatures are furnished would seem to be an effectual defence against the efforts of open foes, while the sensitive nervous nature with which they are gifted would appear to secure them from insidious attacks. Yet the hard, stony shells, that turn the edge of a steel knife, are constantly found to be perforated by creatures that can be squeezed flat between the fingers, and whose bodies are no harder than the human tongue. Formerly, the external characters of shells were the only object of the collector; and the conchologist, as he was termed, might have, and very often did have, a large collection of valuable shells, without the least idea of the form, food, habits, or development of the creature that secreted them. Now, however, those who examine a shell are not satisfied unless they know something of the creature that inhabited it, and from whose substance it was formed: and so this branch of Natural History has leaped at once out of the mere childish toy of conchology into the maturer science of malacology. The former treated merely of shells, and therefore excluded the vast army of molluscs, that wear no shells at all; but the latter treats especially of the animal, considering the shell to be of secondary importance.

And yet, even though the shell is considered to be inferior to the animal by whom it was secreted, much more attention is paid to the shell itself than was the case in the old conchological times. In those days the mere shape and colours of the shell were the characteristics by which its name and place in the system were determined; but now we submit the shell to the searching powers of the microscope, and find that various kinds of shells are characterised by various arrangements of particles, and are acted upon by polarized light in various ways. It is, therefore, quite possible to fix the character of a shell from a single fragment no larger than a pin's head. There are few things more curious than this wonderful arrangement of the particles; which, by the way, are brought within the scope of the microscope, by making very thin sections of the shells, by the aid of saw, file, and hone.

One of the commonest shells found on the sea-shore is the Limpet (*Patella vulgata*). See plate B, fig. 3.

In its living state it may be found adhering closely to rocks or other substances, that give it a firm basis of support. The adhesion is caused by atmospheric pressure, for the limpet is enabled to raise the centre of that part of the body that rests on the rock, while the edge is closely pressed upon it. This movement causes a vacuum; and so firmly does the air hold the limpet in its place, that the unaided fingers will find great difficulty even in stirring it. The firmness of adhesion is also increased by the fact that after the animal has remained for some time in one

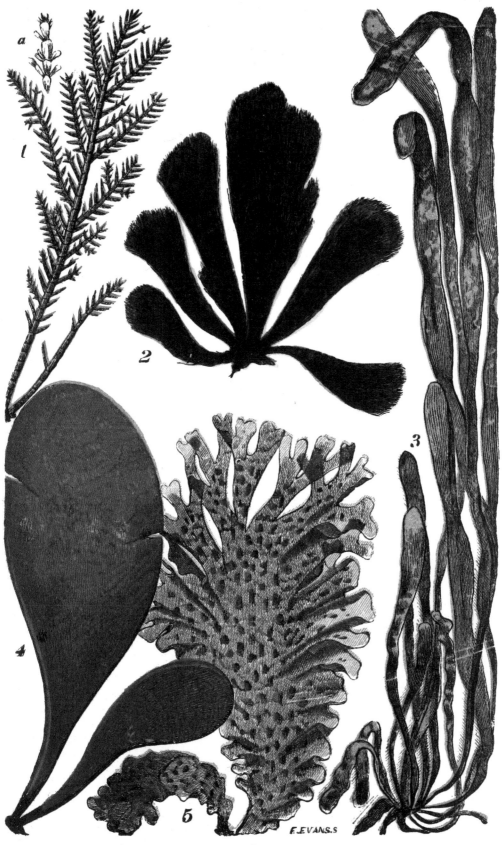

spot, it forms a hollow in the place where it rests, corresponding in size with the shape of the shell. Into this depression the shell sinks, and consequently there is no possibility of reaching its edge, where alone it is vulnerable. When, however, it is not warned, and prepared for resistance, it can be easily detached by a sharp movement of the hand.

In general, it is not a migratory creature, and, consequently, is often seen to be so covered with parasites of various kinds, that its form can hardly be recognised. I have now in my aquarium a limpet-shell, on which a specimen of the common laver (*Ulva latissima*) and another of *Porphyra laciniata* have affixed themselves, and are growing luxuriantly. There was also in the same tank another limpet-shell, on which was growing a whole forest of sea-grass (*Enteromorpha compressa*), expanding as widely as the crown of a man's hat. The acorn barnacle, too, often takes possession of the limpets, and it frequently happens that, in some dark cavity of rock, a colony of limpets may be found, each so covered with these sessile barnacles, that not a particle of the original shell is visible. Of this, however, we will speak hereafter.

The figure, plate B, fig. 3 *a*, represents the

appearance of the limpet as it is generally seen on the rocks; 3 *b* represents the under-surface of the same object, and shows the animal itself. The limpet may easily be thus seen, if it is placed in a vessel of sea-water with flat glass sides, for it soon crawls up the side, and so exhibits itself very perfectly.

The Common Whelk (*Buccinum undatum*) is another shell that is sure to be found on the sands. This is so well known a shell that no particular description is here necessary, but mention will be made of it on a succeeding page.

COWRY.   WHELK

The little shell, figured on the left side of the whelk, is one of the cowries, of which there are almost innumerable varieties. Some of them, found in the tropical seas, are of very large size, while others are much smaller than the specimen represented. One species of this shell is used as a medium of exchange in some countries. Money that can be picked up on the sea-shore is, however, of very small value, fifteen hundred cowries being considered as an equivalent to one English shilling—hardly reimbursing the collector for the trouble of stooping so often.

There is another shell allied or distinctly related to the whelk which is very common on our coasts, and which is well worthy of notice. This is the *Purpura lapillus* (plate B, fig. 4), a shell that is sometimes found nearly white, but mostly banded with brownish orange, as is represented in the figure. Now, the creature that inhabits this shell is one of those animals that furnished the famous purple of the ancients, and from that property it derives its name of Purpura. The colour is not particularly beautiful, and it is rather remarkable that the ancients, who had very good taste in colours, should have placed so high a value on this purple, which, according to their own account and our observations, closely resembled clotted blood.

Perhaps, however, its rarity constituted its value; for there is so little in each shell, that an enormous number of victims must have perished before a sufficiency of the dye for one robe could have been obtained.

The ancients seemed to have managed the extracting process in rather a clumsy manner; but it is easy enough to procure the dye without mixing it with the juices of the animal, as seems to have been the case in the olden times. If the reader would like to try the experiment, it may be done as follows:—

Let him look among the rocks at low water, and plenty of the shells may be found tolerably close together. When a sufficient number are collected, they should be killed by placing them in fresh water, after the shell has been pierced or broken, as otherwise the animal shuts itself up so tightly that the water cannot gain admittance. When the creatures are quite dead, the colouring matter may be found in a yellowish-looking vessel, that derives its colour from the substance contained within. There is very little of this colouring matter in the vessel. Now, if this yellow substance be spread on white paper and placed in the sunshine, a blue tinge enters the yellow, making it green. The blue gradually conquers the yellow, and the green soon becomes blue. Another colour, red, now makes its appearance in the blue, and turns it into purple. The red becomes gradually stronger, and in its turn almost vanquishes the blue, but does not quite succeed in doing so; for the blue, having taken so much pains to turn out the yellow, will not entirely vacate the premises, and, coalescing with the red, forms a deep purple, the red very much predominating. So we have here all the primary colours fighting for the dominion, and yellow, the most powerful of the three, forced to retire before its complementaries.

There are great numbers of little shells, called Tops from their form, which are found plentifully on every coast, either empty and cast ashore by the waves, or living, and found adhering to the sea-weeds that are laid bare at low water. It is not often that these shells are found quite perfect, for the shell is generally worn away at the apex, so that the colouring substance is removed and the point of the shell is white. One of the most beautiful of these shells, the Livid Top (*Trochus ziziphinus*), is represented on plate B, fig. 1.

The tongue of this species is remarkable for its structure. Many molluscs are furnished with very wonderful tongues, the true beauty of which can only be seen by placing them under a microscope of moderate power. Their tongue is easily extracted by drawing it out from its hiding-place with a needle, and cutting it off — the owner being, of course, previously killed.

When this organ is properly displayed, it will appear furnished with one array of teeth, very minute, but very strong, and quite adequate to the work which they have to perform. In fact, the tongue is a miniature file, and is used not so much for tasting the food, as for a rasp, wherewith to cut it off. The top, therefore, is an useful inhabitant of an aquarium, for he saves an immensity of trouble in keeping the glass sides clean. After an aquarium has fairly settled itself, the algæ pour out their spores, and these, adhering to the glass, there affix themselves, so that in a few weeks the glass becomes dimmed by the mass of minute vegetation. Here the tops and periwinkles come to our aid, and by means of the natural scythes with which they are armed, soon mow away the greater part of this vegetable growth. They seem to do their work as composedly and regularly as if they were paid by the day for it. The Livid Top may be found alive among the rocks at low water.

I have already stated that the periwinkles are useful inhabitants of an aquarium, and such is the case as long as they can be kept alive. But they are often very perverse in disposition, and show greater predilection for dying than for mowing. The Common Periwinkle, so familiar in our streets, is tolerably hardy in confinement, and may be kept for some time. But the handsomer Yellow Periwinkle (*Littorina littoralis*), which is represented on plate B, fig. 2, is still more delicate in constitution, and seldom survives for many weeks. But even the Common Periwinkle is a pretty creature, as it exhibits itself when crawling upon the glass of the aquarium, or on the sea-weeds where it finds its food. The body is prettily banded with multitudes of narrow dark markings, and the mode in which the creature slides itself over the glass is very curious.

There is a very pretty shell found in tolerable profusion on our sands, and which will be recognised at once from its portrait (plate B, fig. 5). This is the Common Wentletrap (*Scalaria communis*). It is not only a pretty shell, but holds relationship with a very aristocratic connexion. The Wentletraps are divided into two great sections; the false Wentletraps, the whorls of whose spires touch each other; and the true Wentletraps, whose whorls are disjointed from each other. Of the former section our little friend is a good example; and of the latter, the aristocratic relative alluded to. This is the Royal Staircase Wentletrap, a shell formerly of such rarity that a specimen only two inches and a quarter in height would fetch eighty or ninety pounds.

The next shell which I shall mention is the Common Cockle (*Cardium edule*), represented on plate B, fig. 6.

Perhaps this is the most abundant of all the littorine shells; for if a handful of shells be gathered at random from the sands, nearly one-third will be cockles. When living, the animals find a home under the sand, in which they lie buried. The cockle is a capital delver, and, armed with his natural spade, digs for himself a hole in the sand nearly as fast as a man can dig with a spade of metal. As for the wooden spades, so much in vogue on sandy coasts, they have hardly a chance against the cockle.

Many an observer has been perplexed at the little jets of mingled sand and water which are so often seen issuing from the sand when the waves have retired. These tiny geysirs are occasioned by the cockles that lie buried beneath the sand, and which are still in the water below the sand level, although the surface is tolerably dry.

Our cockle, however, is not only a digger, but a jumper, and the same instrument which serves him as a spade to dig a hole in the sand also serves him as a foot by means of which to spring into the air.

There is another burrowing shell, that is found on most sandy beaches. This is the Razor-Shell (*Solen ensis*), for a representation of which, see plate B, fig. 7.

This creature burrows even deeper than the cockle, being often found at the depth of two feet. It does not, however, seem fond of sinking thus low, but generally remains sufficiently near the surface to permit the tube just to project from the sand. The burrow in which the animal lives is nearly perpendicular, and in it the Solen passes its entire life, sometimes ascending to the surface, and sometimes descending to the bottom of its burrow, for it has none of the locomotive faculties of its fellow-miner, the cockle. But although its range of travel is circumscribed, the narrowness of its habitation is compensated by the activity of its movements therein. The fisherman who wishes to capture the creature is aware of its agility, and takes measures accordingly. As the tide retreats he watches for the jet of sand and water which the animal throws into the air when alarmed by its hunter's footstep. Into the hole from which the jet ascended the fisherman plunges a slender iron rod, which having a barbed, harpoon-like head, pierces the animal, and retains it while it is dragged from its hole. If, however, the fisherman takes a bad aim, and misses his cast, he does not try a second with the same creature, knowing that it will have retreated to the termination of its burrow, whence it cannot be extracted.

Yet another burrowing shell. In most chalky rocks, such as those of which the white cliffs of old England are composed, many portions run well out to sea. If these are examined at low water they will be found to be perforated with numerous holes, running to some depth, and varying considerably in dimensions. These holes are made by the *Pholas dactylus*, plate B, fig. 9, one of the most remarkable animals in creaturedom.

Hard rocks and timber are constantly found perforated by this curious shell, but how the operation is performed no one knows. It is the more wonderful, because the shell is by no means hard, and cannot act as a file. Indeed, in some species, the external shell is almost smooth. And, moreover, if the shell were used as the boring-tool, the hole would be nearly circular, instead of being accommodated to the shape of the shell, as is seen to be the case. However they get into the stone, there they may be everywhere found, and it does not seem to be of much importance whether their habitation be limestone, sandstone, chalk, or oak. Even the Plymouth breakwater, solid stone as it is, was very soon attacked by these creatures.

They are especially obnoxious to the builders of wooden piers, for they seize on the submerged portion of the piles on which the pier rests, and do their utmost to reduce them to a honeycombed state with the least possible delay. Lately, however, the Pholades have been conquered; for they cannot pierce iron, and it is found that if iron nails are closely driven into the submerged portion of wooden piles, they bid defiance to the Pholas.

The specimen represented in our figure is shown resting in its rocky bed, and seen edgeways. At each side may be seen the furrowed shells; the foot appears in the centre, surrounded by the mantle, and the tube is seen projecting far beyond the shell. Very many good specimens may be obtained by splitting open the piece of rock, and thus the shells extracted without injury from the rocky home where they have lived and died. In the interior of a perfect shell may be seen a very curious projection, formed something like a spoon. Its object does not seem to be very clearly ascertained. The tube, which has been so often mentioned, is generally a composite organ, composed of two tubes or siphons, as they are called, which are placed closely together, something on the principle of a double-barrelled gun, or an elephant's trunk. Through these tubes passes the water which is necessary for respiration, being received into one tube, drawn from thence over the gills, and finally expelled from the other tube.

There is another boring mollusc, which is on many accounts worthy of notice. This is the so-called Shipworm (*Teredo navalis*), a representation of which may be found on plate F, fig. 3. It has been placed on the same page with some of the worms, in order to show its very great external resemblance to some animals of that class, and especially its similitude to the Serpula. So closely, indeed, does it resemble the last named creature, that even Linnæus placed the Teredo between Serpula and Sabella in his "System of Nature."

But this is really one of the molluscs, and a

very curious one. It is called the Ship-worm because it has so powerful an appetite for submerged wood, and especially for ship-timber. I have now by me a large piece of oak, the remains of some wreck, which I found entangled among the rocks at low water. It is so completely devoured by the Teredo, that it is almost impossible to find any portion of the wood that is thicker than the sheet of paper upon which this account is printed. Timber, however, can be protected from the Teredo by a closely-studded surface of broad-headed iron nails. These nails soon rust through the action of the salt water, and the whole of the timber is rapidly covered with a thick coating of iron rust, a substance to which the Teredo seems to have a strong objection.

The *Teredo navalis* is not a very large animal, but it has a huge overgrown relation, the Giant Teredo, whose diameter at the thickest part is three inches, and its length nearly six feet.

On plate B, fig. 8, may be seen a shell, which will probably be recognised at once as the Common Mussel (*Mytilus edulis*). The specimen figured is a young one, and is shown as it appears when adhering to the rock by means of the natural cable—or byssus, as it is scientifically named—with which these creatures are furnished.

These shells are exceedingly common, and large masses of them may be found clinging to any rocks or stairs where they can anchor themselves. This mussel is called Edulis, or eatable, because it is largely used as an article of food. But it is by no means a safe edible, as at certain times, or to certain constitutions, it acts as a poison, producing most alarming and sometimes fatal effects.

The byssus is an assemblage of delicate, silky, and excessively strong fibres, the origin of which seems to be at present rather obscure. Many shells are furnished with this substance, which is shown in perfection in the great Mediterranean Pinna, some specimens of which measure nearly two feet in length. The byssus of these creatures is often spun and woven like silk, and in many places may be seen gloves, purses, and other objects, which have been made from this substance. It is, however, too rare to be put to any practical use.

The Common Scallop (*Pecten Jacobœa*), generally known in connexion with oysters, may be found abundantly on our shores. Even the empty shells are pretty enough to attract observation; but the animals are more beautiful than their shelly habitation. A living scallop is well worthy of notice, if it were only for the row of eye-like points which are seen peeping out from the very margin of the shell, when the creature holds the valves partially open. Whether

SCALLOP.

these brilliant spots are really eyes or not has not been clearly ascertained, but at all events there appears no reason why they should not be eyes; and so to us eyes they shall be.

The scallop is capable of changing its position, and does so by the forcible ejection of water from a given point. This mode of progress is analogous to that employed by the larva of the dragon-fly. The title Jacobæa is given to the scallop, from the shrine of St. James (Lat.

Jacobus), at Compostella; to which spot journeys were made by pilgrims, who, in token of having paid their devotions at St. James's shrine, wore a scallop-shell in their hats for the admiration of their contemporaries, and bore it on their coats-of-arms for the information of their posterity.

The story which connects the scallop-shell with St. James is very curious, but too long for insertion.

The last shell-bearing mollusc which I shall mention is one which does not at first appear to be a mollusc at all. This is the curious little Chiton, a creature which, instead of a tubular shell like the Teredo, a single whorled shell like the whelk, or a double shell like the scallop, bears an array of eight shelly plates on his back, and thus gives to the observer an idea of a tiny marine armadillo.

The entire back of the Chiton is covered with a strong leathery coat, much larger than the living centre of the animal. Upon this leathery mantle are placed eight shell-plates, which overlap each other just as do the tiles of a house: They are not very large on our English coasts, but some foreign species are found which exceed four inches in length.

If the shell-bearing molluscs are remarkable

D

NUDIBRANCHS—DORIS.

for the elegant form and brilliant colouring of their habitations, they seem to be equalled, if not eclipsed, in beauty by a race of molluscs which possess no shell at all, and whose chief beauty is derived from the singular peculiarity of formation from which their name is derived. These are the Nudibranchs, or Naked-gilled Molluscs, so named because their respiratory apparatus, instead of being concealed within their bodies, or defended by shells, is placed upon the exterior, in apparently heedless defiance of surrounding objects. And the more that the delicate construction of these branchiæ is seen, the more wonderful does it appear that these organs should be placed in the position which they occupy without suffering serious injury. If the lungs of one of the mammalia were to be attached to its sides, and permitted to hang loosely therefrom, exposed to the invasions and collisions to which they would probably be liable, the owner of the said lungs would hardly feel comfortable. But the lungs, gills, or branchiæ, of the mollusc are so exceedingly delicate, that the mammalian lung appears quite coarse by their side.

There are many species of Nudibranchs found on our coasts, one of the commonest of which (*Doris ptilosa*) is represented on plate N, fig. 4.

The gills may be seen spreading like a feathery plume, or a radiating flower, on the upper surface of the creature. The position of the branchiæ is by no means uniform, for indeed the most fertile imagination would hardly venture to depict such fantastic forms as are found among the Nudibranchs, or, if they were depicted, could hope that such wondrous shapes should be received by men as existing in the same world with themselves.

Some species, like those whose shape has been already alluded to, are nearly flat, and wear their lungs much as a gentleman wears a bouquet, in his button-hole. Others have their lungs neatly arranged round their bodies in little spreading tufts, so that the creature has something the aspect of a floriated coronet. Some have their whole dorsal surface thickly studded with lungs, so that it would bear a decided resemblance to a hedgehog, were it not that the spikes must be semitransparent, and tinged with the most exquisite colours. Again, there are some species which carry their lungs at a distance from their bodies, and present them to the waves as if they were holding the branchiæ in the hands of their outstretched arms; while there are some whose forms are so utterly unique and grotesque, that a description would be useless except it were accompanied by a drawing.

As to the colouring of these creatures, there is hardly a tint, from blackish grey to the most brilliant carmine, that is not found in some member of this strange family. They all belong to that division of the molluscs that go by the name of Gasteropoda, because the lower surface of the body forms the foot by which they move from place to place. By the aid of this foot they often float on the surface of the water, as has been already recorded of other molluscs. This action, however, has been well described, as creeping on the superincumbent stratum of air. Many species of the genera Doris and Eolis, together with others, may be found, at low water, clinging to the rocks and stones. They will hardly be recognised as Nudibranchs at a hasty glance, for they subside into shapeless gelatinous knobs as soon as the waves leave them, and do not resume their expanded form until the surging sea returns.

The Nudibranchs, although most lovely creatures, are very unsafe inhabitants of an aquarium, in spite of their delicate and dainty looks; and a wolf would be about as appropriate an inmate of a sheepfold, as a Nudibranch of an aquarium where sea-anemones live. Even the giant crassicornis, or Thick-horned Anemone, has fallen a victim to the insatiable appetite of these greedy creatures. In closing this short description of the Nudibranch, let me strongly recommend the reader to examine, if possible, the beautiful work on these creatures by Messrs. Alder and Hancock, published by the Ray Society.

CHAPTER III.

MARINE ALGÆ, OR SEA-WEEDS.

SEA and land are, after all, wonderfully like each other. The surface of the land has its mountains, its valleys, its fire-vomiting volcanoes, its mountains of eternal cold. So the bed of the sea is delved into vast valleys, as yet unfathomable by human plummet; and these valleys we of the upper world call depths. Also, it has its precipitous mountains, some towering above the watery surface, and others lifting their heads until they are dangerous neighbours to those that go in ships upon the waters; and these we call by various insulting names according to their degree of elevation. And there are volcanoes of the sea as well as of the land; while the Polar islands, which are, in fact, the tops of submarine mountains, are covered with snows as eternal as those which crown the Monarch of mountains himself.

Then, the sea-bed has its Table Mountains, its vast Saharas, its undulating prairies, its luxuriant forests, and its verdant pasture-lands. And as the sandy tracts or shingly beds are bare and devoid of vegetative life on the upper earth, so are they also in the sea below; while submarine forests lift their branches towards the light of the sun, and submarine herbage waves its many-coloured leaves in the rolling sea, just as flowers and leaves bend to the breezes above. For in the kingdom of Ocean, water is the atmosphere, and, like its more ethereal relative, is ever rolling, and ever changing.

Let us now visit the boundary line of the two great kingdoms, Earth and Water, and though belonging to the former, extend our researches as far as possible into the latter.

Throughout the preceding pages it will be noticed that the expression "at low water" is constantly used. Now, this expression is quite necessary; for were the sea always to remain at the same height, our knowledge of its wonders would be wofully circumscribed. It is little enough even now, but that little would be almost reduced to nothing were there no alternations of high and low water.

Of the theory of tides there is here no opportunity to speak, for it is a most complex subject, and even to give a hasty sketch would occupy many pages, and require many diagrams. Suffice it to say, that the grand exciting cause of the tides is the force called attraction, or gravitation; the moon being the chief among the many agents through which it acts. It matters not whether the water is salt or fresh, whether as an ocean it fills the bed of the Atlantic, or as a drop of dew trembles on a violet leaf. The tide-force still acts on it, and tides there are, although we are incapable of perceiving them. It is the same with the upper sea, namely, the atmospheric air of our earth. In the aërial ocean there are waves, whirlpools, calms, and storms, although our eyes are too dull to perceive them, and can only be made aware of their existence by seeing their effects.

Twice in every day of twenty-four hours the water advances and recedes, and thus at least one opportunity is given daily for the observer to follow the retiring waves, and to discover some small portion of the wonders of the sea. Some of its living and breathing inhabitants have been mentioned in the preceding chapter, and in the following pages will be briefly described some few of its vegetative inhabitants, that breathe not, but yet live.

If we walk on the sea-shore, vast masses of

dark olive vegetation meet our eyes; if we wait until the tide has retreated, and examine the pools of water that are left among the rocks, there we find miniature forests, and gardens of gorgeous foliage, some of which are scarlet, others pink, others bright green, others purple, while some there are that play with all the prismatic colours, each leaf a rainbow in itself. If we take a boat, and rowing well out to sea, cast overboard a hooked drag, we shall find adhering to the iron claws new kinds of vegetation, and probably among them will be found a veritable flowering plant,—apparently as much out of its place at the bottom of the sea as a codfish in a birdcage. Now all this luxuriant, graceful, and magnificent foliage, we dedecorate with the title of sea-*weed*. It is a miserable appellation; but as it is a term in general use, I shall employ it, although under protest.

Those sea-weeds, then, which first strike our eyes, are usually those denominated Wracks, the Common Bladder-wrack (*Fucus vesiculosus*) being the most common. For a figure of this plant, see plate J, fig 6.

There is little difficulty in distinguishing this conspicuous alga; for the double series of round air-vessels with which the fronds are studded,

DANGERS OF ROCKS. 45

and the mid-rib running up the centre of each frond, point it out at once. This plant, together with one or two others of the same genus, is still used in the manufacture of kelp, but not to such an extent as was formerly the case. There is a variety of this plant found in salt marshes, where it congregates in dense masses : this variety is very small, being only an inch or two in height, and the eighth of an inch, or even less, in width. The plant is at all times very variable, according to its locality, both in colour and form.

When trodden on, or otherwise suddenly compressed, the air-vessels explode with a slight report, and seem to afford much gratification to juveniles. This and other fuci grow in the greatest abundance on rocks that are covered by the waves at high water, and left bare when the tide retires. Now on, under, and among these rocks, the great zoological or botanical harvest is to be collected, and therefore among these rocks the collector must walk.

I make mention of this circumstance, because it is necessary to warn the enthusiastic but inexperienced naturalist, that the slimy and slippery fuci make the rock-walking exceedingly dangerous; for the masses of fuci are so heavy and thick, that they veil many a deep hollow, or slightly cover many a sharp point,—in the former of which a limb may be easily broken, and by the latter a serious wound inflicted,—and there is special reason for avoiding any such mishap. Proverbially, time and tide wait for no man; and should a disabling accident occur when no one was near to help, the returning waters would bring death in their train—a death the more terrible from its slow but relentless advance.

Now, the reader must be careful not to confound with *Fucus vesiculosus* another species of somewhat similar appearance, namely *Fucus nodosus;* see plate J, fig. 1.

This plant may at once be distinguished from the Common Bladder-wrack, by the absence of a midrib; it is of a tough consistence, and it grows to a large size, being sometimes nearly six feet in length.

About half-way between high and low water another species of fucus may be found: this is destitute of air-vessels, it lacks the sliminess of the bladder-wrack, and its edges are toothed, like the edge of a saw. It is much about the same size as the bladder-wrack, but perhaps rather longer; see plate D, fig. 2.

This is a very useful plant indeed. It is a

capital manure for land, it can be preserved, and used as food for cattle, it can be made into kelp, and it is an excellent substance in which to pack lobsters, and other marine productions, that are sent inland. The bladder-wrack is much used for the same purpose, but its sliminess renders it liable to heat and to ferment; while *Fucus serratus*, being comparatively free from slime, retains its cool dampness, and preserves the fish sweet. It is really of importance, for a tainted lobster is not only nauseous to the palate, but even dangerous to the whole system.

There is a very tiny fucus, some four or six inches long at the most, that is to be found near high water-mark, chiefly in summer and autumn; it may be recognised by a number of small channels that furrow one side of the frond. It has no air-vessels.

All these plants, together with all the algæ comprised in this chapter, belong to the class of algæ called MELANOSPERMS, or black-seeded; so called from the dark olive tint of the spores, or tiny seeds, from which they spring. They all seem to be exclusively marine.

At spring-tides the waters recede considerably below their usual mark, and these seasons are the harvest-times of the shore-naturalist. As

LAMINARIA.

nearly as possible at six hours after the high tide the waters will have retired to their lowest boundary, and near that boundary will be found myriads of new forms, both animal and vegetable. Indeed, so prolific is the spring-tide harvest, that an hour or two of careful investigation will sometimes produce as good results as several hours' hard work with a dredge. It is better to go down to the shore about half an hour or so before the lowest tide, so as to follow the receding waters, and to save time.

When the naturalist has gained the spots below the usual low-water mark, he will find himself in the midst of a new set of vegetation, contrasting as strongly with the productions of the higher grounds, as forest trees with herbage and brushwood. Huge plants, measuring some eleven feet or so in length, and nearly a yard in width, are firmly anchored among the rocks by roots rivalling in comparative size and strength those of the oak-tree. This plant is commonly known by the name of Oar-weed, and may be easily recognised from the drawing in plate D, fig. 1. Its scientific name is *Laminaria digitata*. It is called "Laminaria" on account of the flat thin plates, or laminæ, of the frond, and "digitata," or fingered, because the frond is split into segments, something like the fingers of a hand.

I may as well mention here, that the sea-weeds have no real root, and do not derive their nourishment from the soil, as do the plants of earth; they adhere to the rocks or stones by simple discs, and draw their whole subsistence from the water that surrounds and sustains them. In the so-called root of the Laminaria there are no root-fibres, but a succession of discs, each connected with the main stem of the plant by a woody cable.

The stem of the Laminaria is very strong, and is used for making handles to knives and other implements. When fresh, this stem is soft enough to permit the tang of a knife-blade to be thrust longitudinally into it. A portion of the stem sufficiently long for the knife-handle is cut off, and in a few months it dries, contracting with such force as to fix the blade immovably; and having much the consistency and appearance of stag's horn. One good stem will furnish more than a dozen of these handles.

Among the Laminariæ may be seen growing a singular plant, more like a rope than a vegetable. It consists of one long, cylindrical, tubular frond, hardly thicker than an ordinary pin at the base, but swelling to the size of a swan's quill in the centre. When the plant is handled, it slips from the grasp as if it were oiled; this effect is produced by a natural sliminess, aided by a dense covering of very fine hairs.

The name of this plant is *Chorda filum*. Its length varies extremely, some specimens being found to measure barely one foot, while others run from twenty to thirty, and even to forty feet. It tapers gradually from the middle to the point, where it is about the same thickness as at the base.

I here make an exception to my general rule of excluding all but the commonest objects, in favour of one sea-weed, which, although not very common, yet may be found quite unexpectedly. It owes its introduction to its very singular form. The name of it is the Peacock's-Tail, deriving its title from its shape. Its scientific name is *Padina pavonia;* see plate A, fig. 3.

The habitation of this plant is midway between high and low water-mark, where it may occasionally be found adhering to the rocks. It is not a large plant, as it is generally only two or three inches in height, but occasionally reaches the height of five inches.

In the same order as the Padina is another

little algæ, which, I think, is one of the prettiest of the Melanosperms. I do not know whether it possesses any popular name, but its scientific title is *Dictyota dichotoma*. For a figure of it, see plate A, fig. 5. It is a very delicate-looking plant, and, unlike the Melanosperms in general, lives tolerably well in an aquarium. The name Dictyota is derived from a Greek word, signifying a net; and it will be seen, on examination, that the surface of the frond appears as if woven into a tiny network, with square, or rather slightly oblong meshes. Its specific appellation of "dichotoma" is also of Greek derivation, and signifies "cut in pairs," in allusion to the shape of the frond.

Failing space permits only one more plant belonging to the class, or rather, to speak accurately, the sub-class Melanospermeæ. This is the plant known to botanists by the title of *Ectocarpus siliculosus*, and which I mention here because it is liable to be confused with other algæ that much resemble it in form, though not in constitution; see plate A, fig. 2. It is called Ectocarpus from two Greek words, signifying "external fruit," and its specific title "siliculosus" is given to it on account of the silicules, or little pod-like bodies, that are found on the branches. These details are of very minute size, and cannot be made out without the assistance of a magnifying glass.

These dark-spored vegetables are very variable in colour, as indeed are all the algæ, without reference to the colour of their spores; sometimes, indeed, even trespassing on the colour of another sub-class. These changes are mostly due to the varying depths of the sea where the plants grow, and to the amount of light and shade which falls to their lot. Even the hardy, rough, and coarse bladder-wrack, which is usually of a very dark olive-green, more approaching to black than to green, becomes of a rich yellow tint when found at any depth of water.

When dried the green vanishes totally, the colour changing to dark-brown, and in many cases to black. Most of this sub-class of algæ require alternations of water and air, the best specimens being found where they are exposed to the heat of the sun and to the force of the winds for some hours daily.

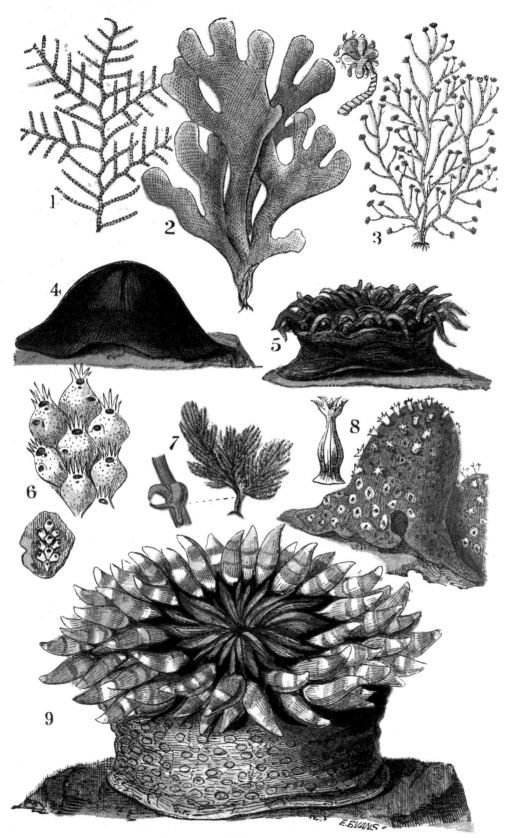

E

CHAPTER IV.

RED-SPORED AND GREEN-SPORED ALGÆ.

WITH this chapter we begin the account of another sub-class of algæ, the RHODOSPERMS, or red-seeded. The plants belonging to this class are among the most beautiful of the algæ, that is, when they are placed in favourable situations; for they also change their colours, and as their most beautiful colour is their natural tint, any change is for the worse. Some of them even become brown when there is too much light for them.

About low water-mark may be found growing largish masses of a dense, thread-like, reddish foliage, sometimes adhering to the rock, or sometimes even fixed to the stems of the Laminaria. When removed from the water the plant does not collapse, like many of its relatives, but each thread and branch preserves its own individuality. This is one of the large genus Polysiphonia, and the specific name is "urceolata." See plate K, fig. 2.

By the side of the plant itself is represented a little object that explains the latter title. This little jar-shaped object is one of the fruits, or ceramidia, as they are learnedly called, much magnified. The word "urceolata" signifies pitchered, if we may be permitted to coin an English word corresponding to the Latin. The name Polysiphonia is Greek, and signifies "many siphons," or tubes. The reason for the name is evident on cutting any of the branches transversely. It will be then seen that the plant is composed of six tubes arranged round a central aperture; the branches are jointed, the length of each joint being several times its own width.

There are twenty-six known British species of this single genus.

That popular author and extensive traveller, Baron Munchausen, tells us that in one of his journeys he met with a tree that bore a fruit filled interiorly with the best of gin. Had he travelled along our own sea-coasts, or indeed along any sea-coasts, and inspected the vegetation of the waves there, he would have found a plant that might have furnished him with the groundwork of a story respecting a jointed tree, composed of wine-bottles, each joint being a separate bottle, filled with claret. It is true that the plant is not very large, as it seldom exceeds nine or ten inches in height; but if examined through a microscope, it might be enlarged to any convenient size.

The name of this plant is rather a long one, but very appropriate, *Chylocladia articulata,* i. e. the "jointed juice-branch." See plate A, fig. 1.

It may be found adhering to rocks, or sometimes parasitically depending on some of the larger algæ, and really does resemble a jointed series of transparent bottles filled with claret or other red wine. The colour is remarkably delicate and beautiful, but is rather apt to fade after a time; when it is preserved, dried, and pressed, the gelatinous juice that filled the interior disappears, and the plant can be flattened until it hardly presents any thickness, even to the touch. There is now before me a dried specimen of another species of chylocladia, which adheres so firmly to the paper on which it is laid, and is so delicate in substance, that several persons to whom I have shown it have mistaken it for a well-executed drawing.

If now the reader will refer to plate C, fig. 1, he will there see depicted one of the most remarkable of the algæ; remarkable in itself, and for the great battles which have been fought over it by scientific individuals. This plant is the Common Coralline (*Corallina officinalis*), which may be found most abundantly on any of our coasts, growing in greatest perfection near low water-mark.

It is well enough known that many creatures, formerly supposed to be vegetable, such as the corals and the zoophytes, have since found their proper place in the animal kingdom; and one consequence of this reformation was, that several real plants were supposed to be animals, because they possessed some of the characteristics which had distinguished those animals that had been placed in their proper position. Of these plants the coralline is a good example; for until a comparatively late period, it was placed among the animals in company with the true corals.

There was reason for this error, for the coralline is a very curious plant indeed, gathering from the sea-water, and depositing in its own substance, so large an amount of carbonate of lime, that when the purely vegetable part of the alga dies, and is decomposed, the chalky portion remains, retaining the same shape as the entire plant, and very much resembling those zoophytes with which it has been confounded. While growing, the coralline is of a dark purple colour;

CORALLINE.

but when removed from the water, the purple tint vanishes, and the white stony skeleton remains. It is, however, a true vegetable, as may be seen by dissolving away the chalky portions in acid: there is then left a vegetable framework, precisely like that of other algæ belonging to the same sub-class.

The coralline is a small plant, seldom exceeding five or six inches in height, and not often even reaching that size. However, it compensates for its low stature by its luxuriant growth, being usually found in dense masses wherever it can find a convenient shelter.

If a dried branch of coralline be inserted into the flame of a candle, it exhibits a most brilliant white light just at the point where it meets the flame. The light is exhibited better by the flame of a spirit-lamp than by that of a candle, and for obvious reasons.

It will live well in an aquarium, and, if tastefully disposed, is an elegant ornament to the vase or tank. There is now in my own aquarium a moderate tuft of coralline, which seems in good health, although the water has lately been assuming an unpleasant milky appearance, from some cause which I cannot as yet detect.

We now come to a most magnificent sea-plant,

magnificent both on account of its gorgeous colouring, and on account of its luxuriance. This is the *Delesseria sanguinea*, represented, about half its usual size, in plate J, fig. 3.

The shape of the leaf, or rather of the frond, so closely resembles that of terrestrial trees, that at first sight few would attribute the beautiful scarlet leaf, with its decided midrib and bold nervures, to an alga. Yet an alga it is, and may be found in its most perfect state about June or July: later in the year it becomes very ragged, the broad flat frond giving way to the fruit. In this state, although interesting to the botanist, it is hardly suitable for the cabinet, as little of the plant is left except the midrib, and a few flapping raglets. When spread on paper and preserved, it retains its colour well, and adheres very firmly.

The fronds are generally from two to seven or eight inches in length, but they are not often found exceeding five or six inches. A branch containing eight or ten fronds, averaging five inches in length, may be considered a good specimen, and worth preserving, if the edges are entire. There is a very peculiar marine scent about this plant, an "ancient and fishlike smell," quite indescribable, but not to be forgotten. A large branch will retain this scent for months. I have by me a tuft of this plant, which I gathered in July last, and its peculiar smell is now (April) very perceptible.

There are five British species of this beautiful genus, none of them very rare. *Delesseria hypoglossum* (plate D, fig. 4) may be found in the summer months growing on almost every coast. It is a very pretty plant, although not so gorgeous as its predecessor. The fronds are generally of small size, being hardly a quarter of an inch in length.

In the little sea-weed landscapes, that are sold so abundantly at the fashionable sea-side towns, there is one species of sea-weed in great request for trees and bushes. It is of a bright pinky red colour, and is thickly branched, so as to afford a tolerable representation of a forest tree, or of a thick bush. This is the *Plocamium coccineum*, a plant sufficiently beautiful to the unassisted eye, but especially so when submitted to a magnifying lens. When examined through a glass of moderate power, it will be seen that even the tiny branchlets, each hardly thicker than a hair, are again furnished with a row of smaller ramifications, somewhat resembling a very finely-toothed comb.

On plate K, fig. 3, may be seen a specimen of the Plocamium of the natural size, and near it a single branch magnified, in order to show the tiny combs.

Many of the marine algæ are used as articles of food; some eaten uncooked, and others after a long course of boiling. To the former of these categories belongs the Dulse, Dillisk, or Dillosk (*Rhodymenia palmata*), although it is sometimes cooked. The species, however, which is here illustrated, is *Rhodymenia bifida*, a plant of a very fine rosy red when fresh, found in tolerable profusion adhering to rocks or on the larger algæ. The fronds are generally two inches or so in length, and about a quarter of an inch in width. For a figure of this plant, see plate K, fig. 4.

The Carrageen Moss, so well known in the form of jelly and size, is one of the Rhodosperm Algæ, by name *Chondrus crispus*. (Plate J, fig. 5.)

It may be found growing on the rocks in large quantities, where its shape will be the best guide to its detection, for its colour is exceedingly variable. Although one of the Rhodosperms, it is very frequently of a greenish tint, and in many places it assumes a yellow jaundiced complexion, not at all of a healthy nature.

To preserve it for esculent purposes, it must be washed in fresh water and then left to dry, when it soon becomes horny to the touch, and resists pressure. If boiled, it subsides into a thick colourless jelly, that is thought to be very nutritive, and is employed for many purposes. Invalids take it in their tea, or epicures in their blanc-mange. Calico printers boil it down into size, and use it in their manufactures. It is said to be a good fattening substance for calves, if boiled in milk; and, lastly, pigs are very fond of it when it is mixed with potatoes or meal. It is sometimes known by the name of "Irish Moss." It will grow in an aquarium.

A plant is represented on plate J, fig. 4, that is found plentifully between tide-marks. It is rather a conspicuous plant, and is appropriately named *Furcellaria fastigiata*, the generic title being derived from a Latin word signifying a little fork. It is of a dark-brown colour with an obscure dash of purple, but in drying the purple departs, and the brown becomes nearly black.

I have already mentioned that some of the algæ reflect prismatic colours. This is occasionally the case with *Chondrus crispus*, and there is one genus which is so resplendent that the name *Iridæa* is given to it; Iris signifying a rainbow.

The species represented at plate C, fig. 4, is *Iridæa edulis*, a plant which is sometimes eaten raw, and sometimes fried by unpoetical gastronomists. I do believe that some people would fry the rainbow itself if it were eatable.

The frond of this species is generally about nine or ten inches in length, and five inches in width, although it sometimes nearly doubles these dimensions. Its colour is an uniform deep red, and its shape somewhat resembling a battledore.

A particularly elegant species of alga, making a good figure when spread on paper, is seen figured on plate K, fig. 5. The fronds are sometimes more than a foot in length, but do not often exceed ten or eleven inches, some being only three or four inches long. The colour is rather apt to fly, unless care be taken; but it is a beautiful plant, were it only for the elegance of its form. Its name is *Ptilota plumosa*, both words having a like signification, and meaning "winged," or feathery.

There is a pretty little alga, called *Griffithsia setacea*, which has the property of staining paper with a fine pinkish-scarlet hue, when the enclosing membrane bursts. Contact with fresh water will usually cause the membrane to yield, and then the colouring matter is shot out with a slight crackling noise.

Its length is generally about four or five inches. A drawing of the plant of the natural size, together with a magnified sketch of the fruit, may be seen on plate K, fig. 1.

The last of the Rhodosperms that will be noticed in this volume is a very delicate species, entitled *Nitophyllum punctatum;* see plate c, fig. 5. This plant will easily be recognised from the drawing. Its usual size is six or ten inches in length, and nearly as wide; but it is not uncommon to find specimens that exceed a foot in length, while some huge monsters have been found that measured five feet in length and a yard in width. It is easy enough to distinguish this plant from the Delesseria, as it has no midrib.

The CHLOROSPERMS, or Green-seeded Alga, are the best friends of those who keep marine aquaria, for they are endowed with the power of pouring out oxygen in very large quantities when placed in favourable circumstances. If any of my readers wish to preserve alive the creatures that they find on the sea-shore, they can do so without difficulty, by imitating as nearly as possible the natural state and accompaniments of the animals which they have captured.

If even one or two fish, crabs, or indeed any living animals, be placed in a jar of sea-water, they speedily exhaust the free oxygen of the water, and, as the water cannot absorb fresh oxygen from the atmosphere so rapidly as the animals consume it, the water soon becomes unfit to support animal life, and its inhabitants die as surely as a man would who was enclosed in an air-tight box. It is possible to renew the oxygen by dashing water into the jar from a height, or even by pumping fresh air into it; but such a process would be very fatiguing, as it must be continually carried on day and night. But it is found that plants have the property of pouring out oxygen when they are in a healthy state and acted on by light. So, if we can procure plants that will thrive in a confined space, and keep them in a light room, we shall find that each plant acts as a natural pump, and not only supplies continually fresh oxygen, but consumes the carbonic acid gas that loads the water with its stifling influence. The Chlorosperms are peculiarly useful for this purpose, as many of them will live for an unlimited time in confinement, continually regenerating the water in which they are placed. I have now an aquarium containing water that I brought from the sea last August, and by the untiring exertions of a few green sea-weeds the water has been preserved bright and pure, even though inhabited by all kinds of marine animals.

Among the most useful, as well as the most elegant of the sea-weeds used for this purpose, is the little *Bryopsis plumosa;* see plate D, fig. 3. This brilliant and delicate little plant is common enough, and may be found in the pools left by the retiring tide, where it adheres to their rocky walls. The colour of the plant is a very bright green, and its form is so feathery, or rather fan-like, that it well deserves its name of " plumosa."

In almost any little pool, between tide-marks or even hanging from rocks that have been left quite dry, may be seen thick tufts of a coarsish horsehair-like plant, of a dull green colour, often dashed with black. This is the *Cladophora rupestris,* one of the commonest species out of the twenty that are exclusively marine. There are two species that inhabit ditches and lakes where the sea occasionally obtains admission, and several others that prefer water entirely fresh. The length of the tufts is about four or five inches, often less, but seldom more.

Another species of the same genus, *Cladophora arcta,* is of a brighter green than the preceding.

and altogether a prettier plant. It grows in a radiating manner from a very broad disc. This plant is represented on plate c, fig. 2.

But the most useful of the Chlorosperms may be found almost at the very margin of high water, where they live rather more in the open air than under water. These are the *Ulvæ* and *Enteromorphæ*, the first being known by the popular title of Laver, and the second of Sea-grass. There is another plant that is also called Sea-grass; but it is not an alga, and will be mentioned at the end of this chapter.

The Common Sea-grass (*Enteromorpha compressa*) may be seen in abundance on the stones and rocks that are even for a few hours submerged daily. The leaf, or rather frond, of this species is variable in width, sometimes being hardly wider than common sewing thread, and sometimes so wide as to resemble a very narrow ulva. It is this variety which is represented in the engraving, plate c, fig. 3. When the waves retire, leaving sundry pools fringed with this and other sea-weeds, their fronds form hiding-places for innumerable living beings of very many species; and by gathering masses of the wet weed into a basket, and then putting it into a large vessel filled with sea-water, myriads of animals may be captured with hardly any trouble. They will live perfectly well in the vessel if it is kept in a light spot with a free circulation of air.

The Common Green Laver (*Ulva latissima*), plate K, fig. 6, sometimes called the Sea Lettuce, is found most abundantly on the same spots as the preceding plant. Of all the sea-weeds for an aquarium, the Green Laver is perhaps the very best. It is very pretty, from its delicate green colour, and the various folds and puckers into which it throws itself. Its power of expiring oxygen seems to be almost unlimited. I have in my aquarium a large plant of this species, which generally lives very contentedly in the place where it had been deposited. But, a few days ago, the sun shone brightly enough to pierce through the veil of smoke with which the metropolis is generally hidden from his presence, and consequently there was a greater abundance of light than usual. On looking at the aquarium, I found that the ulva had risen in the water, and was hanging in most elegant festoons from the surface, forming emerald caves and grottos such as the sea-nymphs would love. Even at a little distance it was a pretty sight, but a closer inspection revealed still more beauties; for being excited by the unwonted light, the plant had poured forth so much oxygen that its entire surface was thickly studded with tiny sparkling beads, that had buoyed up the whole plant, each bubble acting as a miniature balloon. When, however, a black cloud came over the sun, the bubbles soon detached themselves, ascended to the surface, and as there were no more to take their place, down dropped the plant to the bottom.

On a bright day the little oxygen bubbles are so rapidly exuded, that they quite fill the water, rising to the surface, and there dissipating, very much like the sparkling air-bells in champagne.

The Purple Laver, as it is called in England, or the Sloke, as it is termed in Ireland, is another of these useful plants. In external appearance it very much resembles the ulva, save that the colour, instead of being light green, is purple. From this peculiarity of colour it is called by botanists, Porphyra, which word signifies "purple." There appear to be only two species of this genus belonging to our coasts, the one *P. laciniata*, and the other *P. vulgaris*. The former of these plants is engraved in plate A, fig. 4. Only a portion of the frond is given. It

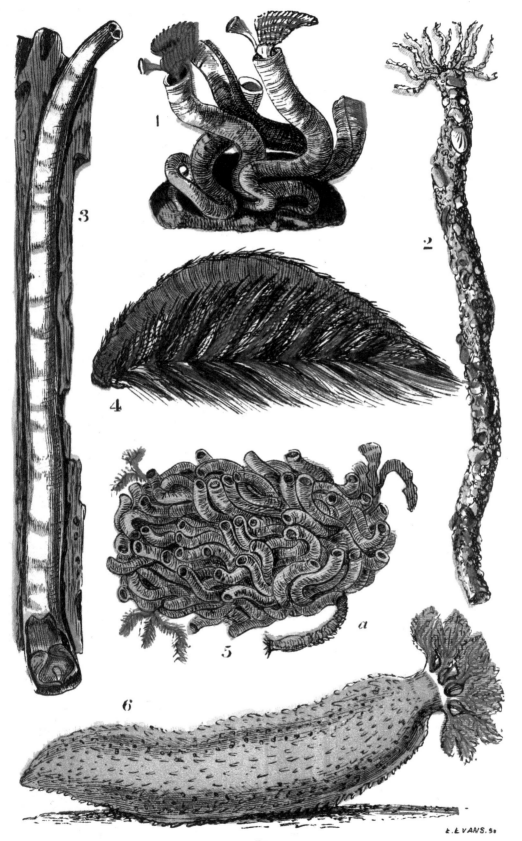

F

is hardly inferior to the preceding plant in value to the aquarium-keeper, and flourishes wonderfully. *P. vulgaris* may be used for the same purpose. I have seen one of these plants in an aquarium, which had increased to such an enormous size, that it was aptly compared by a bystander to a lady's purple silk apron.

The ulva and porphyra, if intended to be eaten, must be gathered in the winter, or, at all events, the very earliest of the spring months. The purple laver is said to be much superior to its green companion, but I cannot speak from personal experience. If any of my readers would like to try the experiment for themselves, they may easily do so; the laver should be stewed for several hours, until it is reduced to a pulpy mass, which, with the addition of lemon juice, is considered by some persons a dainty.

I may here mention that, although both ulva and porphyra will live in an aquarium, when floating freely through the water without any attachment, yet it is better that they should be adherent to some stone or shell, by which they can be anchored in a convenient spot. Now these plants are very constant, for they never have but one attachment during their whole lives, and if torn from that one object they never affix themselves to any other: so it is necessary to use a chisel and mallet, or at all events a geologist's hammer, for the purpose of detaching the portion of rock or stone to which the plant is adherent. Generally the geologist's hammer, if properly chosen, answers every purpose. Almost at the commencement of my last shore-season I dropped both my chisels into a rock-pool, and not being able to find them again, brought the hammer into play; and so useful was that hammer, that I did not find it requisite to procure a fresh set of chisels during the four weeks of shore-searching.

A plant has been mentioned, which does not belong to the sea-weeds, although from its residence at the bottom of the sea it is often thought to be of that family; this is the *Zostera marina*, for a drawing of which see plate J, fig. 2, a true flowering plant, growing with a real root at the bottom of the sea. Its entire character is so completely terrestrial, that it can at once be distinguished from the alga.

The zostera is an useful plant to the zoologist, for it grows in great numbers, or rather in great fields, affording pasture to innumerable living beings, which he captures in his net or dredge. It will live well enough in an aquarium, and gives a decided character to that portion of the tank in which it is placed. Again. when dried, it is largely manufactured into bed-stuffing, under the name of Alva, and is used instead of hay or straw for packing glass, china, and other fragile wares. On many coasts this plant is known by the name of Grass-wrack, and is cast up in great quantities on the shore, where it soon turns black, rots, and presents a very unsightly aspect.

If the naturalist wishes to dry and preserve the algæ which he finds, he may generally do so without much difficulty, although some plants give much more trouble than others. It is necessary that they should be well washed in fresh water, in order to get rid of the salt, which being deliquescent, would attract the moisture on a damp day, or in a damp situation, and soon ruin the entire collection. When they are thoroughly washed the finest specimens should be separated from the rest, and placed in a wide shallow vessel, filled with clean fresh water. Portions of white card, cut to the requisite size, should then be slipped under the specimens, which can be readily arranged as they float over the immersed card. The fingers alone ought to answer every purpose, but a camel's-hair brush and a needle will often be useful. When the specimen is properly arranged, the card is lifted from the water, carrying upon it the piece of seaweed.

There is little difficulty in getting the plants to adhere to the paper, as most of the algæ are furnished with a gelatinous substance, which acts like glue, and fixes them firmly down. Where they do not readily adhere, the use of hot water will generally compel them to do so; and if they still remain obstinate, the gelatine obtained by boiling the carrageen (*Chondrus crispus*)—see p. 39—will be an unfailing remedy. This is a much better cement than animal glue, or even gum-water, as it approaches nearer to the natural glue of the plant. *Furcellaria fastigiata, Cladophora arcta,* and others, are not easily affixed to the paper, and will often require the aid of some adventitious substance.

This sketch of the British marine algæ is necessarily very imperfect; but even as it is, they have occupied rather more than their proper share of paper. Still, there are sufficient different genera here mentioned to prevent the inexperienced marine botanist from erring very widely, and the specimens chosen have been selected for two reasons; the one being that they may be found on almost every coast; and the other, that they form a series of landmarks, by means of which the observer can be directed in the right course.

CHAPTER V.

EGGS OF MARINE ANIMALS—CUTTLES AND THEIR HABITS.

It is impossible to walk on the sea-shore without being struck by the strangely-shaped objects that are cast up by the waves, and left high and dry until swept away by the next tide, which in its turn brings new and varied forms; and these objects are continually changing according to the season of the year. One week may pass, and the observer will see nothing on the sand with which he is not thoroughly acquainted; and in the course of the next week new and grotesque objects will be found profusely scattered at his feet. Some of these objects are purely natural, and their presence is occasioned by the development of nature, while others are but a mixture of the natural and artificial.

For example, when green peas come into general use, the empty pods are thrown into the sea, and are after a space washed up by the waves, having been so chemically acted upon by the salt water, so abraded by sand and pebbles, and so nibbled by various marine animals, that they can hardly be recognised even by the very persons who have consumed the peas that were once enshrined in these metamorphosed husks. Nut-shells, gooseberry husks, currant stalks, cherry stones, and many similar objects, assume, after a temporary sojourn in the ocean, very singular forms, and may easily deceive an unaccustomed eye, especially as they often resemble vegetable and animal remains that properly belong to the sea. I mention this, because I have seen many instances of such deception.

Only a year or two ago, I commissioned a friend to procure for me any marine curiosities that could be found, and to forward them when a sufficient quantity had been amassed. In due time the parcel arrived, having a pleasant marine smell about it, and on being opened, was found to contain some very curious objects. Among them was one on which the collector especially prided himself, and in chase of which he had bravely waded into the sea and undergone a complete wetting. It was apparently a kind of sponge, about eight inches in length, of a light brownish yellow colour, and hollow at one end, as are most sponges. But, on a nearer examination, this sponge proved to be a cabbage-stalk, of which only the fibrous portion remained, and which had probably been tossing about for many months in the sea; sometimes soaked by the waters, sometimes lying on a rock and bleaching in the sun, until the next high tide carried it back again; and at other times entangled among heavy sea-weeds, and anchored by them under water. I still preserve it as an example of marine curiosities.

It is quite necessary, therefore, to exercise much caution in selecting objects. The surest mode of obtaining success is to gather indiscriminately everything that presents itself, and having conveyed the cargo to a place of shelter, deliberately to examine the heap. By so doing, the valuable objects will be retained, and the useless rejected, without so much danger of passing over the one or preserving the other, as if the choice were made immediately. After a time, the eye will become so accustomed to note distinguishing characters, that such a process will be no longer required, and the eye will make its selection at once.

Among the singularly shaped substances that are found thrown on the sands, are the eggs of various marine creatures. Many of these eggs are so curiously formed, that they would hardly

EGGS OF WHELK. 77

be recognised as such by one who was not acquainted with the animals to which they belonged. Very many eggs are found on the shores; but as most of them may be referred to one out of four or five classes of animals, I will only mention those that are, as it were, the general types.

Plate H is specially dedicated to eggs, and, as will be seen on referring to the plate, some of them have anything but an egg-like aspect. The commonest of all the eggs, masses of which are to be found at almost all seasons of the year, are those of the common whelk, the shell of which is represented on p. 22, and its eggs on plate H, fig. 3. The egg mass from which the drawing was made, is now before me. It contains many eggs as yet unhatched, many which are addled, and some which have already discharged their inhabitants. It was taken out of the sea at the very beginning of April, but if it had been permitted to remain in its habitation until the summer, all the egg-sacs would have been found empty. I am keeping it in the aquarium in faint hopes that some of the young whelks will be hatched; but it is very doubtful whether the surrounding conditions are sufficiently favourable. The enormous size of some of these egg-clusters is remarkable; for the whelk itself is by no means a large shell, and so it often happens that those persons who are practically acquainted with the whelk, but not with the eggs, entirely refuse to believe that there is any connexion between objects so dissimilar. The empty egg-cluster bears some resemblance to a rather dingy honey-comb, partially squeezed between the hands. But when the membranous egg-sacs contain their living inhabitants, the animal nature is evident from the presence of the young whelks, whose forms can be plainly seen through the semi-transparent substance that envelopes them until they are sufficiently strong to lead an independent life.

A description of the purpura has already been given on p. 24, and it will be seen that this creature is interesting, not only on account of the beautiful dye which it contains, but also for the singular shape of its eggs, a cluster of which is represented in plate H, fig. 2. Sometimes these curious eggs are found affixed to little stones, and, indeed, when first deposited, some of them seem always to be thus anchored as it were, and to afford support to the others, who stand on each other's shoulders, something like the human pyramid that is occasionally formed at Astley's and similar establishments. The cluster from which this sketch was made I found lying among the rocks, and put it carefully away. But in the course of travel, the box in which it was placed gave way, and my poor little egg-cluster was thrown among a large boxful of shells. There it rested for some seven or eight months, and when discovered was dry, shrivelled, and hardly to be recognised. But, when placed in hot water, it absorbed the liquid as if it had been composed of blotting paper, and in two minutes had completely resumed its natural aspect.

Often may be found, lying on the shore, masses of dark soft substances, not unlike purple grapes, both in size and in shape. If you ask a fisherman the name of them, he will tell you that they are sea-grapes, but for any further information you may usually ask in vain. Indeed, as a general fact, those who live on the sea-shore are hopelessly ignorant of its treasures. I knew a person of intellect, education, and ordinary observing powers, who had resided within a stone's throw of the sea for a period of thirty years, who had been accustomed to walk on the sands almost daily, and yet had never in his life seen, and hardly ever heard of, a common sea anemone, although the shores were studded with them as the sky with stars. And one of those strange amphibious humanities, who get their living by collecting shells, curious pebbles, sea-weed, zoophytes, and other saleable curiosities, persisted in declaring that the hermit crab was the young of the common edible crab, and that when it grew old enough, and was too large for its shell, it abandoned the useless adjunct, and commenced another course of life.

But to return to our sea-grapes, of which a sketch may be seen on plate H. fig. 5. These are the eggs of a cuttle-fish, and curious eggs they are. Each is produced into a flexible stalk, by means of which the mass is held together, and affixed to any convenient object. The egg-cluster from which the sketch was taken was one of four or five which I preserved at different times, in order to watch their progress. Here and there, among the dark mass of eggs, appeared one nearly white, and semi-transparent, through whose delicate walls might be seen the little cuttle within, very lively and seemingly anxious for his emancipation. At the bottom of the egg-cluster may be seen one of the young creatures escaping from the prison that had confined him, and, as will be seen, the young cuttle is rather a comical-looking little animal.

I was much amused with the perfect self-possession of the first that was hatched in my presence. It had not been free from the egg-shell for one minute before it began a leisurely tour of the vessel in which it first saw the light, examining it on all sides, as if to find out what

CUTTLE.

kind of a place the world was, after all. It then rose and sank many times in succession over different spots, and after balancing itself for a moment or two over one especial patch of sand, blew out a round hole in the sand, into which it lowered itself, and there lay quite at its ease. It executed this movement with as much address, as if it had practised the art for twenty years.

The mode by which the creature forms this little burrow is sufficiently curious. Its siphon is a slightly projecting tube, and by bending this towards the spot selected, and then forcibly ejecting a column of water, the sand is displaced apparently by magic. This siphon is also useful as a means of progression; and, in one of the cephalopods, as these creatures are called, because their feet are situated on their head, has not only cast out water, but also the mistaken notions with which careless observers had obscured its history.

Some cephalopods have bodies soft and naked, while others are protected by a shell secreted by themselves. Among these shelly cuttle-fish is the well-known nautilus, who was once said to row himself over the sea with his legs, and to stretch out his wing-like arms as sails to catch the breeze. But it is now known that these sails are kept closely wrapped round the shell, which, indeed, they secrete originally, and can mend if injured, while the legs are suffered to trail loosely. Successive jets of water are then ejected from the siphon, by which the creature is driven in a contrary direction. By this water-power the nautilus is urged through the waves, but when it wishes to move about on the bottom of the sea, it just crawls exactly as a large spider might be supposed to do.

On the arms, legs, feet, or tentacles of the cuttles, are arranged rows of suckers, which are capable of taking a very firm hold of any object to which they are applied, aided in some species by sharp hooks. If any one of these suckers is examined, it will be found to be the living type of the air-pump an exhausting syringe that was in full operation thousands of ages before man worked in metals, and more perfect than the best air-pump ever made. For there are extant many specimens of fossil cuttle-fish, the relics of one of which, by the way, are familiar to most people by the title of "thunderbolts"— long cylindrical bodies, composed of calcareous spar, pointed at one end, and slightly hollowed at the other, not unlike an elongated Minié bullet, and which, when cut or broken across, display a radiated structure.

If an arm of a cuttle be taken, and any one sucker examined, it will be seen to consist of a thick muscular membranous cup, having a cavity at the bottom, something like the chamber at the bottom of a mortar. The sucker should now be divided longitudinally, and then at the end of this "chamber" will be seen a soft muscular piston, exactly fitting the cavity. Now, if the circumference of the sucker be closely pressed against any substance of sufficient size and consistency, and the piston withdrawn, a vacuum is at once created, and powerful adhesion takes place. As, on an average, each cuttle is furnished with nine hundred suckers, the force of its hold may be imagined.

There is a substance that is often to be picked up on the shore, and oftener to be purchased at the perfumer's shops, known by the name of cuttle-bone, and when reduced to powder, used for various purposes. This so-called cuttle-bone is not bone at all, but a very wonderful structure, consisting almost entirely of pure chalk, and having been at one time embedded loosely in the substance of some departed individual of the species called *Sepia officinalis*. The "bone" is enclosed within a membranous sac within the body of the cuttle, by which sac it is secreted, and with which it has no other connexion, dropping out when the animal is opened. On taking one of these objects into the hand, its extreme lightness is very evident, and if it be cut across and examined through a lens, the cause of the lightness will be perceived. The plate is not solid, but is formed of a succession of excessively

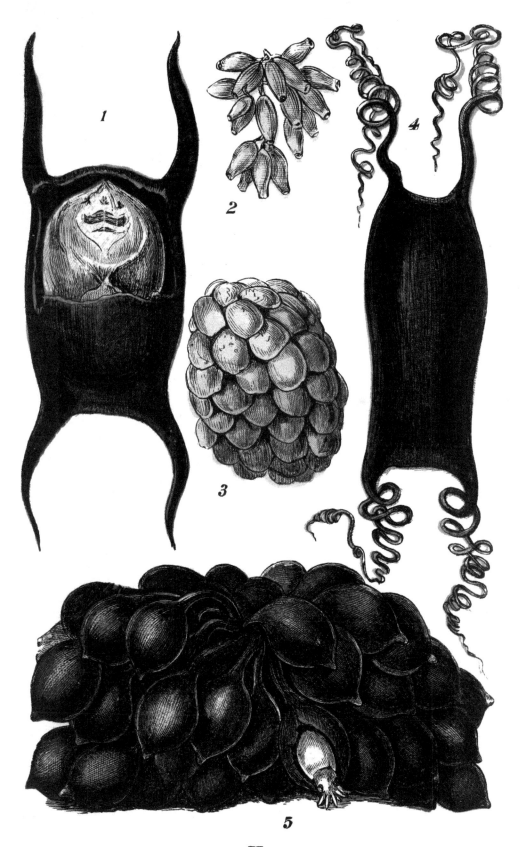

H

SEPIA.

thin laminæ or floors of chalk, each connected with each by myriads of the tiniest imaginable chalky pillars. When the cuttle is living, this structure runs through the entire length of the abdomen, being of equal length with it, and occupying about one-third of its breadth. In the Calamary the analogue of this object is of a horny consistence, semi-transparent, something resembling in shape the head of a spear, or the feather of a large pen, from which latter resemblance it is sometimes called the Sea-pen.

The well-known colour, sepia, is or ought to be manufactured from a black liquid, which is possessed by most of these creatures, and which can be ejected at will, probably with the view of darkening the water, and so temporarily baffling their enemies, of whom they have many, even including their own species. And it can also be employed while the animal is out of the water, as was once rather amusingly exemplified.

There was an officer employed, as I hope my readers have been and often will be occupied, in searching the coast for objects of marine natural history. After a while he came unexpectedly on a cuttle, who had taken up his abode in a convenient recess. The cuttle has a pair of very prominent eyes, and for a short time the cuttle looked at the officer, and the officer at the cuttle. Presently, the cuttle became uneasy, and taking a good aim at his military visitor, shot his charge of black ink with so true a range, that a pair of snowy white trowsers were covered with the sable fluid, and rendered entirely unpresentable. Even in many of the fossil cuttles this ink has been discovered dry and hard in its proper place within the creature. This most ancient substance has been removed, and ground down like very hard paint, and has been found to produce so beautiful a sepia tint, that an artist to whom it was shown inquired the name of the colourman who prepared it. And, in order to prove the character of the colour, a drawing of the fossil animal was made, and a description of it written, with its own ink.

Some of the Cephalopods are gifted with great powers of locomotion, and of those so gifted the Flying Squid is a good example. One of these creatures has been known to spring from the sea clear over the bulwarks of a ship and to fall on the deck, where it was captured. This specimen was six inches in length, and its habitation was the Pacific Ocean, lat. **34 N.**

The eye of the cuttle is a most singular organ,

its anatomy having long perplexed the dissectors, who could not conceive by what means the creature could see at all. It would be impossible here to describe this beautiful structure, and therefore I will content myself with observing that the celebrated Coddington Lens is merely a reproduction, though unwittingly so, of the lens belonging to the cuttle's eye. So here we have a pair of achromatic lenses and a series of air-pumps, contained in the structure of one creature belonging to the lower orders of the animal kingdom. Many other analogies exist, but space suffices not for them here.

The eggs of many fish are small and globular, being generally known under the name of spawn. The "hard-roe" of a herring furnishes a good type of this class of eggs. These, however, are the eggs of fish that are destined to be produced in countless myriads, and to serve as food for the other inhabitants of the deep, as well as for man. Uncounted thousands of these eggs perish before their maturity, being devoured by other fish which watch for them; and even when the young fry are born, comparatively few of them escape destruction. In order to compensate for such a loss of animal, the number of eggs is proportionably increased; one single cod-fish having

88   EGGS OF FISH.

been known to cast forth into the waves, in one single season, nine millions of eggs—equalling eight times the population of London.

But the destroying fish are not multiplied to such an extent, or the ocean would for a season teem with battles, and after a time be utterly

THE COD-FISH.

depopulated. The eggs of such a creature as a shark, for example, are singly committed to the ocean; and in order to prevent them from being carried about at the mercy of the waves, or thrown to perish on the shore, they are of a

DOG-FISH.

most singular form. An egg of one of our British sharks, the Common Dog-fish, is represented on plate H, fig. 4.

The egg is of a softish, horn-like consistency, so that it is not liable to be broken, or easily to be penetrated. The general shape of the egg has been aptly compared to a pillow-case, with strings tied to the corners; the enclosed pillow being the young shark. The long, curling, tendrilous appendages speedily affix themselves to sea-weeds, or other appropriate substances, and from their form and consistence anchor the egg

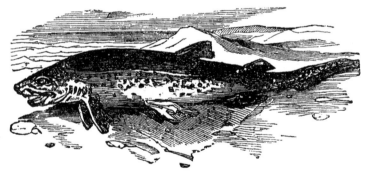

THE DOG-FISH.

firmly. In order to enable the little shark to breathe, there is an aperture at each end of the shell, through which the water passes in sufficient quantity to renovate the blood. And in order to permit the enclosed fish to make its escape when sufficiently developed, the end of the egg nearest to the shark's head is formed so as to open by the slightest pressure from within. After the newly-born shark has left the egg-shell, there is no perceptible external change in its shape, for the sides are elastic, and immediately close up as before.

These eggs are popularly known by the poetical title of Mermaid's Purses—alas, empty!

There is another and more common Mermaid's Purse, that is found of various sizes and of various tints, although it is usually of a very dark brown, closely approximating to black, and in length from three to five inches. It will be at once recognised by the figure on plate H fig. 1.

This egg is the production of one of the skates, and harmonises well with the strange, weird-like aspect of the creature from which it was produced. The accompanying figure represents the Thornback Skate (*Raia clavata*), a species that often attains a very great size, measuring some ten or eleven feet in length, and very nearly as much in breadth. From the four short arms with which these eggs are furnished, they are thought to bear some resemblance to hand-barrows; and if a fisherman is asked the name of the object, he will generally call it "Skate-barrer." If the egg is picked up in the early part of the year, it will usually be found to contain the young animal—not a very prepossessing creature—as may be seen by reference to the engraving, where a portion of the egg is represented as removed. Perhaps the reader may

SKATE.

remember Hogarth's "Gate of Calais," where a fisherwoman has upon her knees a huge skate, into whose countenance the painter has wickedly infused an expression precisely like that of the weather-beaten, withered old dame who holds it.

I was once talking about these eggs to some fishermen, who told me that in the spring they often found these eggs before the young were hatched, and were accustomed to boil and eat them just as hens' eggs are eaten. Whether to believe them or not I could not make up my mind, for fishermen are wonderfully loose in their details. However, as they gave me the information, I present it to the reader, and leave it to his own discretion to judge, or haply to his own energy to prove or disprove by actual experiment. I trust the latter.

In the summer months the eggs are invariably empty, or only filled with sand, little pebbles, and other shore *débris;* so that, unless the experiment can be tried in the earlier portions of the year, it cannot be made at all.

CHAPTER V¹.

SEA ANEMONES AND OTHER ZOOPHYTES.

THERE is a singular class of animals, called by the scientific name of *Anthozoa*, or living-flowers, because their formation and external appearance seem to partake much of the vegetative nature. Among the most conspicuous of the Anthozoa, are the creatures called Sea-Anemones, although they are not in the least like anemones, but do bear a very decided resemblance to China-asters, daisies, or dahlias. Very many species of these curious beings may be found on various portions of the British coasts, especially those which lie quite to the south; some of these creatures are only to be found in certain localities, but there are two species which are to be found on every coast where there is the best shelter. These two species I shall describe.

On plate E, figs. 4 and 5, is represented the commonest of the Sea-Anemones, or Actinias, as they ought to be called; one figure showing the animal as it appears when covered with water and all its tentacles expanded in search of prey; and the other, as it appears when closed, and at rest. I regret to say that it has an exceedingly long title, and one which takes some time to say as well as to write, "*Actinia mesembryanthemum,*" and which is generally curtailed in conversation or notes to "Mes," just as Mephistopheles was contracted to Mesty, by Mr. Easy, junior. Its popular name is the Smooth Anemone, it being so called from the smooth and slippery surface of its skin. When I say the "popular" name, I must be understood to mean the name popular among naturalists; for as to the people who live in close vicinity to it, they have no name for it at all. Very often, the fisherman has never seen such a thing as an actinia, or if he has been sufficiently observant to note it, he has no name by which he could describe it to another fisherman. Now, indeed, that so many enthusiasts crowd the sea-shores in their search after these actiniæ, and press fishermen into their service, a little knowledge on the subject is being gradually diffused; but some six or seven years ago, no one troubled himself about creatures which he did not catch with a net or line, and which he could not take to market.

The Smooth Anemone is generally the first

that meets an observer's eye. As he wanders about among the rocks and stones that are left dry by the receding tide, he will see on many of them certain little lumps of green or red jelly, varying in size from a pea to a plum. On touching them, they do not feel soft, like jelly, but are very smooth, slippery, and much firmer in consistence than would be imagined from their aspect. In this state, some of them are entirely green or red, but the greater part are marked longitudinally with lines of different colours according to the variety. For there is but one species of "Mes," although the variety of colour pass through several tints of green and red.

If they are to be examined, they can be detached without injury, by slipping the thumbnail or an ivory paper-knife under the base, and so gradually peeling them away from the support. Great care must be taken of the sucking-base by which the animal affixes itself, for a wounded base often causes death; and if the anemone has any nerves at all, as is but natural to suppose, their situation is on the base. Still, the process of peeling occupies some time, and if the tide is coming in, every minute is of importance. In this case, they can be removed instantaneously, by a judiciously aimed blow with the sharp end of the geological hammer, so that the portion of rock on which the animal is fixed is detached by the stroke. In this way a cargo may be collected in a very short time. They are generally found in colonies, consisting of five or six up to twenty or thirty in number; but now and then a giant may be found by himself or herself—for with them it is all the same thing—living in solitary state.

The smooth anemone seems to lead almost an amphibious life, for numbers of them are found affixed to stones and rocks in such situations, that they spend rather more time out of the water than in it. So that the rock is tolerably shady, the creature seems to be perfectly satisfied; but there are some hardy individuals who expose themselves to the full blaze of the sun. Even when they cannot be seen, owing to failing light, or rising tide, the touch will easily detect them, although in the latter case there is a slight chance of getting the finger nipped by a bad-tempered crab. I once filled a small basket with these creatures in a quarter of an hour or so, although the tide was driving me gradually to the shore, and there was not sufficient light to distinguish the hands of a watch.

The chief beauties of this, or indeed of any

ANATOMY OF ACTINIA.

other of its relatives, cannot be seen until it is removed from the sea, and placed in a vessel where it can be subjected to light and exposed to close examination. While it remains in its native rock-basin, or adheres to its opaque support, the light cannot shine through the creature, and its base, one of the most remarkable features of its structure, is concealed by the stone on which it rests. If, then, the observer wishes to examine the anemone more closely, he may easily do so by acting as follows:—

Let him procure any glass jar—a large one is not at all requisite—and having chipped off a morsel of stone on which is growing a frond of ulva or porphyra, place it conveniently in the jar, and fill up with sea-water. After an hour or two, the sea-weed will be studded with tiny air-bells; and when that is the case the water is fit for the reception of the anemones, of which there should be only one or two. If the anemones are merely dropped into the water, they will sink to the bottom, and after a little while begin to crawl up the sides of the jar. Now is the time for examination, and with a lens of moderate power much of the structure may be made out.

Perhaps the "Mes" is the least attractive of all the anemones, but yet it possesses some ex-

H

ANATOMY OF ACTINIA.

tremely beautiful features which are peculiar to itself. When the creature is fully expanded, a row of little globules will be seen round the edge, among the tentacles. They are usually about the size of a No. 5 shot, and are blue and bright as turquoise, to which jewel they bear some resemblance. Indeed, if there is any distinction between the two, the animal turquoise is more beautiful than the mineral.

If a dissection is contemplated (and it will well repay the trouble), the creature can be killed by placing it in fresh water. Its anatomical structure is very remarkable, and even the muscular structure, that enables the creature to expand or contract at pleasure, is worthy of observation on its own account, independently of the fact that the presence of muscle is a proof that there must be nerves by which the muscles are excited. Although the details are complicated enough, yet the general notion of the creature is sufficiently simple, and may be readily imitated. Let a linen bag be made, and the mouth of it sewn up. Then let the closed mouth be pushed inwards, until the bag assumes the form of a cup with double walls, and it is the type of an actinia; the outer wall being the exterior of the animal, and the inner its stomach. The tentacles are hollow, and communicate with the space between the walls. This space is always filled with water, and by the contraction of the walls water is driven into the tentacles, and so expands them.

This is a capital species for an aquarium, as it will bear travelling, and is very hardy, enduring extremes of heat and cold bravely, but perishing immediately in pure fresh water. A rather remarkable circumstance connected with these creatures occurred some little time ago. A gentleman had brought some of them to town with him, and had been examining them in company with a friend. After the examination supper was brought by an unsophisticated servant, and removed by the same individual. While the table was being cleared the servant asked what was to be done with the anemones, and was told to put them carefully away in a jug. Now the only jug at that time on the table was a jug containing porter, and into that jug the anemones were severally dropped. About a fortnight afterwards, the anemones were again called into requisition, and the jug demanded. Great was the astonishment of their owner to see the porter-jug produced, and still greater when he found the creatures were still living. They have also been known to live in soapsuds for a considerable time.

If specimens are gathered for an aquarium, they should be chosen of a moderate size, for the larger actiniæ require a considerable bulk of water. They are in general very migratory in their character, marching at their own sweet wills over the sides of the vessel in which they are confined, and now and then paying a visit to the rocks, stones, and shells in its centre. Sometimes they will attach themselves to a frond of ulva, whose broad leaf affords a good hold for their base; and there they will stay for weeks. Sometimes they will not take the trouble of crawling down the sides, and then over the bottom, but turn themselves into boats by hollowing the base, and thus floating with the base upwards, until they think fit to contract themselves, and to sink. They generally remain for some time partially supported by one edge of the base against the side of the vase, but after a while they commit themselves freely to the water. Fresh-water snails may be seen floating in a similar manner.

The base of the actinia, by which it moves, and of which it is so careful, is a very pretty object when seen pressed against the glass side of

a tank, and the light shining through its substance, showing the dark-green lines that radiate from the centre, and are so well contrasted with the azure gems that surround the disc. The mode by which the anemone travels is simple enough; it pushes forward one portion of the base, and, having fixed it firmly, draws the remaining portion after it.

There is a delicate gelatinous membrane that covers the entire animal, and which it frequently throws off. After an actinia has been sojourning in one spot for some time, and then moves away, it generally leaves a cast coat behind, as if to mark the exact locality of its habitation. Sometimes the creature appears to find a difficulty in getting rid of this membrane, which generally adheres strongly to the mouth. In such a case it is useful to assist nature, and a camel's-hair brush will generally give great aid to the actinia; who, in gratitude for help, expands itself immediately on being freed. Numbers of their cast membranes will be soon found in the aquarium, and should be removed.

If these creatures are kept merely for their beauty, they should be treated as greyhounds are treated; that is, kept almost entirely without food. They will live very well for many months without requiring food, but when it comes within their reach, they can devour and digest unlimited quantities. Perhaps the best plan is to give them a very tiny portion, which they eat, and which only stimulates them to protrude their hungry arms in the hope of getting more. Whenever I fed my own specimens, I generally gave them small pieces of beef, which they swallowed, and after a few days rejected, having extracted all the nourishment, and only left a white fibrous mass.

They were also very fond of flies, and ate many of them. Once a great bluebottle fly came buzzing into the room, and made such a disturbance that I immolated him, and gave him to the largest and hungriest anemone. About three or four days afterwards I saw the bluebottle floating on the surface of the water. On lifting it out of the tank it fell to pieces, being in fact the mere shell of the fly; all the interior having been, by some mysterious process, extracted in the stomach of the anemone. It is amusing enough to see the way in which an actinia eats. If a fly or a piece of meat be presented to its tentacles, it is instantly seized by them, and drawn to the mouth, the tentacle closing upon it on all sides. The animal then literally tucks in the morsel, and with it all the tentacles and the upper portions of its body, until they reach the stomach. Digestion then goes on very quietly, the presence of its own arms in its stomach being of no consequence at all to the animal; and in due time it untucks itself, and tosses away the indigestible portions of its food.

The "mes" multiplies readily in a tank. At the beginning of last September I had fifteen specimens in the aquarium, and by February in the present year there were between forty and fifty, many of them very minute, and very transparent. Indeed, scarcely a day passed without the discovery of a nursery full of little pink or green actiniæ, contained in an empty shell, or studding the surface of a shady pebble. Funny little things they are, and have a most consequential look as they spread out their tiny tentacles in search of food.

When the "mes" is caught for the purpose of stocking an aquarium, it will travel well if wrapped in wet sea-weed. If care is taken, the ulvæ, or enteromorphæ, that are used for this purpose, can be transplanted into the tank, and so two objects secured at once. The best plan is to put some wet green algæ at the bottom of a basket, and then to lay on them the anemones.

each wrapped in wet sea-grass; over them another layer of green algæ should be placed, and so will they be quite comfortable.

The second species of Sea Anemone which I shall describe is, in my opinion, the most magnificent of the family, whether size or colour be the criterion. It is possessed of a scientific name hardly inferior in length to that of the smooth anemone, being called by the learned, *Bunodes crassicornis*. It is much too long a title for every-day use, and so it is contracted into "Crass." A portrait of a half-expanded crass may be seen on plate E, fig. 9.

This creature does not bear exposure to air and heat so well as the smooth anemone, and must be sought for in the shallows at low water. Sometimes at spring-tide a solitary specimen is seen, high, dry, and discontented; but such is an exception; usually it may be found just beyond ordinary low water-mark, expanding its gorgeous tentacles, and waiting for the numerous crustaceans or fish that are always left in the tide-pools.

The colours of this animal are very varied, hardly any two specimens being found of exactly the same tint, and the magnitude of its fully expanded disc is nearly equal to that of the crown of an ordinary hat. Some of them are scarlet, some pink, some lilac, some delicate grey, some of an olive-green, and all so delicately transparent that no colour can faithfully represent their beauty. On referring to the engraving, the reader will see that the base of each tentacle is surrounded with a pear-shaped, dark line. It is on this line that the depth of colour is chiefly lavished, the tentacle itself being always much fainter in tint, and as transparent as if formed from gelatine. The tentacles are very thick in proportion to their length, and it is from that peculiarity that the creature derives its name of "crassicornis," signifying "thick-horned."

The voracity of these animals is quite surprising. I have often amused myself by watching them in their native haunts, and experimenting upon their powers of digestion. One single "crass," measuring barely three inches in diameter, required two crabs, each the size of a penny-piece, and a large limpet, before it ceased to beg with extended arms.

It is evident by the fact of the crab-eating that the crass must possess great powers of grasp, or it could never hold, retain, drag to its mouth, and finally devour, a creature of such strength as a crab of the size above-mentioned. Such a crab struggles with great violence, and requires a very firm grasp of the human hand to prevent it from making its escape. And yet the anemone, whose entire body is not larger than the closed hand, and whose substance is quite soft, can seize and retain the crab, if it is unfortunate enough even to thrust one of its legs within reach of the tentacles. There must, therefore, be some strange power by which this object is achieved; and the mode by which it is accomplished I now proceed to describe.

It has been long known that many water-inhabiting animals, of a class so low as to be scarcely more than water animated, can throw out long fishing-lines, of a substance so delicate as only to be descried by the sparkle of light upon them as they float about in the water, and that when these delicate lines even touch a little fish or crustacean, some power destroys that fish as effectually as if it had been struck by lightning. But this is not the mode of attack with the anemone, which resorts to means purely mechanical.

If the finger is brought into contact with the outspread tentacles of a healthy anemone, it will adhere to them with a peculiar, almost indescribable sensation. It is not in the least the adhesion of gum, glue, or any such substance, but bears some resemblance to that which is felt when the finger is thrust into the mouth of the common ringed snake, or of several fish. The mode of adhesion is only to be discovered by the aid of a tolerably powerful microscope, and when discovered is found to be equal in beauty to any structure that is yet known.

Scattered at intervals over various portions of the body, and especially crowded on the tentacles, are found tiny organs that are called by the name of thread-capsules. These are little oval vesicles, imbedded in the substance of the anemone, and containing within them a long delicate thread, closely coiled, in forms varying according to the description of capsule. The extreme tenacity of the thread may be imagined from the fact that the largest capsules are not more than the three-hundredth of an inch in length, and that within so small a compass the thread is coiled like a watch-spring in the barrel. Indeed, the simile of a watch-spring will nearly express the object, for the thread is so strong, in spite of its tenuity, that it has aptly been compared to the hair-spring of a watch. When the tentacles are irritated or compressed, myriads of these capsules start forward, become everted, and shoot forth their tiny spears. The length and shape of these wonderful filaments are very various, some being of a very great length, and so fine that a microscope of high power can hardly distinguish them; while others are only two or three times the length of the capsule that contained them, and covered with an armature of short hairs even more minute than themselves.

It is easy enough to see these singular organs, even with the aid of a good hand-lens; but with a microscope nothing more is required than to cut off part of a tentacle, get it well in the field of the microscope, and then apply pressure: some of the capsules will dart forth their threads almost immediately, but others will require greater force before they will evert themselves.

The crass may often be passed by without observation if it does not choose to display itself; for when closed nothing appears but a round mass of sand and broken shells, about the size of a penny-piece, and projecting so slightly above the sand that an inexperienced eye would see nothing remarkable in it. This unpromising aspect is assumed when the creature has had a large dinner, or when it is alarmed. It is rather curious to see how suddenly a magnificent specimen of living flower collapses into a shapeless, and not at all pleasing knot of sand, stone, and

RAPID CHANGE OF FORM. 109

shell. The crass is a delicate animal to preserve, for it seems to require a large body of pure water for respiration, and if in the least injured does not recover from the wound like the smooth anemone. If it were as easily detached as the "mes" there would be less difficulty in preserving it; but it has an unpleasant habit of forming its base into some six or seven lobes, attaching four or five of them to separate stones or portions of rock, and pushing the others into any crevices that may be convenient.

Very seldom, indeed, are the fingers alone sufficient to extricate the creature without injuring the base, and unless that important part be preserved in its integrity the crass assumes various wonderful shapes, and very soon dies. In such a case it generally begins by puffing out several striped lobes from its mouth, which project, and soon assume so enormous a size that they quite overshadow the tentacles, and render it a matter of some small difficulty to ascertain clearly whether the animal is not reversed. After it has thus displayed its own temper, it proceeds to try that of its possessor, by assuming various forms with such rapidity, that to draw it with any accuracy is quite out of the question. In the morning it may present a splendid and regular disc of tentacles, expanded to their utmost, and glowing with scarlet, pink, or lilac: just as a rough sketch of the creature is taken, it half withdraws the tentacles, and begins to puff out the striped lobes or lips. Another sketch is now taken of the crass, as it appears when in a pouting humour; but there is no time for any colouring, because the perverse animal now begins to take in water to an alarming extent, and soon succeeds in turning itself into an animated hour-glass, hardly a vestige of tentacle or pouting lips being visible. Even this form does not seem to please it, and is speedily exchanged for another, which in its turn gives place to a fifth, and so on *ad infinitum*.

In order to avoid the danger of tearing the very sensitive base, I was in the habit of digging out all the stones and pieces of rock to which the crass had affixed itself, and permitting the creature to free itself in the tank; a feat which it generally soon accomplished. By taking these precautions I succeeded in bringing a very large and gorgeous specimen to town, and preserved it alive and healthy for upwards of six weeks, in one of the most crowded parts of the city. It would probably have lived much longer, had not the density of the water in which it resided been corrected too suddenly. The animal was in a large vase of sea-water, which necessarily presented a wide surface to the air; and so the water rapidly evaporated, leaving all its salts behind, and rendering the remaining water much too dense. In the correction of this evil the fresh water was added too rapidly, and inflicted on the nervous system of my poor crass a shock from which it never recovered.

In such a case the tentacles begin to droop, they then lose their beautiful transparent colouring, and become dull, opaque, flaccid, and exceedingly small. Whenever the creature is in this state it is not easily restored to health and brilliancy.

This species does not travel in wet sea-weed nearly so well as the smooth anemone, and besides, pours out a vast amount of mucus, which makes the whole of its neighbourhood exceedingly unpleasant. In order to transport it in perfection it should be indulged with a jar of sea-water, and have no travelling companions. Its tentacles are so strong, and the animal is so voracious, that it will frequently destroy any other creature that happens to be placed in the same vessel. The very specimen which I have described, killed, during the few hours of its journey, a beautiful Æsop prawn, on which I rather prided myself, and a young gurnard. It did not eat either of these creatures, but simply caught them, and retained them until they were dead. How they were killed I do not exactly know; but it is suggested with some reason, that the capsular threads convey with them a kind of poison, that is effectual enough among the smaller animals, but not sufficiently powerful to affect human beings.

The coating of little stones and broken shell has already been mentioned. These substances are evidently chosen instinctively with a view to concealment, and are fastened to the body by little sucker-like protuberances, with which the greater part of its surface is studded. If the animal is used for culinary purposes, for which it seems to be adapted as well as the oyster or the periwinkle, this shelly coating must be removed; an operation which is easily enough effected by the fingers, and not so tedious as plucking a fowl. There is no difficulty in finding a sufficient number to form a respectable dish, for any one who knows where to search, and what to see, may capture an unlimited supply in an hour or so.

I conclude this short history of the creature by

observing that the reader must not expect to find the crass presenting precisely the appearance of the specimen here depicted, for the whole of the body is generally invisible when the creature is offended, part being buried in the sand, and the remainder overshadowed by the thick and close-set tentacles. If, however, it is carefully removed, and takes up its residence in a jar or tank where there is no sand, it then assumes the shape of the figure, at all events for a time; but after a little experience of prison it changes its shape so frequently that it almost realizes the fable of Proteus, and his efforts in a similar difficulty.

On plate L, fig. 1, is a representation of another British zoophyte, allied to the anemones, but yet distinct from them. It is the Common Madrepore (*Caryophyllia Smithii*), a member of that wondrous family that produces the coral and similar substances.

When removed from the water, or even when alarmed, the animal portion of the madrepore suddenly shrinks, and little is visible save a series of thin calcareous plates, standing on edge, and radiating from a centre. But when the creature has recovered its self-possession, and begins to feel rather hungry, a beautiful semi-transparent living substance emerges from the chalky plates, and after a while puts forth a number of tentacles, tinted with most delicate hues, and much resembling those of the anemones, except that each tentacle is terminated by a little globular head. These tentacles, like those of the anemones, are covered with filiferous capsules, and adhere to the hand in much the same manner, though not so strongly. This madrepore is voracious enough in its own way, but does not seem to care very much for food supplied artificially. I had a specimen alive for some months, but could not get it to eat any but the minutest portions of meat, while it would have nothing to do with a small fly; yet it was healthy, and almost constantly protruded its transparent tentacles. The proximate cause of its death appeared to be attributable to a bad-tempered Daisy Anemone (*Actinia bellis*), which lived in a cave like a hermit, and did not approve of intrusion. I am accustomed to stir the water daily, in order to imitate as far as possible the natural stir of the sea, and in so doing this madrepore was washed into the cave where resided the daisy anemone. Although it was speedily replaced in its own little home, which had been specially chiselled in the rock, it never properly recovered;

and after leading a dull, inactive, colourless existence for a week or two, fairly died, and left me nothing but its skeleton as a memorial.

The madrepore may be found adhering to, and in fact almost forming part of, the rocks, requiring the aid of a strong knife to detach it without injury. The specimen represented in the engraving is rather larger than the general run, and only exhibits the tips of the tentacles, and the expanded membrane that edges the calcareous plates. When the creature is dead, or alarmed, the arrangement of these plates is plainly visible, reminding the bystanders of a reversed mushroom suddenly petrified.

There is a certain substance often found on the shore, tough, soft, fleshy, and unprepossessing, well deserving the popular name that is given to it, namely "dead man's fingers," or "dead man's toes," as the case may be. For, just as the smooth anemone when touched collapses into a shapeless lump of green jelly; as the "crass" shrinks into the sand, an almost undistinguishable and unrecognisable excrescence; as the madrepore retires within its own chalky walls, so does this zoophyte withdraw all its beauties in the uncongenial conditions of heat and drought, and only present an exterior anything but agreeable to sight or touch. The scientific name of this zoophyte is *Alcyonium digitatum*, or the Finger-shaped Alcyonium, and a representation of it may be found in plate E, fig. 8, accompanied by a magnified sketch of one polyp.

In this and other zoophytes, the qualities, fraternity and equality, are exhibited in a manner far superior to any republic, ancient or modern; but there is very little liberty in the case. In these curious creatures communism prevails to its fullest extent, one for all and all for one. There is one body, so to speak, the polypidom as it is called, and from this body protrude, under favourable circumstances, innumerable polyps, each one gathering its nutriment from the surrounding water, and conveying that nutriment not to its own body only, and for its own aggrandisement, but into the general polypidom, affording to each of its thousand relatives a portion of its own nourishment, and receiving from each of them some modicum of their own.

When placed in clear sea-water, the alcyonium soon begins to put forth a few crystalline columnar polyps, each standing boldly out, and bearing a mouth or head, composed of eight radiating, slender, pointed petals, fringed with delicate hairs. The internal anatomy of this

K

creature, or rather of this creature-mass, is very interesting, and worth studying practically.

On plate E, fig. 1, is shown an example of a plant-like, compound animal, very common on the coasts, either thrown on the shore dead or dying, or affixed to large algæ, near low water-mark. This is the Sertularia, a beautiful family of zoophytes, of which some sixteen or seventeen species are found on our own coasts. The species represented is perhaps the commonest of all, *Sertularia filicula*.

If one of these creatures is examined by the aid of a moderately powerful lens, it will be seen to consist of a horny, many-branched stem, each branch being studded with a double row of little cells, open at the mouth, which is much smaller than the base. If the creature is placed in clear sea-water, and still watched through the lens, each cell will be seen to protrude a tiny polyp, whose star-like head is all that is visible externally. The polyps are easily alarmed, and in such a case withdraw themselves wholly within their cells. There is a very similar zoophyte called by the name of *Plumularia*, which may, however, be easily distinguished from Sertularia, by the position of the polyp cells, which only occupy one side of the branches;

whereas those of Sertularia are to be found on each side equally, sometimes in pairs, sometimes alternately, according to the species. When dried, both these creatures retain their chief characteristics, and if any of the sea-weed landscapes be examined, both Plumularia and Sertularia will generally be found among the algæ. They are delicate in constitution, and not easy to keep in an aquarium, unless under singularly favourable circumstances.

There is another pretty, plant-resembling zoophyte, found plentifully enough near low water-mark, but at a rather higher elevation than the preceding. This is known among zoologists as *Coryne pusilla,* and is chiefly remarkable on account of the peculiarity from which it derives its name. "Coryne" is a Greek word, signifying a club, or knobbed stick, or more properly a mace, such as the steel-headed, spike-armed weapons with which our gentle ancestors were in the habit of exciting the brains of their adversaries.

The stalk of this zoophyte is about as thick as ordinary sewing thread, and it clings to the seaweeds among which it resides as much as the cotton in question would do. Even when seen with the unassisted eye it is rather an elegant creature, but when a lens is brought to bear upon it, sundry hidden beauties become obvious, and among them that peculiar formation of the polyp from which it has derived its name of Coryne. Its club-shaped head is studded with numerous tentacles, that are arranged in a manner somewhat similar to the steel spikes of the war mace. Each of the tentacles is furnished with a globular head, and if submitted to a higher power of the microscope, appears covered with minute knobs, at the extremity of each of which is a short straight bristle. The stalk is merely a horny tube, ringed in structure, and increasing in size towards each polyp head, so as to allow room for them to change their position. The movements of these creatures are not very rapid, but can be clearly seen. A magnified representation of a single polyp head accompanies the figure of the zoophyte.

We now come to one of the most remarkable objects in the whole range of animated creation, and which requires a microscope of some power to develop. On plate E, fig. 7, may be seen a kind of miniature tree; this is the representation of an elegant little zoophyte that is found plentifully near low water-mark. It is small, seldom exceeding two or three inches in height, and is of a delicate and feathery texture. To the naked eye there is nothing of any great importance in this creature, but if a portion of a branch be detached and brought under the field of a microscope, a very strange sight meets the eye. As is the case in all these zoophytes, the branches are studded with cells, in which live little polyps, at one time expanding wide their feathery tentacles, and at others sulkily gathering them up like an unsuccessful fisherman gathering his net on his arm. But to each of the cells is attached a most singular appendage, precisely resembling the head of a bird and one joint of its neck, by which it is attached as a point to the cell, and on which it works. There is certainly no eye in the head, but there is a most decided beak, which opens and shuts precisely like the beak of a bird, while the entire head keeps up a continual nodding backwards and forwards on its joint. If the portion selected is tolerably healthy, it will contain from ten to fifteen, or even more, of these birds'-heads, bowing to each other in the most oppressively polite style, and every now and then shutting their beaks with a sharp snap.

By the side of the zoophyte itself is shown one of the bird-head appendages, as it appears when fastened to the cell. The object of these strange organs is not at all clearly ascertained, for they seem to have but little connexion with the polyps that inhabit the cells, and bow with as much perseverance when the cell is empty as when it is occupied by its living inhabitant. Many zoophytes possess the bird's-head; but as that species which I have mentioned is perhaps the most common, and is easily detected, it has been admitted as the representative specimen. It must be understood that when these creatures are subjected to the action of a microscope, they must be well supplied with water, or they will die speedily. If the power is not very high, the entire creature may be placed in a flat glass cell, and slightly compressed against the side by a glass plate. The microscope should then be set horizontally, and will show the structures tolerably well. But if a higher power is requisite, a portion must be removed, and placed in the animalcule cage with which every good microscope is furnished. A flat watch-glass, and a piece of the thin microscopical glass used for covering preparations, will make a good extemporised animalcule cage, if the regular machine is not at hand.

Among the other members of the animal kingdom that are popularly ranked as vegetables are the *Flustræ*, for an example of which see plate E, fig. 2. These creatures do, indeed, much resemble the leaf of some plant, and so closely that uneducated people can hardly be made to believe their animal origin. If, however, the fingers are passed over the surface of the flustra, and especially if they are passed from the point of the leaf towards the base, a peculiar rough, harsh, and stony sensation will be perceived, and this sensation is caused by the innumerable spine-crowned cells with which the leaf is covered, and, indeed, of which it is composed. A close examination with the naked eye shows that there is a curious structure not usually met with in plants; but if an ordinary pocket-lens is brought to bear on it, the entire surface will be seen to be composed of little oval cells, arranged in rows something like the scales of a fish, or tiles on a house-top. Each cell is armed with four short, sharp spines, that project from the upper portion of the edge, two at each side, and these spines are the cause of the peculiar rough sensation that is communicated to the finger. The cells are placed back to back, like those of a honeycomb, so that there is no right or wrong side to the flustra leaf; for leaf it must be called, although its proper title is polypidom.

The species given in the plate is *Flustra foliacea*, a very common zoophyte, and often found on the shore, thrown up by the sea in large masses.

On many sea-weeds may be found a kind of stony scurf that spreads over their leaves or stems, and often destroys the beauty of the specimen. It is true that the appearance of the alga may be injured, but the creature that injures it is of so curious and beautiful a form that it ought to be preserved. This stony scurf is called Lepralia, and consists of innumerable cells, not unlike those of the flustra, spread evenly over the surface of the substance to which it adheres, and often so thickly that there is hardly a spot left uncovered. I have now a remarkably fine specimen of *Delesseria sanguinea*, measuring eighteen inches across, the whole of whose stem, and great part of the leaves, is overrun by several species of Lepralia, which, although they certainly rather disfigure the plant as a dried specimen, look so lovely under the microscope that I would not on any account have them away.

A common species of Lepralia is depicted on plate E, fig. 6. The upper figure represents the zoophyte as it appears when magnified about thirty diameters, equalling nine hundred times superficially; and the lower figure shows its appearance when slightly magnified by a simple pocket-lens. Almost every sea-weed of any dimensions, and especially the *Laminariæ*, will be partially covered with the cells of this zoophyte, and it may also be found on shells and other objects which have been submerged in the sea for any length of time. There are nearly forty British species of this single genus, and if half a dozen specimens are examined, it will probably happen that three or four will be distinct species.

CHAPTER VII.

STAR-FISHES AND SEA-URCHINS.

PEOPLE seem to have a strange love for comprehending various descriptions of objects, whether animal, vegetable, or mineral, under a single term, and generally contrive to hit upon a word which could not be rightly applied to any of them. Take, for example, the word "fish," as we are speaking of marine objects. Not to mention the whale, and other cetaceans, which are popularly called by the name of fish, we have lobsters, crabs, shrimps, oysters, limpets, mussels, &c., all comprised in the term "shell-fish." Then, we talk of cray-fish, cuttle-fish, jelly-fish, and starfish, not one of the whole party having the very smallest right to the title of fish. Still, custom has so inextricably woven the name into the idea, that they cannot well be separated, and therefore must be retained until the same power shall unweave its own web. These prefatorial remarks must be my excuse for employing the word "star-fish" in the present case, and "jelly-fish" in a succeeding chapter.

Every one has heard of star-fish, and most people have seen them, either in a preserved state, or as they appear when thrown up by the waves. There are very many British species of star-fishes, but out of them I have chosen three, as types, to which, indeed, most of the species can be referred. The commonest of the British star-fishes is the Five-finger (*Uraster rubens*); for a figure of which see plate L, fig. 4. There are few days when some of these creatures are not cast on the shore, and there left by the retiring tide, so that their habits and anatomy may be easily studied. Generally they appear to be dead, and, indeed, sometimes are so; but their apparent death is often but quiescence, and if they are placed in sea-water they become lively in a very short time.

If a star-fish is thus rescued, and laid in a shallow rock-pool, where its movements can be watched, it will give ample food for contemplation, were it only for the mode in which it moves from one place to another. This movement is very slow, gentle, and so regular, that the eye cannot detect any motive power at work. Should a stone, a ridge of rock, or any other

MODE OF PROGRESSION. 127

impediment, be in the path of its progress, the star-fish does not seem to trouble itself in the least, but continues its still, gliding movement, as quietly as if it were moving on level ground. As the stone is reached, one ray of the star-fish is gently pushed upwards, and seems to adhere to the stone; another follows, then a third, and presently the creature is seen to climb the stone with quite as much ease as if it were walking on level sand. The rays accommodate themselves, in a very curious manner, to the shape of the substance over which the animal is crawling; so that if it is passing over a sandy spot, interspersed with furrows and pebbles, the arms of the star-fish never bridge over the furrows, but pass down one side and up the other, while precisely an opposite process takes place with regard to the pebbles. The star-fish can thus climb rocks that are perpendicular, and clings firmly even when they overhang. How this process is conducted we shall presently see.

I may as well remark here, that if the star-fish were dead when put into the water, the observer would find his patience well tried if he waited to watch for these movements, so that he ought to be quite sure whether he has hit upon a specimen that is living. Now, however dead a star-fish may appear to be, if it is of a tolerably firm consistence to the touch, it is a living being, even though there should be no perceptible movement. If, however, in taking it up, it hangs loose and limp, life has departed, and it can only be used as a specimen for preparation, or for anatomical purposes. Any doubt will soon be settled, by placing the dubiously vital animal in clear sea-water for a few minutes, and then suddenly turning it over. If there is any life remaining, the numerous feet that occupy the under surface will move about, and the creature will soon recover its wonted activity.

When a living star-fish is laid on its back, a number of semi-transparent, globular organs will be seen in constant movement, being thrust forward and then withdrawn, moving from side to side, as if feeling for something, as indeed they are. These are the ambulacral organs, as they are scientifically called, but I prefer to call them feet, on account of their office. These feet are, in fact, suckers, and can be protruded or withdrawn by a very curious piece of mechanism, which is not easily described without the use of diagrams, but which I will endeavour to explain, as far as possible, without them. The feet are hollow tubes, each passing into the interior of the animal through a circular aperture, and being furnished with a globular, membranous head, just within the skin, filled with a fluid; so that, in fact, each foot or sucker bears some resemblance to a brass-headed nail driven through the skin, the head remaining within the animal, and the nail itself projecting. Now, if the creature compresses the membranous head, the fluid contained within it, being comparatively incompressible, seeks an exit, and finds none except the hollow of the tube. Through this tube it accordingly runs, and so pushes forward the sucker which terminates the foot. When the pressure is removed, the fluid returns into the head, and the sucker is retracted.

The mouth of the star-fish is placed underneath, and in the very centre of the body, the stomach being immediately beyond the mouth, as is the case with the sea-anemones. The stomach does not seem to occupy very much space, but it is capable of accommodating a large amount of nutriment, in which it is assisted by certain supplementary stomachs, which run through each ray, nearly to its extremity. These supplementary stomachs, or cæca, as they are called, may be seen by slitting up the skin of the upper surface of the rays, when the cæca will be seen lying immediately beneath, looking more like dark, loose, unformed masses of liver, than mere appendages to the stomach. And it is a very remarkable fact, that although these large and very important organs exist in each ray, the star-fish appears to be indifferent to the loss of one or more rays, and fills up the wounded space so perfectly that it is hardly possible to distinguish the spot where the ray once was. This circumstance accounts for the fact that star-fishes, apparently perfect, but only possessing four rays, are sometimes met with, whereas the minimum of number is five. When the star-fish has been seen to cast away its rays in captivity, the amputated rays still continued to move their feet-suckers as when they were attached to the body, but they did not appear to be capable of continuing their march.

Small as the mouth of a star-fish appears to be, small as is its stomach, and feeble as are its muscular powers, it can swallow a bivalve mollusc entire, or if needful, open it, and suck out the contents in some mysterious way; a feat that no man could accomplish without tools. Even with the proper knife, oysters are not very easy to open without some practice; but if a man's food were restricted to oysters, which he must open without the assistance of any tool, he would run considerable risk of starvation. The ancient naturalists were well aware that the star-fish possessed the power of eating oysters, but they thought that the creature accomplished its design by watching until an oyster opened its shell, and then poking one of its rays between the shells as a wedge; then, having once gained a partial admission, it slowly insinuated itself, and finished by devouring the inhabitant. It, appears, however, by the reports of careful observers, that the oyster-eating is true as to the fact, but false as to the mode. The star-fish seems to bring its mouth in contact with the edge of the shell, and then from some delicate vesicles never protruded at any other time, to pour into the oyster some drops of a poisonous fluid, which forces the animal to open the shells, and finally kills it. Such is the account as it stands at present.

The skeleton of the star-fish is one of the most complicated structures imaginable, much too complicated for description here. It may easily be obtained by any one who wishes to possess such an object, if he takes a perfect specimen of the creature, and places it near an ants' nest. In a very few days, the ants will nibble it to pieces with their sharp, sickle-like jaws, and eat away every particle of the soft portions, leaving only the skeleton, which will then look like a singularly beautiful specimen of carved ivory. Ants, by the way, are very useful insects to the naturalist, and are capital skeleton developers. Only they do *not* store up the food in their subterranean mansions, as is popularly imagined; for as they feed on animal substances, and not on corn, their stores would soon be exhaled in the form of gas. There are always plenty of ants' nests near the coast, and it would be useful to look out for them as soon as possible, taking care to choose those that are not exposed to the public gaze, or near a public path. I would recommend the use of a box, perforated with many holes, as a convenient mode of keeping the specimens from dust, and at the same time of permitting free access to the ants. I had, until lately, an exquisite skeleton of a lark that had been prepared in a similar manner.

The colour of the five-finger star is generally a dusky red on the upper surface, the colouring matter of which is sometimes irritating to those who possess delicate skins. Sometimes, however, specimens are found of a purple or violet hue and are by some authors considered to be a distinct species, although they are probably but

L

a variety of the common red five-finger. This species, especially if large, is not very suitable for an aquarium, and seldom survives for any length of time. One individual that I tried to domesticate had, for the last few days of its life, a curious habit of resting merely on the points of its rays, and elevating the disc in the centre, so that it presented somewhat the aspect of a five-legged table, the rays forming the legs and the disc the table itself.

There is another tolerably common species, that is found on the shores, and is very different from the Five-finger, being composed of a large disc, with twelve short pointed rays proceeding therefrom; so that when a large specimen is seen in scarlet splendour on a rock, it seems to blaze out like the sun, and has accordingly been called the Sun-star. It is not easily mistaken for any other creature, but in order to make its recognition easier, a figure of it will be found on plate L, fig. 5. Its scientific name is *Solaster papposa*. Its usual colour is a bright red, but it is often seen to be tinged with violet, while in some specimens the rays are very much paler than the disc, and one most singular example has been recorded of a mixture of bright green.

Occasionally near low water-mark may be found specimens of a very curious star-fish, differing as entirely from the sun-star, as it does from the five-finger. This is the Brittle-star, of which there are several British species: the commonest of them, *Ophiocoma rosula*, is given on plate L, fig. 3, and, as will be seen, is a very curious creature; its form has been well described by the image of a little sea-urchin, surrounded by five very lively centipedes. Indeed, it hardly resembles the sun-star at all; but these creatures assume such singular shapes, that forms the most dissimilar are found actually to be closely united; this we shall see presently when we come to the urchin. The *Ophiocoma* is called the brittle-star on account of its inexplicable custom of breaking itself into little bits when alarmed. It is really a matter of some little difficulty to secure an entire specimen, so that a really perfect brittle-star is rather a valuable acquisition, though the creatures are so common that a dredge will haul them up by pailfuls. One of the largest British species of star-fish, *Luidia fragilissima*, a creature measuring some two feet across, possesses this suicidal property in a high degree. In Forbes's British Star-fish, a work which I strongly advise all naturalists to obtain, or at least to read, is a most ludicrous account of an adventure with a *Luidia* and a bucket.

When undisturbed in their own element, the brittle-stars are well worthy of observation; for their long fringed arms wriggle about with great vivacity, and well carry out the simile of the centipede. In order to destroy these creatures without damage, a vessel of fresh water should be brought to them, and if they are rapidly submerged, the saltless fluid destroys them before they have time to discover that there is anything wrong. In the case of the *Luidia*, however, the sight (if star-fishes see) of the fresh water was so alarming that the precautions were useless.

If the star-fishes are needed for the cabinet, they must be dried; an easy process enough, but requiring patience. They must first be thoroughly washed in fresh water, in order to get rid of the salt, and then carefully spread out on a clean smooth board, and dried in the open air. They should not be placed in the cabinet until they are thoroughly dry, or there will be sad damage done.

I said just now that we should soon find that forms, apparently dissimilar, were in reality closely connected with each other; and this fact we shall see exemplified in the creature that next comes before our notice, the Common Sea-egg or Sea-urchin (*Echinus sphære*), whose external appearance is shown on plate L, fig. 2. There does seem to be some slight connexion between the three star-fishes which we have just examined; but that there should be any connexion at all, or any relationship, between the brittle-star and the sea-urchin, appears too preposterous an assertion for credibility: yet such is really the case, as will soon be seen. The specimen from which the drawing was made is a tolerably perfect one, being still furnished with its array of spines from which it derives its name of sea-urchin, the urchin being a popular name for the hedge-hog. But if these spines are rubbed away, a smooth surface will be left, on which are numerous tubercles, marking the spots on which the spines formerly rested. If now the reader will take a damaged urchin, plenty of which are to be found on the shores, and examine its external appearance, he will see that it has a very close relationship indeed with the star-fishes. Let a common five-finger star-fish be laid on its back, and the points of its rays stitched round a little disc of leather, it will then assume very much the aspect of a skeleton urchin. Let then the spaces between the rays be filled up with a substance of the same

SHELL OF ECHINUS. 137

structure as the rays themselves, and then we have a complete Echinus, complete, at all events, as to its general external appearance.

The specimens that are found cast up by the waves are generally destitute of the spiny armature that is found upon them in their living state, and thus permit the eye to perceive the formation of the shell. Close by the figure of the Echinus itself will be seen a little diagram composed of several angular forms studded with little tubercles. These represent the pentagonal plates of which the shell consists, and which are most wonderful instances of animal economy. In the shell of every Echinus are hundreds of these plates, varying in size according to their position, and so closely connected with each other that externally their marks of junction are not perceptible; but if the shell is broken, and examined from the interior, the shape of these plates becomes tolerably well defined. It will be observed also, that when the shell is broken, the serrated edges of the fractured portions show the angular form of the plates. As the shell is composed of these plates, it may well be asked how the creature can possibly increase in size, because it cannot, like the lobster and other crustaceans, throw off its old coat when too small, and take to itself a better; and to add to the difficulty, there is no supply of arteries and veins ramifying through these plates, as is the case with the bones of a vertebrate animal, but each plate is dense and dead.

In order to overcome these apparently insuperable difficulties, a very beautiful arrangement takes place. The delicate living membrane with which the entire surface of the body is covered insinuates itself between the edges of these plates, and continually deposits round the margin of each particles of calcareous matter; so that each plate simultaneously increases round its edge, and the original form of the shell is preserved.

If we still keep before our eyes the image of the rolled-up star-fish, we shall see that as the mouth is precisely in the centre of the disc, it would also be found in the centre of the Echinus shell. And such a mouth as it is could hardly be conceived. If a human being, say a man of six feet in height, were to be possessed of a similar mouth, it would be about the size, and very much the shape, of an ordinary wooden pail, the teeth being as long as the staves of the pail, only they must be made very sharp at the top, and but five in number. The teeth of the Echinus may be seen protruding from the mouth, and their extreme hardness may be tested by the finger without any danger. The entire arrangement of teeth and muscles, and bony scaffolding, is so exceedingly complicated, that even with the help of diagrams it would be difficult to explain the structure, and without their aid quite impossible. I may, however, mention, that there is some resemblance between the teeth of the Echinus and those of rodent quadrupeds, there being a provision for adding fresh substance to the tooth as fast as it is worn away by use.

If the reader will now examine the interior of an Echinus shell, he will see that it is marked out into five equal parts, by five double rows of perforated plates, containing many hundreds of very minute apertures. Through these apertures protrude sucker-feet, just like those of the starfish, which have already been described, and worked in the same manner.

Reverting now from the interior to the exterior, we shall find its surface thickly studded with spines, which, as well as the suckers, are employed as a means of locomotion, and therefore must be freely movable. If a single spine be removed, and note taken of the part which it previously occupied, it will be seen that on the shell is placed a rounded tubercle, and that the base of the spine is furnished with a hollow socket into which the tubercle fits, so that the spine has perfect facility of movement. The spine is bound to the tubercle by a short tendinous ligament, connecting the centre of each, much as is the case with the larger joints of vertebrate animals. The power of motion is communicated by the membranous covering that envelopes the body during the life of the animal. The spines of some foreign species of Echinus are very delicate and sharp, piercing the unwary hand like so many needles.

Besides all these multiplied means of progression, there are other very tiny organs, which may possibly be used as assistants for the same purpose, or they may possibly perform some office at present unknown. Among the spines there may be seen, with the assistance of a lens, a very great number of little three-headed pincers, standing each upon a flexible footstalk. These are called *pedicellariæ*, and are also found on several of the star-fishes. Some naturalists have regarded them as distinct animals, residing parasitically upon the Echinus. The Echinus is often boiled and eaten, just like eggs; from which circumstance it is sometimes called the

Sea-egg. All these creatures are called by the general name of *Echinodermata*, signifying " urchin-skinned." We have already seen how great is the apparent distinction between the several creatures that are classed together under this title, and this outward distinction is quite as great in the last example that will be here mentioned.

There is a curiously-shaped creature represented on plate F, fig. 6, looking something like a cucumber, with a feathery fringe attached to one of its ends. This is popularly called, on account of its shape, the Sea-cucumber, and is scientifically termed *Holothuria*. The derivation of the word is Greek, but its signification is very uncertain. It was used by Aristotle in his " History of Animals;" but the reason why he so named the creature no one can tell.

The holothuriæ are very curious creatures, for they possess some organs of the Echinodermata, by virtue of which they rank as Echinoderms, and they also have other organs, which seem to imply a connexion with the sea-anemones, while the shape somewhat approximates to the annulate form of the worms. Like the star-fishes and the Echinus, they possess five rows of sucker-feet along the body, although in some species these sucker-feet are scattered over the entire surface. If a holothuria be opened, almost the entire cavity of the body is filled with small white tubes, which are apt to tumble out, and become inextricably confused, if care is not taken. Indeed, at first sight, a freshly-opened sea-cucumber reminds one of the famous cucumber in the "Arabian Nights," which was stuffed with pearls by command of the talking bird, only that the sea-cucumber appears to have been stuffed with white bobbin. These white threads are the egg tubes. Altogether, it does not present the most inviting aspect to the eye, nor does it appear to be a very suitable object for the table. Yet it is one of the favourite dishes of that omnivorous nation, the Chinese, who pay large sums for fine specimens.

When the holothuria feels unwell, or is displeased, it has a very remarkable habit of dispensing with its teeth, stomach, and entire digestive apparatus, and so converting itself into a mere empty bag, with an useless mouth at one end of it. However, animals of this order are not easily killed, and before very long a fresh set begins to grow, and in a few months the holothuria is as perfect as ever. The beautiful feathery plume that surrounds the head, or rather the mouth, is the organ that has caused some naturalists to class the creature with the actiniæ or anemones, to whose tentacles the plume bears so close a resemblance. In general, the body of these animals is too thick, and the skin too tough, for the adoption of the suicidal habits of the star-fishes; but in some of the species, whose diameter is very small in proportion to their length, the creature actually does succeed in breaking its body into several fragments. The reader may compare with this habit the similar custom that prevails among many lizards, of snapping their tails off if they are touched or suddenly alarmed; as is especially exemplified in the case of the Common Blindworm, which is often known to break itself across, as if it had been made of glass. If any of these creatures are found in a living state, they will not at first put forth their tentacular crown, and the owner must be content to wait. But if they are properly supplied with clear and pure seawater, they will generally exhibit a large portion of the tentacles, if not the whole. They may usually be found clinging firmly to stones and pieces of rock, in situations where they are not exposed to light; for the influence of light seems to be exceedingly painful to them.

CHAPTER VIII.

ANNELIDS—BARNACLES, AND JELLY-FISH.

On plate F may be seen some figures of strange-looking creatures, having a kind of general resemblance to each other, but belonging to very different ranks in the animal kingdom. They are, however, placed near each other, in order to show how the same idea of form runs through different genera. The figure on the left is no worm, although it bears some resemblance to the creatures whose portraits occupy the top and right of the same plate. No. 3 is a mollusc, ranking with the periwinkle, mussels, nudibranchs, and other creatures, which have already been described in Chap. II. Then, again, the figure occupying the bottom of the plate has rather a worm-like aspect, and would bear even a close likeness, if it were much longer in proportion to its diameter, as is the case with some of its congeners. This animal, however, belongs to the star-fishes. The central figure, which is really one of the worms, looks much more like a common garden slug than an earth-worm; to which latter creature, however, it is in near relationship. One may therefore easily pardon the errors of the earlier naturalists, who were deceived by the external form, and classed the creatures according to shape, and not according to anatomical structure.

To begin, then, with the worms, or annelids, as they are called, being composed of a series of rings bound together by muscular and tendinous substances. The insects and many other creatures, by the way, are also composed of a series of rings; but they possess jointed limbs, and by virtue of those limbs occupy another place in the system of living beings.

The commonest of all the terrestrial annelids is the earth-worm; and there is a marine earth-worm that corresponds with its terrestrial relative in habits and uses. On the sand may often be seen little heaps of contorted sandy strings, possessed of no compactness, but dispersing when touched, and looking as if Michael Scott's familiar were hard at work at his task of twisting ropes from sea-sand, and throwing down his abortive attempts. These ropes or strings are the sand-casts of the lug-worm, a creature that is possessed of no particular beauty, but is very useful to the

L

fishermen, who use it as bait, much as the earth-worm is used by fresh-water anglers. Parties of boys may be seen, armed with spades and boxes, trudging knee-deep in the muddy sand-flats, as soon as the tide goes out, in full search after logs, as they call the worms. Although the shape of this worm is not very beautiful, yet it is not utterly devoid of some beautiful features; for the double row of scarlet branchiæ, or lung tufts that fringe the central portions of the creature, are remarkable for their brilliant tints.

While speaking of this worm, and its representative, the earth-worm, I may as well mention that the popular idea of the multiplication of th earth-worm by division is quite erroneous. The general notion on this subject is that if an earth-worm be cut in two near the middle, the divided portions reproduce those organs which they have lost, and so in a short time the earth is richer by one more worm than before. This notion, however, is untrue. The severed worm seldom seems to recover in the least from its wound, although the portion on which is placed the head survives longer than that to which the tail is attached. The ring next to the wound very soon dies, contracts, withers, and drops off by mortification. The next ring is then attacked in the same way and dies in its turn. And so on in both portions, the anterior perishing from the wound towards the head, and the posterior portion from the wound towards the tail. The latter portion, indeed, loses the power of locomotion altogether, and can only twist and wriggle about on the spot where it is placed. If only a portion of the tail end be cut off, the remaining part has sometimes sufficient strength to heal the wound, and the creature survives; but the wounded portion is not capable of producing a fresh tail, or even of forming a single fresh ring.

If the sand or stones be carefully examined at low water, certain curious objects will often be found between tide-marks, sometimes existing singly, but generally living in societies. One of these objects is represented on plate F, fig. 2. It is a tube composed of innumerable fragments of shell, or sometimes of entire shells, if they are sufficiently minute, grains of sand, and other similar substances, agglutinated together by a secretion that is poured from the surface of the body, and which soon hardens into a tough membranous substance. The mouth of the tube is adorned with a fringe, which is composed of a number of much smaller tubes formed from the same substances, and in the same manner, as the principal tube. The creature that inhabits this dwelling goes by the name of Terebella. Its empty tubes are sometimes torn away from their attachments by the power of the waves, and in this case are thrown upon the sea-shore together with the algæ, shells, and other *débris* that mark the line beyond which the proud waves can no further go. Generally, however, they are fixed with such firmness, that to procure an entire specimen is a matter of some difficulty. There is no connexion between the tube and its inhabitant, who seems on occasion to be able to take little journeys among the rocks, and even to swim on the surface of the water by spreading its numerous tentacles abroad as the molluscs spread their foot. Sometimes the terebella becomes ambitious, and instead of contenting himself with sand and tiny stones, affixes a stone of some size to his tube. One that I possessed for some time had fastened the centre of its tube to a pebble more than an inch in length, and very nearly the same in width.

The reader will not fail to remark the analogy between the tube-inhabiting annelid and the larvæ of the common caddis, or stone-fly of anglers, which build tubes in a very similar manner, pressing into the service all sorts of

SABELLA. 149

substances, and which, like the terebella and others, has no organic connexion with the tubes in which their soft bodies are sheltered from danger. The tube that is represented in the engraving is of the natural size, and was drawn, as indeed were most of the figures, from an actual specimen.

On the same plate as that which is occupied by the terebella, and at fig 5, may be seen a group composed of innumerable tubes, massed together as if they had been a handful of earthworms compressed into a small space, and then suddenly liberated. These are the tubes of another annelid, called the Sabella, and, like those of the terebella, are built up from the particles of sand on and among which the worm lives. At the bottom of the mass may be seen one of the worms crawling from its tube. These tube-masses may be found in abundance at low water-mark, especially where the corallines are plenty; and the size of the masses is very various, some being only composed of a few tubes twisted together, while others are several feet in diameter. It is not often that a fragment is found where the tubes are so plainly shown as in the specimen depicted. Generally the surface of the mass somewhat resembles a sponge with circular apertures, with here and there a tube, or a portion of a tube, twining itself into the substance. Various algæ are often found affixed to the tubes of these creatures.

Another of these tube-inhabiting worms, or Tubicolous Annelids, to use the correct term, forms a shelly tube so closely resembling that of the ship-worm, *Teredo navalis*, that the two are often confounded with each other, especially if a portion only is in question. This is the Serpula, a group of which is given on plate F, fig. 1, the species being *Serpula contortuplicata*. There are several species of this curious and beautiful worm, one of which, and perhaps the most common, possesses a bayonet-shaped shell, which twists about on the surface of stones or other convenient substances, and does not erect itself freely. But the species that will be more particularly noticed here, after taking a turn or two upon its support, as if to obtain a firmer basis, and at the same time to determine its direction, shoots boldly upwards.

Now if a group of these tubes, situated, we will say, on an oyster-shell, be taken into the hand, they will all appear to be empty and useless; but if the tube is not very much contorted, a something scarlet may be seen at some little distance down the tube, and by that sign the living state of the inhabitant may be known. When the serpula is placed in clean sea-water, it generally remains quiet for a few hours, as if to make itself acquainted with the atmosphere of its new home; by very slow degrees the scarlet object is pushed nearer and nearer to the mouth of the tube, and at last emerges, when it is seen to be an exquisitely formed, conically shaped cork or stopper, its small end being prolonged into a kind of footstalk. Two of these stoppers may be seen on reference to the engraving, but they are hardly represented of sufficient size in proportion to the diameter of the tubes. After a little time, a row of scarlet feathery objects slowly follow the stopper, and in a few minutes spread themselves out into a most elegantly shaped plume.

Slowly as the serpula protrudes itself from the tube, it is by no means slow in retreating. If one of these creatures is fully extended in an aquarium, and the hand is rapidly moved outside without even touching the glass, the worm pops back into its tube with marvellous rapidity, so rapidly, indeed, that the eye fails to follow the movement, and the creature vanishes as if by magic. A cloud passing over the sun, or even the shadow of a person passing by, will have the same effect. It seems evident, therefore, that the serpula must be able to see, although, as yet, no eyes seem to have been discovered. After awhile, however, the creature appears to become partially tame, so to speak, and is less alarmed at a casual movement or shadow. Such at all events was the case with my own specimens, which at first were painfully shy, and avoided all close inspection, but, after a fortnight or so, permitted me to place a lens sufficiently near them to examine the beautiful plumes and stopper.

This last-mentioned organ is the developed one out of a pair which the creature possesses, the other being very small and not put forward to view. This may remind the reader of an analogous arrangement in the tusks of the Narwhal; in which cetacean there are really two tusks, but one only is fully developed, the other lying concealed in the jaw. The beautiful fan-shaped plume is composed of that part of the breathing apparatus which separates the oxygen from the water, and is analogous to the gills of fishes, or the lungs of man.

If serpulæ are kept in an aquarium, they should be closely watched, as they, in common with the sabella and others, have a bad habit of dying when they are not suspected, and so tainting the water, to the destruction of animal life. Most of the tubicolous worms come out of their houses before they die, but the serpula often excepts himself from the general rule, retreats into his shell as far as he can go, and there dies. It is very difficult to discover whether the animal is really dead or only sulky—if the latter, he recovers his temper in a day or so, and waves his plumes as usual; but if the former, a white film begins to collect over the mouth of the tube: this must be accepted as a hint for instant removal. In general, if a serpula does not spread its fans boldly and decidedly from the tube, and permits the stopper to droop over the mouth, it should be gently touched with a camel's-hair brush; if it does not smartly shoot back into its tube, that serpula is in a bad state of health, and must be looked after. It is always better to remove the creature at once, than to run the risk of tainting the water with the unpleasant smell that immediately follows upon the death of any marine animal.

On plate F, fig. 4, may be seen a figure of a creature that does not look at all prepossessing, yet this very animal is one of the most gorgeous creatures that can be imagined, the metallic brilliancy of whose colouring would not suffer in comparison with the plumage of the brightest humming-bird. This animal is popularly known by the name of Sea-mouse, its scientific title being *Halithœa*, or *Aphrodite aculeata*. Edging the body may be seen rows of bristles or hairs which, when simply printed in black and white, give no idea at all of the iridescent colouring of their surfaces; while even, if coloured, the resemblance is but faint, because it wants the changing tints which flash along the hairs whenever they are moved.

It is a strange thing, and one that shows the lavish beauty of creation, that an animal endowed with such glorious colours, that can only be exhibited by a full supply of light, should have its habitation in the mud. When kept in an aquarium, they generally appear to avoid the light rather than to seek it, and keep themselves so hidden among the weeds and stones, that it is not always an easy matter to find them. They are rather migratory in their habits, but not erratic, for they seem to go over the same course week after week; so that, having seen them on one day, it is not difficult to predict their locality on the next.

The bristles of the aphrodite are not only

worthy of notice on account of their wonderful colouring, but also on account of their shape. Among other offices, they seem to play the part of weapons, like the spines of the porcupine or hedgehog. But as they surpass the hedgehog's quills in external beauty, so do they in form. There are certain islands, called "Friendly," whose amicable inhabitants are famous for the ingenuity of their clubs wherewith to knock out another friend's brains, and of spears wherewith to perforate him. Many of these spears are made with rows of several barbs, one placed immediately above the other, in order to add more destructive power to the weapon. Now, if the Friendly islanders had possessed microscopes, we might with some justice have said that they took their idea of the many-barbed spear from the bristles of the Sea-mouse; for a magnified representation of one of these bristles would give a very fair idea of the Friendly lance.

All these lances can be withdrawn into the body of the sea-mouse at the will of their owner, and it would therefore be a most unpleasant circumstance, if the creature were to wound itself with its own weapons. In order, therefore, to obviate this difficulty, each spear or bristle is furnished with a double sheath, which closes

156 ACORN-SHELL.

when it is retracted into the body, and opens again when protruded. It is hardly possible to conceive a more wonderful structure in the whole of the animal kingdom, and certainly not possible to conceive one more beautiful, when the changing tints of orange, scarlet, or azure are taken into consideration.

There is a kind of slimy muddiness about the back of a sea-mouse that rather counteracts the beautiful effect of its hairs. This is caused by the muddy soil in which it loves to reside, and which is strained through a dense mass of fine hairs that interlace with each other, and arrest the muddy particles, while they permit the water for respiration to pass between them. The whalebone plates that fringe the mouth of the Greenland whale have a somewhat similar office, only the fringe of the whale catches molluscs, and that of the aphrodite catches mud.

Wherever rocks are found between tide-marks, their surfaces are usually selected as resting-places by some very curious animals, known by the name of Acorn-shells, which will at once be recognised by the sketch on plate N, fig. 3, where is represented a group of these creatures which have affixed themselves to the shell of a limpet. In the original specimen, the entire surface of the limpet was covered with acorn-shells; but one or two were removed, in order to show the nature of the substance on which they rested. Their scientific name is *Balanus balanoides*, and they belong to a class of molluscs that are called Cirrhopoda, on account of the cirrhi, or ciliated arms, which form their chief characteristic.

The first acquaintance that is usually made with these animals is seldom of an agreeable nature, and generally takes place after the following manner:—An inexperienced but earnest observer is picking his way among the rocks and stones at low water, his eyes being more engaged in searching for curiosities than in looking after his own feet. Suddenly, he puts his foot on a sloping rock, rendered slippery as ice by the slimy algæ that cover it, or is inadvertently caught in a rocky pitfall, whose orifice was concealed by the heavy masses of wrack or tangle that are flung over it by the tide, and at the bottom of which is a pool of water just deep enough to wet his feet, and to irrigate his body by spirting up along the sides of the cavity. In either case he catches frantically at the nearest piece of rock, and finds his fingers cut in several places by the sharp edges of the acorn-shells that have there affixed themselves, and present as uncomfortable a hold for the undefended hand as a wall tipped with broken bottles.

While they are left in the open air, there is nothing attractive in the balani, which seem rather to disfigure the rock than to improve its appearance. But when the sea returns and brings back the welcome supply of nourishment, these dull, lifeless objects suddenly start into activity, and begin to fish as industriously as if they knew that they had only a limited time for eating, and must, during that time, procure a sufficiency of food to employ their digestive organs while the tide is out. The manner in which the Cirrhopoda fish is very remarkable. Some animals, like the sea-anemones, hang out a net and await the approach of prey, who unwarily come within the scope of their power, and so rush to their own destruction. Other creatures hang out fishing lines, like the common fresh-water Hydra, or the beautiful marine Beroe, which will be described on a future page. Others, again, chase their prey through the water, and capture it by virtue of superior swiftness or cunning. But neither of these modes is employed by the balanus, which is furnished with a veritable casting-net, which it ever and anon throws expanded into the water, and then retracts when closed. The action of a man throwing an ordinary casting-net is much the same as that of the acorn-shell. If the reader will refer to the plate, he will see two of these creatures in the act of making their cast, and the net is formed from the cirrhi from which the entire class derives its name.

Each cirrhus is found, on examination, to be double, the pair springing from a single footstalk. They are of a partially horny consistence, and separated into numerous joints, each joint being furnished with long stiff hairs. These hairs stand out boldly from the centre, and the consequence is, that when the whole apparatus is fully extended, it forms a kind of network of hairs, permitting none but the smallest substances to pass between them. In the balanus, the cirrhi are of a delicate white colour, and have a singularly elegant appearance as they are alternately thrown abroad and gathered together again. There is hardly a prettier sight than a large stone, or piece of rock, that is covered with balani, and immersed in clear sea-water. Each little conical shell opens at the tip, and from the aperture a fairy-like little hand is constantly thrust, grasping at some coveted object, and then closed and withdrawn. There is a grace and elegance about the whole movement that is not easily described. This sight may often be witnessed in the rock-pools, when they are of sufficient depth to cover the balani, and are not exposed to the action of the wind.

With all their beauty, however, the balani are uncongenial inhabitants of an aquarium, although they add much to its appearance at first. They soon become languid, their graceful cirrhi remain half protruded from the shell, they then die, and shortly exude such a detestably-scented gas, that the surrounding water soon becomes unfit for the respiration of the other inhabitants—they in their turn die, and the whole aquarium is ruined. So, present beauty must be sacrificed to ulterior service; and if any balani are growing on a rock intended for the aquarium, they must be removed before it is placed in the tank.

They do not seem to be particular as to their place of residence, for they may be found on rocks, wooden piles, stones, and even on living shells, of which they most affect the limpet, because it is not of migratory habits.

It is a very remarkable fact, that although the balanus never moves from the spot on which it has taken up its habitation, and, indeed, is incapable of any kind of locomotion, yet when very young it was an active, wandering little creature, furnished with jointed limbs, much resembling a young shrimp or crab, and swimming freely through the water with a succession of bounds. When first discovered, the young balani were thought to be veritable crustaceans; but after careful observation they were seen to affix themselves to the sides of the vessel in which they were placed, and straightway to change their roving life for an existence of settled quiet. Similar strange developments take place in many marine animals, but there will not be sufficient space for their discussion.

The balanus has a very near relative going by the popular name of Ship-barnacle, and the

BARNACLE.

scientific title of *Pentalasmis anatifera*, the latter title signifying "the five-plated goose-bearer." It is called Pentalasmis, or five-plated, because its shell is composed of five distinct portions, curiously arranged, and between them the cirrhi are protruded. The word "goose-bearing" is given to it because an old writer named Gerard, who lived in 1636, discovered that the Bernicle-

BERNICLE-GOOSE.

goose (*Bernicla leucopsis*) was produced from the ship-barnacle; and in order to prove his own rather startling account, he gives drawings of the creatures in all their stages, from the mollusc to the bird. Whether the worthy man intended to deceive, or was himself the victim of others, it is impossible to say. His account is so quaint that I here give an extract:—

"What our eyes have seen, and hands have touched, we shall declare. There is a small island in Lancashire called the Pile of Foulders, wherein are found the broken pieces of old and bruised ships, some whereof have been cast thither by shipwracke, and also the trunks and bodies with the branches of old and rotten trees, cast up there likewise; wherein is found a certain spume or froth, that in time breedeth into certaine shels, in shape like those of the muskle, but sharper pointed, and of a whitish colour; one end whereof is fastened into the inside of the shell, even as the fish of oisters and muskles, the other end is made fast into the belly of a rude masse or lumpe, which in time commeth to the shape and form of a bird: when it is perfectly formed the shell gapeth open, and the first thing that appeareth is the aforesaid lace or string; next come the legs of the bird hanging out, and as it groweth greater it openeth the shell by degrees, till at length it is all come forth and hangeth only by the bill: in short space after it commeth to full maturitie, and falleth into the sea, where it gathereth feathers, and groweth to a fowle."

The anatomy of this barnacle is curious, and will repay examination. The shell should be removed, and the animal carefully displayed with the assistance of a pair of scissors and a needle or two: all dissections of small animals should be made under water; or if the dissected creature is intended to be permanently preserved, it should be immersed in the fluid which is used as a preservative.

These creatures are often found clinging in great numbers to the bottoms and keels of vessels, sometimes interfering with their speed. Their growth is very rapid, and it has often happened that a ship has started upon a short voyage without a single barnacle adhering to her planks, and yet has come back encumbered with a whole army of them. They are often found adhering to pieces of wreck, or to floating spars that are cast upon the shore by the waves. The stalks or tubes of the individual represented in the engraving are not sufficiently long in proportion to the dimensions of the shells themselves, and ought to have been drawn nearly double their length.

After a gale, especially if the wind sets landwards, the shores afford a great harvest to the naturalist ; and if the gale and the spring-tides coincide, he inwardly wishes for the hundred eyes of Argus to look after the objects that lie scattered on the shore, and for the hundred arms of Briaræus wherewith to pick them up. Among the strange things that are cast on the shore will be seen many lumps of a jelly-like substance, varying much in size, called popularly by the

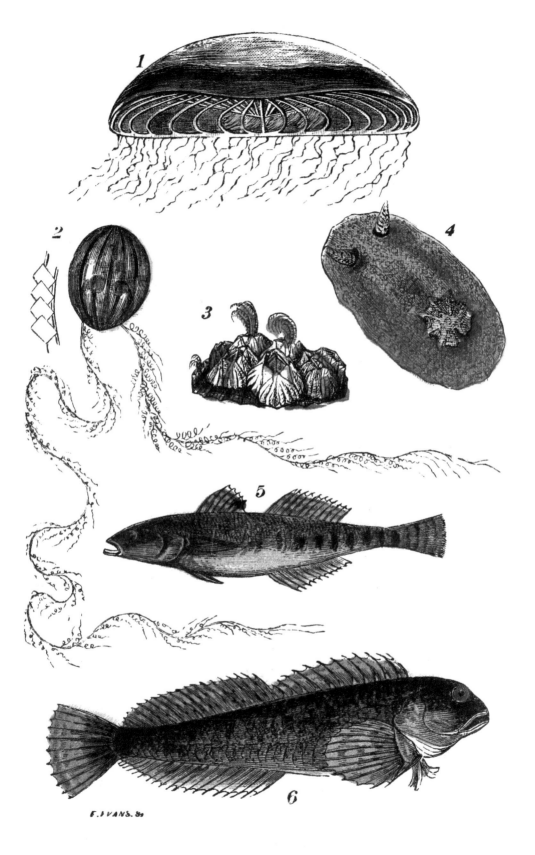

name of Jelly-fish. These are strange creatures, of wondrously low organization, that thickly populate the ocean, and are anything but shapeless when living and in health. Scientifically they are called *Acalephæ*, from a Greek word signifying a nettle, because many of the species have the power of stinging the hand that incautiously touches them.

Sometimes they lie on the shore in vast numbers; and there is a story on one of our coasts, that a farmer ordered down his carts to the sea, and carried away several cartloads of jelly-fish to serve as manure for his fields; but by the next morning the heaps of jelly-fish had disappeared, leaving behind them a few lumps of membranous threads. In fact, all the real animal matter that the carts had carried to the fields might have been conveyed in the farmer's own hand, for jelly-fish are really little but animated sea-water.

On plate N. fig. 2, will be seen a singularly shaped creature, bearing two long threads covered with spiral tendrils; this is one of the jelly-fishes, called by the name of Beroe or Cydippe, and a wonderful creature it is. If on a calm day a gauze net is passed gently through the water, there will often be found adhering to its sides sundry little gelatinous knobs, perfectly transparent, and apparently lifeless. Now, if the net be lowered into a glass vessel of pure sea-water, and slightly agitated, the lump of jelly will be loosened, and left in the water. For a moment the eye fails to perceive that the water has any inhabitant at all; for the beroe, as the gelatinous knob turns out to be, is itself little but sea-water, but may soon be recognised by the flashes of light that appear on its surface. It is a creature that can hardly be drawn, for it ought to have no outline, and only to be shown by the brilliancy of its surface, which surpasses that of the water around. Presently, as the creature begins to feel more at home in its new habitation, it swims about with an easy gliding movement, and an iridescent light shows itself on one part of the surface. The iridescence continues to increase, and at last is seen to reside in eight longitudinal bands that completely encircle the animal; over these bands the light plays, and at last all the colours of the rainbow ripple over its surface with indescribable beauty.

These iridescent bands are the organs of locomotion, and it is to their form and mode of use that the beautiful colour is owing. By the side of the beroe may be seen a magnified portion of one of these bands. Its surface is covered with little square scales, disposed in a manner somewhat similar to the boards of a water-wheel. Each of these steps, so to speak, is capable of motion backward and forwards, and by their rapid and successive motion a series of prisms are formed, and by them the light is decomposed to the prismatic colours; this iridescence is best seen when the sun shines upon the creature.

When the beroe has been watched for a little time as it swims about in its glass prison, two long and most delicate threads will be seen depending from its exterior, and falling into graceful curves as the creature ascends or descends in the water. The threads are so exceedingly delicate that they are not observed at a first glance; and when they are seen, rather convey to the spectator's mind the idea of spun glass, than of any animated structure. Indeed, the whole creature looks as if it were formed of crystal, cut and polished, and the threads almost seem to be spun from its substance as it moves about. These threads are called the fishing-lines, and if closely watched are found to be fringed with smaller tendril-like threads, that are dispersed along the chief line, just as a fisherman attaches several baits to his line by supplementary strings. The fishing-lines can be entirely withdrawn into the body of the animal, or they can be shot out to lengths that appear wonderful, considering the size of the creature to which they belong. The supplementary tendrils elongate themselves when the fishing-line is drawn out to its full length, and become more tightly twisted as the line is retracted into the body. Many people, and especially those who live on the sea-shore, imagine that the beroe is the egg of the sea-urchin.

It must be remembered that the creature can alter its shape by expansion and contraction; which circumstance accounts for the fact, that if several artists sketch this creature, each figure may have a different form. The species represented in the engraving is *Cydippe pileus*, shown as it appears when fully expanded. The life of the creature is fragile as its form; and if it is kept in a vessel of water, it plays about rapidly for a time, then dies, and disappears as if it had melted into nothing. Yet, if it be cut into pieces while lively, or broken up by the force of the waves, as is often the case, its ciliated bands still continue to perform their work, and the iridescent light plays over the fragments as beautifully as when the creature was entire. It is seldom or never found in an entire state at the surface of the water when the wind is rough, but sinks below into the calmer regions, where its delicate organization is not exposed to the rude collision of wave and wind.

The beroe glides along by means of the ciliated bands; but this is only one of the means of progression employed by the Acalephs. Some move themselves about with cirrhi, and are therefore called Cirrhigrade; the beroe being called a ciliograde. Others again are named Physograde, because they are buoyed up by a kind of bubble, or bladder filled with air. The well-known Portuguese Man-of-war is a good example of this order. There is another order which moves through the water by a series of regular pulsations like those of the lungs, and the species belonging to it are called, in consequence, Pulmonigrade. An example of a pulmonigrade Acaleph is given in plate N, fig. 1. It belongs to the genus Ægerea, of which many species may be found on our coasts, and, with many others, passes under the general title of Medusa. The size of these creatures varies excessively, and there are strange peculiarities in their growth and structure, which should be learned from some one of the books devoted exclusively to the Acalephs. The movements of a little Medusa in a clear glass vessel are exceedingly graceful, and may easily enough be witnessed; as if a vessel is filled with water drawn from the surface of the sea on a calm day, there are generally a few Medusæ in it. But if there should be none, a little work with a gauze net will secure plenty.

Very many of these Acalephs are phosphorescent, and through their instrumentality the sea appears at night as if filled with fire. This property is especially shown when a little breeze ripples the surface, or a boat dashes the water aside. In the latter case the oars appear to throw from them torrents of fire, and a blazing line marks the direction which the boat has taken. The chief cause of this phosphorescence is a very tiny creature, called *Noctiluca miliaris*. If a vase be filled with sea-water and placed in the dark, it will emit small sparkles of light whenever it is tapped, or even when the foot is stamped on the ground. This phenomenon is more exemplified at the sides of the jar, and but a very few sparks are perceptible at the bottom. Each spark is caused by a Noctiluca; and if it is wanted for examination, it may be caught in a glass tube, in the manner employed for microscopic animalcules, and brought under the necessary magnifying power. Its natural size is about half that of a common mustard-seed. When submitted to magnifying power, it appears to be a creature of form nearly round, but with a notch or depression on one part of its circumference. Close by that depression is a little knot of muddy matter, from the centre of which springs a kind of tail, or perhaps proboscis, by the agitation of which the creature rows itself about in the water. The entire form of a Noctiluca is not unlike that of a melon, the proboscis being the stalk.

CHAPTER IX.

CRABS—LOBSTERS—SHRIMPS—PRAWNS, AND FISH.

Among the living creatures that force themselves on the notice of any one who walks on the borders of the sea, the various crustaceans are perhaps the most conspicuous. Without attempting here to treat of the Crustacea scientifically, I shall mention those creatures that may be seen almost on any day and almost on any shore, leaving the deeper scientific details to be obtained from the very elaborate works that exist on the subject.

It is nearly impossible to walk for more than a few paces on the wet space between tide-marks, without disturbing a host of little crabs, that scuttle about in dire perplexity, either trying to flatten themselves against the ground, hoping to be mistaken for pebbles, or endeavouring to conceal themselves under the shade of a bunch of wrack. Sometimes, on lifting up a heavy mass of seaweed, out comes a crab very unexpectedly, holding up a pair of claws with so ferocious an air

EDIBLE CRAB.

that he often escapes before his discoverer has quite recovered his presence of mind. The former species, that try to escape in such a hurry, are generally the uneatable green crab, although even it is often found of no small size; while the big pugnacious one is of the edible kind, and, from his objection to capture, seems to know it. Sometimes crabs of a very tolerable size may be found concealed in the crannies of the rocks, where they are concealed by the rock itself, and by the fuci

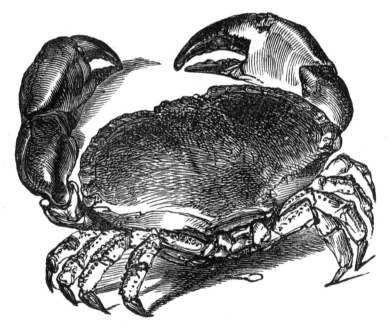

CRAB.

and laminaria that hang about in great profusion. Bare-legged boys may be often seen creeping about among the rocks, and armed with a basket and an iron rod hooked at the end. This latter weapon is ever and anon thrust into the holes and clefts of the rock; and should an unfortunate crab have there concealed himself, he is soon hooked out of his retirement, and, if edible, consigned to the basket.

The great monsters that are brought to market are mostly caught in sunken baskets, much on the principle of the mousetrap, which permits an animal to enter without any difficulty, but opposes an effectual barrier to his egress. It is not always safe to grope for crabs, unless a companion be at hand, for a powerful crab has actually been known to grasp with its claws the hand of its opponent, and to hold him there without the power of moving until a passer-by came to his assistance. Should such a circumstance occur, the best plan for making the animal loosen its hold is said to be by detaching the claw that is unemployed.

The young of the crab is quite unlike the adult animal, and has been described under the name of Zoea. In this state it is a very quaint-looking creature, is possessed of a long tail, two great eyes, something like those of a diver's helmet, and wears a spike on its throat nearly as long as its entire body. It is no marvel that it has been treated of as a separate creature from the crab, for it bears about the same resemblance to the crab that a caterpillar bears to a butterfly, or a wire-worm to a beetle. The long tail of the Zoea formed one of the distinctive points that separated it from the crabs; and yet, if a crab is laid on its back, a tail is seen to be tucked up under its body, in a position something like that assumed by the tail of an alarmed dog.

Many species of crabs may be found on the seashore; that is, if they are sought in the proper localities. The two species already mentioned are totally incapable of swimming. They can crawl upon the shore, half bury themselves in the sand, or push their way among the algæ with much rapidity; but if they are thrown into deep water, they sink helplessly to the bottom, spreading about their limbs in the vain search after some object which they can grasp. There are, however, several species of crabs found on the British shores, which are good swimmers, one of which is given on plate M, fig. 4. This is the Velvet Swimming Crab, or the Velvet Fiddler, as it is sometimes called. If the figure of this animal be compared with that of the common crab, on p. 173, the reader will observe that there is a considerable difference between the two creatures; one of the chief distinctions lying in the shape of the hinder pair of legs, which in the common crab are sharp and rounded, but in the swimming crab are flattened at their extremities. These flattened limbs are used as oars or paddles, and by their repeated strokes the creature is able to urge itself through the water with some velocity. The peculiar movement of the limbs being thought to resemble the action of a violinist's arm, the crab has thence derived its name of Fiddler.

The Velvet Fiddler (*Portunus puber*) is common enough on our coasts, and is a tolerably hardy inhabitant of an aquarium, but not a safe tenant. It is a most ferocious creature, lurking unseen in a corner, and from its den darting forth at any unsuspicious inhabitant that may come near. One of these creatures has been known to attack a moderately-sized hermit crab, and to destroy it with a single snap, despite its shelly habitation and strong claws. After having killed the poor hermit, the fiddler proceeded to eat its body.

There are some crabs, again, which, in consequence of their peculiarly shaped body and long sprawling limbs, are termed Spider Crabs; for an example of which creatures, see plate M, fig. 3. The bodies of these crustaceans are short, wide, and produced into a snout-like form in

SPIDER CRABS.

front. If they will live, they are useful creatures in an aquarium; for they are good scavengers themselves, and, in addition, often carry on their shells a whole army of zoophytes. Two of these creatures, which inhabited a large aquarium belonging to a friend, were perfect treasures to the microscopist; for when a specimen of a living zoophyte was wanted, one of the spider crabs was hunted up, and the requisite portion removed. They were both rather sluggish animals, and generally resided in two dark holes, in different parts of the aquarium, where a practised eye was needed to discover them.

The species represented is *Maia squinado;* but there are several other spider crabs to be found on the shores, some of whom possess limbs so wonderfully elongated, that they seem to have been subjected to the process of wire-drawing. These may be captured at spring-tides, when the water has sunk much below its usual level, and left the unsuspecting crabs on dry land. Various curious fish, and other creatures, may also be taken at the same time.

We now come to a very curious race of creatures, the Soft-tailed Crabs. Those already mentioned are entirely covered with a strong shelly mail; but there are others whose tails are left bare and

HERMIT CRABS.

defenceless, and which are forced to seek an artificial defence in lieu of natural armour. These creatures are generally called Hermit Crabs, because each one lives a solitary life in his own habitation, like Diogenes in his tub; and sometimes go by the name of Soldier Crabs, on account of their very pugnacious habits. The species here given is the Common Hermit Crab (*Pagurus Bernhardus*), and the particular in-

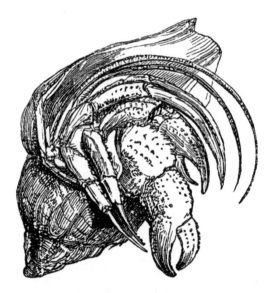

HERMIT CRAB.

dividual is inhabiting a whelk-shell, a domicile that is in great request when the creature grows to any size. Hermit crabs may be found plentifully on the coasts, of all sizes, and inhabiting all kinds of shells, from the trochus to the whelk;

and I have often seen a poor little hermit forced to take up with a huge whelk-shell, of which only the last whorl remained entire, and into which he exactly fitted. He was almost at the mercy of his habitation, for he could not hold it against the power of the waves, which tumbled it over and over most ruthlessly, while the hermit was making futile grasps at stones and sand by which to arrest his progress.

The hermit is furnished with an apparatus of pincers at the extremity of his tail, by which he holds firmly to the shell in which he takes up his temporary habitation, and he flattens himself so firmly against the shell that it is difficult to seize the creature at all; and even when a grasp of any portion can be secured, the hold of the tail is so firm that the animal runs some risk of being torn apart, sooner than leave the shell. Some years since, I was rather anxious to see how the hermit got into his shell, and so, having caught a tolerably large one in a whelk-shell, I tried to pull him out. However, he stuck so close to his shell, that there was no hope of success without inflicting much injury, and I should probably have let him escape, had not an idea then come across my mind. Close by the rock-pool where the hermit had been captured was a colony of very fine sea-anemones (*B. Crassicornis*), and I thought that, probably, by their aid, Mr. Hermit might be enticed out of his shell, even if he would not be dragged out. So I popped the hermit among the wide-expanded tentacles of the crass, which immediately began to contract. The hermit was evidently acquainted with his danger, and, in his hurry to escape from the adhesive tentacles that were twining about him, loosened his hold of the shell, and was instantly plucked out. I let him walk about for a while in the pool, where he looked very woe-begone, trailing his defenceless tail behind him as if he were ashamed of it. After a while I dropped a damaged Purpura shell into the pool, and the crab at once went up to it, and, after a very short examination, stuck the end of his tail into it, for it was not large enough to accommodate the entire tail, and walked about as before. At last, I put the original habitation into the pool, to the very great delight of the hermit, who exchanged shells with a marvellous rapidity, and seemed so much at home again that I could not think of disturbing him. I have frequently tried the same plan of enticing the hermits out of their shells, and never found it to fail.

The combative propensities of these creatures are quite wonderful for their development. If only two hermits of tolerably equal size are placed in an aquarium, they are not content with appropriating different portions of the vessel to themselves, but must needs travel over it and fight whenever they meet. This struggle is constantly renewed, until one of them discovers his inferiority and makes way whenever the victor comes near. When they fight, they do so in earnest, tumbling over each other, and flinging about their legs and claws with singular energy.

They are not at all particular about diet, so that it is of an animal substance, and will eat molluscs, raw meat, or even their own species. More than once, when a hermit has died, I have dropped the body into the water so as to bring it within view of another hermit. The little cannibal caught the descending body in one of his claws very dexterously, and holding it firmly with one claw, he picked it to pieces with the other, and put each morsel into his mouth in a rapid and systematic manner, that was highly amusing It was literally "tucking in." Only the soft abdomen was eaten, and the hard legs, claws, and thorax rejected. Some prawns came and tried to eat the rejected portions, but unsuccessfully, although they dragged them about the water for a few minutes.

When one of these hermits is in captivity, and feels ill, he crawls out of his shell, and generally dies in an hour or two afterwards. It is a very curious propensity, and one that is shared by the tube-inhabiting worms, as has already been mentioned. The habit is the more remarkable, as the usual instinct of animals leads them to the most retired spots that they can find, and there to resign themselves to death.

The formation of the hermit is wonderfully suited to the strange habitation which it adopts, and a hermit when removed from the shell would hardly be recognised for the same creature as that which was snugly curled up within it. Even the claws are modified, for the purpose of lying smoothly in the shell's mouth, one of them being very large, and placed in front, as a shield and weapon united, while the other is very small, and is almost wholly retracted within the shell.

When a hermit desires to change his habitation, he goes through a curious series of performances, which, if he had hands, we should be disposed to call manipulations. A shell lies on the ground, and the hermit seizes it with his claws and feet, twists it about with wonderful dexterity, as if testing its weight; and having examined every portion of its exterior, he proceeds to satisfy himself about the interior. For this purpose, he pushes his fore-legs as far into the shell as they will reach, and probes with their assistance every spot that can be touched. If this examination satisfies him, he whisks himself into the shell with such rapidity, that he seems to have been acted upon by a spring. Such a scene as this will not be witnessed in the sea, unless the hermit is forcibly deprived of his shell, as in the cases above mentioned; but when hermits are placed in a tank or vase, they seem to be rather given to "flitting."

The crab is not always the sole inhabitant of the protecting shell, for one of the curious worms, called Nereides, is very frequently found in joint possession. The fishermen value the Nereis for bait, and turn out the poor hermits in order to obtain this worm that is concealed within the whelk-shell.

The Common Lobster (*Astacus marinus*) is an example of another order of these creatures, the Macroura, or Long-tailed Crustaceans. The crayfish, prawns, and shrimps, all belong to this order. The lobsters are usually taken in basket-traps, after the same manner as the crabs, and when caught they are kept in well-boxes, through which the sea-water runs freely; that is, if there is no immediate sale for them.

Both crabs and lobsters, together with other crustaceans, possess the remarkable faculty of throwing off a limb or two when injured, or even

THE LOBSTER.

if they are alarmed, and of reproducing the lost members. This voluntary amputation often takes place even in the huge claws that lobsters carry, and which are so valued at table. Sailors sometimes hold out a threat to the fishermen, that they will sail to the lobster-grounds, and there discharge heavy guns; the effect of which would be to make the lobsters throw off their claws, and thereby render them unsaleable.

Most crustaceans are pugnacious in character, and it often happens that when they fight, they inflict serious injury upon each other's limbs; and in such cases the maimed limb is detached, not at the wounded spot, but at the joint immediately above, and after awhile a slight protrusion opens itself at the amputated joint, which protrusion becomes more protruded, and in short space develops into a limb. It is a very common circumstance to find a lobster with one very large claw, while the other is of comparatively small size. There is a very wise object in this power. The blood vessels of the crustaceans are but slightly contractile, and, in consequence, if a wound is inflicted, the vessels continue to bleed freely. But by this amputation the wounded surface is reduced to a very small portion, and the substance of the joint contracts with sufficient force to stop the bleeding.

The shell of these creatures is formed of unyielding calcareous substance; and although it may be a most excellent defence for the full-grown crab or lobster, it leaves no room for growth. In order to obviate this difficulty, the crustaceans are possessed of the power of discarding their shells at certain seasons of the year; at which time, also, a new and larger shell is formed. This, in its turn, is cast off; and so continually, until the creature has attained its full growth. Not only are the mere shelly coats of the body and limbs thus changed, but the lobster, for example, when it changes its shell, discards with the shell the following portions of the body:—

The footstalks of the eyes.
The external cornea of the eyes.
The internal thoracic bones.
The membrane of the ear.
The membranous covering of the lungs.
The tendons of all the claws.
The lining of the stomach.
And the stomachic teeth.

The lobster seems to experience some difficulty in casting its skin; and from this list of organs that are changed, it is but little wonder that there should be a difficulty.

There are certain glass toys that were once much in vogue. A reel covered with thread, or a model of a railway engine and tender, were exhibited in a bottle with a very narrow neck. The puzzle was, to discover how the reel got into the bottle. But in the case of the lobster, there is a puzzle of precisely an opposite character: for it is easy enough to understand how a lobster

AGILITY OF LOBSTER.

gets into his shell, or rather how the shell forms over the lobster; but how he gets out of it, and especially how he withdraws his huge pincers without leaving the slightest mark of fracture, is a riddle far more wonderful than that of the bottle. At all events, it is accomplished, and that with such nicety that the cast shell exhibits precisely the same appearance as when it surrounded the living animal.

When the shell is rejected, the creature is almost undefended, its body being only covered with a membranous skin; and is therefore liable to be injured by foes who would be treated with contempt were the shelly armour in its place. So the defenceless animal conceals itself in a quiet spot, and there waits until another shell has been secreted.

The movements of the lobster, and indeed of all its relatives, may be reduced to two kinds, crawling and shooting. Its legs are the means by which it accomplishes the former mode of progression, and the tail by which the latter. I should rather have said, that the tail was the organ of retrogression; for when the lobster shoots through the water by means of the tail, it is in a backward direction. The extremity of the tail is furnished with an array of broad plates, so disposed, that when the lobster **violently bends** its body into the curved shape that it assumes when boiled the force of its action against the water is so strong, that a single stroke will urge the creature to a distance of twenty feet or more, and even enable it to spring out of the water. The natural position of the lobster is straight, and it only curls itself on emergencies, such as a sudden fright, or immersion in hot water. And its sight is so good, or its instinct so wonderful, that it can thus throw itself between two rocks where is barely room for its body to pass. As to the sight, it may well be good, for the lobster possesses compound eyes like the insects, only the shape of the lenses is square instead of hexagonal. Many crustaceans have their eyes hexagonal instead of square.

Beside the crab, the lobster, and the crayfish, two other crustaceans find their way to the table; some of them gaining in liveliness of hue by their passage through the hot water, such as the lobster, the shrimp, and the crayfish; while some, as the crab, make very little change, one way or the other, and some, as the prawns, positively lose their exquisite tints. Heat applied in any way has the same effect, and so has alcohol.

As the shrimps are of more common occurrence

than the prawns, they will have the precedence. Still, many so-called shrimps are really prawns, and among them may be noticed the red shrimp, as distinguished from the brown. The true shrimps are called by the fishermen "sand-raisers," and with reason; for they have a habit of scooping out furrows in the sand, and sinking into them until hardly any portion of their bodies is visible. And in doing so, they raise quite a cloud of sand, which settles down again and assists in obscuring their bodies.

They are caught in peculiarly-shaped nets. which are pushed along the bottom of the sea, and into which the alarmed shrimps rush. The shrimp net, however, contains many objects besides shrimps or prawns, and it is useful to bargain with a shrimper to put all his "rubbish" into a basket and bring it back at the end of his short but hard work. In this way all sorts of marine creatures are captured, especially the smaller fish, and various crustaceans. It is also a charity to the shrimpers, for their work is severe, very trying to the constitution, and badly paid. They are mostly good-natured, honest fellows, always ready to earn a shilling, and can be made into useful assistants as soon as they have got over their astonishment at the value attached to the objects which they have been in the habit of throwing away as rubbish, and which they will still throw away if not looked after.

In the little rock-pools that are left by the tide, numbers of small shrimps may be found, but not easily caught; for they are exceedingly active, darting about with the rapidity of arrows, and, being just the colour of the sand, are difficult to perceive. They cause innumerable furrows in the sand by their little darting flights; and although their marks are usually obliterated by the returning tide, yet there are instances where the furrows, together with the spots produced by rain-drops, and the footmarks of shore birds, have been petrified, and remained as witnesses of events that took place many ages ago. If these little shrimps are desired as specimens, they may be easily captured, by passing a gauze net rapidly through the water. The shrimps are startled at the flash of the net, and as they dart about wildly, here and there, are caught in the very object which they were endeavouring to avoid.

The shrimp is a prolific creature, and produces a large number of eggs, which it carries about until hatching time comes. The young shrimps are comical little creatures, anything but harmonious in their proportions, and bear no more resemblance to their parents than a newly-hatched blind chick does to a gallant game cock. They seem to be gregarious in their habits, and at a little distance look like a cloud of active white particles. They can be easily brought together for observation, or captured, by holding a lighted candle to the side of the vessel which contains them; for they crowd to the light like so many moths or gnats, which latter insects they slightly resemble. Unfortunately, many marine animals are very fond of young shrimps, and a great amount of catching and eating goes on whenever a fresh batch of shrimps comes into existence, so that only a very small per centage attain maturity.

The prawns when living are most exquisite beings, their partially transparent bodies being diversified with delicate tintings, and their radiant eyes glowing like living opals. A boiled prawn loses as much of its living beauty as a trout or a mackerel when subjected to the same process. I have already remarked that the compound eyes of the crustacea are analogous to the corresponding organs in the insects, and the analogy is further carried out by the power of reflecting and refracting light. The eyes of many insects, especially those who fly by night, possess this property; and of all British insects, the **Common Death's-head** Moth is perhaps the most conspicuous in this respect. Either by daylight, or when the rays from a candle fall upon the eyes, they glow as if lighted from within by some hidden fire; a circumstance which adds in no small degree to the terror which is often inspired in the uneducated mind at the sight of one of these insects. In all cases the light departs together with the life of the animal: its origin is not as yet clearly ascertained. Even the eyes of the common Dragon-fly, a diurnal insect, possess a kind of fiery glow when viewed during the life of the creature, but turn to dull, dead hemispheres as soon as it perishes. The light that is reflected from the eyes of cats, &c. is accounted for on principles that do not hold good with regard to the compound eyes of insects or crustacea. The ordinary edible prawns are not found between tide-marks, except occasionally when an unhappy individual has been driven towards the shore, and has not been able to regain the sea before the waves have retired. For this creature, the shrimping net or the dredge is requisite; and as it is tender in constitution, a vessel of sea-water should be ready for its reception when taken out of its native haunts.

There are, however, several species of shor prawns which quite equal, if not excel, those of the deeper waters in beauty. One of these, the Common Æsop Prawn, is given on plate M, fig. 1. It is called the Æsop Prawn because it wears a kind of hunch upon its back, thereby following the example of the great fabulist. Its scientific name is *Pandalus annulicornis*, or the Ring-horned Pandalus. The title "ring-horned" is given to it, because the horns, or antennæ, are exquisitely ringed with scarlet lines at regular distances. These antennæ are most lovely organs; and as the prawn swims through the water in its usual graceful gliding progression, the antennæ wave to and fro, producing elegant and everchanging curves. The whole body of the creature is covered with scarlet lines, which show out exquisitely upon the pellucid groundwork.

These creatures will not be found in the winter, or even in the early spring; but in the summer months they may be seen in abundance in the rock-pools, and captured by means of the gauze net without any difficulty. If the pool is too large, and permits the enclosed animals to escape from the net by means of their extreme activity, the water may generally be drained away by a judiciously cut channel, well guarded by stones and

o

pebbles, or even by the more simple but more tedious mode of baling; the collecting jar makes a very good baling pan. By adopting either of these plans, the surface of water soon becomes contracted, and the imprisoned animals are driven into narrower limits, from which they may be extracted at leisure.

On most sandy shores a curious appearance may be seen bordering the skirts of the waves, an appearance as if innumerable little grasshoppers continually leaped into the air, and in some places so numerous as to fill the air with a sort of misty cloud, to the height of several inches from the ground. Often as the promenader walks along the sea-shore, his footsteps put up whole swarms of these creatures, and induce him to catch them, or rather to attempt their capture. Perhaps one very large individual jumps into the air, and comes down so determinately, that it is marked out for a victim. Down comes the hand upon the spot, but the creature has actively hopped away, and is making off with a succession of agile leaps, that remind one of a kangaroo or a bull-frog. If the pursuer can drive the agile creature from the sea, he may run it down after a smart chase; and when he has caught it he will see that it is a little crustacean, whose form may be recognised on plate M, fig. 2. From its hopping propensities, it goes by the name of Sand-hopper, or Sand-skipper.

It generally lives on the shore, burrowing deep holes in the sand, where it lies concealed until the waves again cover the sands. And if fine specimens are wanted for collection or preservation, they may easily be obtained by digging into the sand with those wooden spades, of which there is no lack wherever there are children, and so pouncing on the sand-skippers before they can recover their alarm at so sudden an entrance into the light of day. They may also be found plentifully swimming about in the rock-pools, or concealed in the masses of ulva or enteromorpha that mostly fringe those miniature ponds. If a basketful of these weeds be plucked at random, and then thrown into a large vessel of sea-water, some twenty or thirty sand-skippers will generally be seen swimming about, and may so be captured.

They feed on the green sea-weeds, and would be hurtful inhabitants of the aquarium did they not serve as food for the anemones, crabs, and other living creatures that are generally kept in such receptacles. It is surprising how soon they vanish from the scene, as soon, indeed, as a stock of carp and roach vanish if placed in a pond where several large pike have taken up their abode. A whole handful of sand-skippers may be transferred to a well-stocked aquarium, and in a week or so hardly one will have survived; there will be plenty of empty shells and rejected limbs at the bottom of the aquarium, but nothing more than their vestiges to tell that sand-skippers once were.

In the same rock-pools where the shrimps, prawns, and sand-skippers are found, there reside also temporarily numbers of little bright-eyed active fish, hardly distinguishable from the shrimps until captured. One of these fishes is shown on plate N, fig. 4, and its name is popularly, the One-spotted Goby, and scientifically, *Gobius unipunctatus*. It derives its name from the single spot that may be seen on the dorsal fin, and which is so conspicuous a mark that by it the creature may be easily distinguished, at all events with sufficient accuracy for ordinary purposes. There is another species of goby, called the Two-spot, that is very common on the coast; so common indeed are these little fish, that I have taken upwards of thirty in as many seconds, merely by sweeping with the gauze net the waters of a rock-pool that had been condensed, as it were, by draining.

The gobies are hardy little fishes, and are able to withstand the prejudicial influences that are inseparable from even the best-regulated aquarium. Some three or four are sufficient in number, and impart to the tank a liveliness that is very pleasing. The ventral fins of the gobies are so formed that they can be pressed together and used as a sucker, by means of which they can adhere firmly to the glass forming the sides of the aquarium, or to the rocks and stones of their native sea. The rapidity, too, with which a goby affixes itself to the glass is quite surprising. These little fish are terrible enemies to the shrimps, for they feed greedily either on the eggs themselves, or on the young shrimps when they have just emerged from the egg. They also feed much on the animalcules of various kinds that throng the alga, and so may be conveniently fed by placing in the tank a handful of freshly gathered ulva, enteromorpha, or indeed any of the sea-weeds whose growth is sufficiently dense to afford shelter to the animalcules. By the aid of a lens, the tiny creatures may be seen coming by thousands out of the floating sea-weed, and snapped up almost as fast as they show themselves.

The Black Goby (*Gobius niger*) may also be

captured as he lies lurking in cunning recesses beneath the stones and rocks, waiting for prey. He is decidedly a fierce fish, and its admission into an aquarium is a doubtful point, inasmuch as he has been known to catch and devour the two-spot goby. It is a larger fish than either of those already mentioned, being about three inches in length.

Another very curious fish is found in much the same locality as the gobies. This is the Shanny, Tansy, or Smooth Blenny, as it is indifferently named, one individual of a large family, whose features are sufficiently remarkable for recognition. The scientific name of this fish is *Blennius pholis*, and a portrait of it is given on plate N, fig 5.

Any one who possesses an aquarium should search after this creature, for it is quite as hardy, if not more so, than the gobies themselves, and is also a bold active fish, making itself very comfortable in its new home, and sparing no opportunity of procuring food, even snatching it from the very jaws of less active fish. The colour of this fish is variable, some specimens being beautifully marked with green and yellow, while there are some almost wholly black or brownish olive. But in all varieties, it has one beauty that never seems to change, and that is the eye, which is decorated with a ring of brilliant crimson. It is a small fish, only a few inches in length, and is to be caught in precisely the same manner as the gobies.

INDEX.

Acorn-shell, 156.
Alcyonium, 115.
Algæ, 41.
Alva, 71.
Anemones, 93.
——— Daisy, 114.
——— smooth, 94.
——— thick-horned, 104.
Annelids, 145.
Aphrodite, 154.
Aquarium, 63.
Auks, 9.

Balanus, 156.
Barnacle, 161.
Bernicle-goose, 162.
Beroe, 165.
Birds, 1.
Bird's-head, 120.
Bladder-wrack, 44.
Blenny, 198.
Bryopsis, 65.

Carrageen, 60.
Cephalopods, 82.
Chiton, 36.
Chlorosperms, 63.
Chorda, 50.
Chylocladia, 55.
Cirrhopoda, 157.
Cladophora, 65.

Cockle, 28.
Cod, 88.
Coralline, 56.
Cormorant, 5.
Coryne, 118.
Cowry, 22.
Crab, edible, 174.
—— green, 173.
—— hermit, 178.
—— spider, 176.
—— swimming, 175.
Crustaceans, 172.
Cuttle, 80.

Dead-man's Fingers, 115
Delesseria, 58, 59.
Dictyota, 51.
Dog-fish, 89.
Doris, 37.
Dulse, 60.
Dunlin, 12.

Echinodermata, 141.
Ectocarpus, 51.
Enteromorpha, 66.

Fish, 196.
Flustra, 122.
Fucus, 44, 46.
Furcellaria 61

INDEX.

Gannet, 8.
Goby, 196.
Grapes, Sea, 79.
Griffithsia, 61.
Guillemot, 9.
Gulls, 1.

Holothuria, 141.

Iridæa, 61.
Irish Moss, 61.

Jelly-fish, 165.

Laminaria, 48.
Laver, 66.
Lepralia, 123.
Limpet, 20.
Lobster, 183.
Lug-worm, 145.

Madrepore, 113.
Medusa, 169.
Melanosperms, 47.
Mermaid's Purse, 90
Mouse, Sea, 154.
Mussel, 33.

Nautilus, 82.
Nereis, 183.
Nitophyllum, 63.
Noctiluca, 170.
Nudibranchs, 37.

Oar-weed, 48.

Peacock's-tail, 50.
Periwinkle, 27.
Pholas, 30.
Plocamium, 59.
Plumularia, 117.
Polysiphonia, 53.

Porpesse, 14.
Porphyra, 68.
Ptilota, 61.
Prawns, 191.
Puffin, 11.
Purpura, 23.
——— eggs of, 78.
Purre, 14.

Razor-shell, 29.
Rhodosperms, 53.

Sabella, 149.
Sand-skipper, 195.
Scallop, 34.
Serpula, 150.
Sertularia, 117.
Shanny, 198.
Shells, 18.
Shrimps, 189.
Skate, 90.
Star-fishes, 125.
——— brittle, 134.
——— five-finger, 126
——— sun, 133.

Tansy, 198.
Teredo, 32.
Terebella, 148.
Terns, 3.
Thread-capsules, 106.
Tides, 42.
Tops, 25.

Ulva, 67.
Urchins, 136.

Wentletrap, 27.
Whelk, 22.
——— eggs of, 77.

Zostera, 70.

INDEX TO PLATES.

A (Front.)
1. **Chylocladia** articulata.
2. **Ectocarpus** siliculosus.
3. **Padina** pavonia.—(Peacock's-tail.)
4. **Porphyra** laciniata.—(Purple Laver, or Sloke.)
5. **Dictyota** dichotoma.

B
1. **Trochus** ziziphinus.—(Pearly Top.)
2. **Littorina** littoralis.—(Periwinkle.)
3. **Patella** vulgaris.—(Limpet.) Ditto, showing under-side.
4. **Purpura** lapillus.
5. **Scalaria** communis.—(Common Wentletrap.)
6. **Cardium** edule.—(Common Cockle.)
7. **Solen** ensis.—(Razor-shell.)
8. **Mytilus** edulis.—(Mussel.)
9. **Pholas** dactylus.

C
1. **Corallina** officinalis.—(Common Coralline.) A portion of frond, with terminal ceramidium.
2. **Cladophora** arcta.
3. **Enteromorpha** compressa.—(Sea Grass.)
4. **Iridæa** edulis.—(Dulse, or dillosk.)
5. **Nitophyllum** punctatum.

D
1. **Laminaria** digitata.—(Tangle.)
2. **Fucus** serratus.—(Notched wrack.)
3. **Bryopsis** plumosa.
4. **Delesseria** hypoglossum.

E
1. **Sertularia** filicula.
2. **Flustra** foliacea.
3. **Coryne** pusilla.
4. **Actinia** mesembryanthemum.
5. Ditto.
6. **Lepralia** ciliata.
7. **Cellularia** avicularis.
8. **Alcyonium** digitatum.
9. **Bunodes** crassicornis.

F
1. **Serpula** contortuplicata.
2. Tube of **Terebella**.
3. **Teredo** navalis.—(Ship-worm.)
4. **Aphrodite**, or **Halithea** aculeata.—(Sea-mouse.)
5. **Sabella**.
6. **Holothuria**.—(Sea-cucumber.)

G
Dunlin.—See p. 9.

INDEX TO PLATES.

H
1. Egg of Skate.
2. Eggs of Purpura.
3. Egg-cluster of Whelk.
4. Egg of Dog-fish.
5. Egg-cluster of Sepia, or Cuttle.

J
1. Fucus nodosus.
2. Zostera marina. — (Grass-wrack, or Alva.)
3. Delesseria sanguinea.
4. Furcellaria fastigiata.
5. Chondrus crispus. — (Irish, or Carrageen Moss.)
6. Fucus vesiculosus. — (Bladder-wrack.)

K
1. Griffithsia setacea. A fruit magnified.
2. Polysiphonia urceolata. A fruit magnified.
3. Plocamium coccineum. A portion magnified.
4. Rhodymenia bifida.
5. Ptilota plumosa.
6. Ulva latissima. — (Green Laver, or Sloke.)

L
1. Caryophyllia Smithii. — (Common Madrepore.)
2. Echinus Sphære. — (Urchin, or Sea-egg.)
3. Ophiocoma rosula. — (Brittle-star.)
4. Uraster rubens. — (Five-finger star.)
5. Solaster papposa. — (Sun-star.)

M
1. Pandalus annulicornis. — (Æsop Prawn.)
2. Sandskipper.
3. Maia squinado. — (Spider-crab.)
4. Portunus puber. — (Velvet swimming Crab, or Velvet Fiddler.)

N
1. Æquorea. — (Medusa, or Jelly-fish.)
2. Cydippe pileus. — (Beroe, or Egg Jelly-fish.)
3. Balanus. — (Acorn-shell, or Acorn-barnacle.)
4. Doris ptilota. — (Naked-gilled Sea-slug.)
5. Gobius unipunctatus. — (One-spotted Goby.)
6. Blennius pholis. — (Shanny.)

THE END.

Woodfall and Kinder, Printers, Milford Lane, Strand, London.

CATALOGUE OF NOVELS, &c.

HAPPY HOME SERIES,

In large fcap. 8vo, fancy covers, 1s. each.

Grimm's Fairy Tales.
Andersen's Fairy Tales.
Edgeworth's Moral Tales.
Edgeworth's Popular Tales.
Sandford and Merton.

Undine, and the Two Captains, by *Fouqué*.
Sintram, and Aslauga's Knight, by *Fouqué*.

BEADLE'S SIXPENNY LIBRARY,

Each 128 pages, fcap. 8vo, fancy covers.

Seth Jones.
Alice Wilde.
Frontier Angel.
Malaeska.
Uncle Ezekiel.
Massasoit's Daughter.
Bill Biddon.
Backwoods Bride.
Natt Todd.
Sybil Chase.
Monowano.
Brethren of the Coast.
King Barnaby.
Forest Spy.
Far West.
Rifleman of the Miami.
Alicia Newcombe.
Hunter's Cabin.
Block House.
The Allens.
Esther.
Ruth Margerie.
Oonomoo.
Gold Hunters.
Two Guards.
Single Eye.
Mabel Meredith.
Ahmo's Plot.

The Scout.
King's Man.
Kent the Ranger.
Peon Prince.
Laughing Eyes.
Mahaska.
Slave Sculptor.
Myrtle.
Indian Jim.
Wreckers' Prize.
Brigantine.
Indian Queen.
Moose Hunter.
Cave Child.
Lost Trail.
Wreck of the Albion.
Joe Davies's Client.
Cuban Heiress.
Hunter's Escape.
Silver Bugle.
Pomfret's Ward.
Quindaro.
Rival Scouts.
Trappers' Pass.
Hermit.
Oronoko Chief.
On the Plains.
Scout's Prize.

Others to follow.

GEORGE ROUTLEDGE & SONS'

ROUTLEDGE'S SIXPENNY NOVELS.

DICKENS, Charles.
  Sketches by Boz.
  Oliver Twist.
  The Pickwick Papers. Part 1, 6d.; Part 2, 6d.
  Nicholas Nickleby. Part 1, 6d.; Part 2, 6d.

MARRYAT, Captain.
  Peter Simple.
  King's Own.
  Newton Forster.
  Jacob Faithful.
  Frank Mildmay.
  Pacha of Many Tales.
  Japhet in Search of a Father.
  Mr. Midshipman Easy.
  Dog Fiend.
  Phantom Ship.
  Olla Podrida.
  Poacher.
  Percival Keene.
  Monsieur Violet.
  Rattlin, the Reefer.
  Valerie.
  The Pirate and the Three Cutters.

COOPER, J. F.
  Waterwitch.
  Pathfinder.
  Deerslayer.
  Last of the Mohicans.
  Pilot.
  Prairie.
  Spy.
  Red Rover.
  Homeward Bound, and Eve Effingham.
  Two Admirals.
  Miles Wallingford, and Afloat and Ashore.
  Pioneers.
  Wyandotte.
  Lionel Lincoln.
  Bravo.
  Sea Lions.
  The Headsman.
  Precaution.
  Oak Openings.
  Heidenmauer.
  Mark's Reef.
  Ned Myers.
  Satanstoe.
  The Borderers.
  Jack Tier.
  Mercedes.

SCOTT, Sir Walter.
  Guy Mannering.
  Antiquary.
  Ivanhoe.
  Fortunes of Nigel.
  Heart of Midlothian.
  Bride of Lammermoor.
  Waverley.
  Rob Roy.
  Kenilworth.
  Pirate.
  Monastery.
  Old Mortality.
  Peveril of the Peak.
  Quentin Durward.
  St. Ronan's Well.
  Abbot.
  Black Dwarf.
  Woodstock.
  Anne of Geierstein.
  Betrothed.
  Fair Maid of Perth.
  Surgeon's Daughter, &c.
  Talisman.
  Count Robert of Paris.
  Redgauntlet.

VARIOUS AUTHORS.
  Artemus Ward, his Book.
  Artemus Ward, his Travels.
  Nasby Papers.
  Major Jack Downing.
  Biglow Papers.

CATALOGUE OF NOVELS, &c.

ROUTLEDGE'S SIXPENNY NOVELS, *continued*.

VARIOUS AUTHORS, contd.
Biglow Papers, 2nd series.
Orpheus C. Kerr.
Robinson Crusoe.
Uncle Tom's Cabin.
Colleen Bawn.
Vicar of Wakefield.
Sketch Book, by Irving.
Sterne's Tristram Shandy.
—— Sentimental Journey.
English Opium Eater.
Essays of Elia.
Notre Dame.
Roderick Random.
Autocrat of the Breakfast Table.
Tom Jones. 2 vols. 6*d*. each.
Gulliver's Travels.
Wandering Jew. Pt. 1. (The Transgression).
—— Pt. 2. (The Chastisement).
—— Pt. 3. (The Redemption).
Mysteries of Paris. Part 1. (Morning).
—— Part 2. (Noon).
—— Part 3. (Night).
Lamplighter.
Professor at the Breakfast Table.
Last Essays of Elia.
Hans Breitmann.
Josh Billings.
Romance of the Forest. *Mrs. Radcliffe.*
The Italian. *Mrs. Radcliffe.*
Mysteries of Udolpho. By ditto. 2 vols. 6*d*. each.
Shadowless Man.
Swiss Family Robinson.
Sayings and Doings of Sam Slick. 1st series.
—— 2nd series.
—— 3rd series.
Baron Trenck.

ROUTLEDGE'S 3/6 STANDARD NOVELS,
In cloth.

LYTTON, Lord.
Eugene Aram.
Night and Morning.
Pelham.
Ernest Maltravers.
Alice.
Last Days of Pompeii.
Harold.
Last of the Barons.
Lucretia.
Caxtons.
Devereux.
My Novel. 2 vols. 3*s*. 6*d*. each.
Disowned.
Coming Race.
Godolphin.
Paul Clifford.
Zanoni.
Rienzi.
A Strange Story.
What will He Do with It? 2 vols. 3*s*. 6*d*. each.
Leila, and the Pilgrims of the Rhine.
Falkland and Zicci.
Kenelm Chillingly.
The Parisians. 2 vols. 3*s*. 6*d*. each.
Pausanias, the Spartan.

Or the Set Complete in 28 *vols., brown cloth, price* £4 18*s*.; *half-roan, gilt edges,* £5 10*s*.; *half-calf or half-morocco,* £9 10*s*.

GEORGE ROUTLEDGE & SONS'

ROUTLEDGE'S 3/6 STANDARD NOVELS, *continued.*

LOVER, Samuel.
  Handy Andy.

MAYNE REID, Captain.
  Scalp Hunters.
  Rifle Rangers.
  Maroon.
  White Chief.
  Wild Huntress.
  White Gauntlet.
  Ocean Waifs.
  Guerilla Chief.
  Half Blood; or, Oceola.
  Headless Horseman.
  Lost Lenore.
  Hunters' Feast.
  Wood Rangers.
  White Squaw.
  Tiger Hunter.
  Boy Slaves.
  Cliff Climbers.
  Giraffe Hunters.
  Afloat in the Forest.
  Fatal Cord.
  War Trail.
  Quadroon.

  Or the Set Complete in brown cloth, £3 17s.

SMEDLEY, F. E.
  Lewis Arundel.
  Frank Fairlegh.
  Harry Coverdale's Courtship.
  Colville Family.

FIELDING, Henry.
  Tom Jones.
  Joseph Andrews.
  Amelia.

SCOTT, Sir Walter.
  With the Author's Notes and the Original Steel Plates by GEORGE CRUIKSHANK, J. M. TURNER, and others.

  Waverley.
  Guy Mannering.
  Old Mortality.
  Heart of Midlothian.
  Rob Roy.
  Antiquary.
  Bride of Lammermoor.
  Black Dwarf and Legend of Montrose.
  Ivanhoe.
  Monastery.
  Abbot.
  Kenilworth.
  Pirate.
  Fortunes of Nigel.
  Peveril of the Peak.
  Quentin Durward.
  St. Ronan's Well.
  Red Gauntlet.
  Betrothed and Highland Widow.
  Talisman and Two Drovers.
  Woodstock.
  Fair Maid of Perth.
  Anne of Geierstein.
  Count Robert of Paris.
  Surgeon's Daughter.

  Or the Set Complete in 25 vols., green or red cloth, price £4 7s. 6d.; half-roan, £5.

MARRYAT, Captain. 3s. 6d.
  Peter Simple.
  King's Own.
  Frank Mildmay.
  Midshipman Easy.
  Jacob Faithful.
  Dog Fiend.
  Percival Keene.
  Japhet.
  Rattlin the Reefer.
  Newton Forster.
  Poacher.
  Pacha of Many Tales.
  Valerie.
  Phantom Ship.
  Monsieur Violet.
  Olla Podrida.

  Or the Set Complete in 16 vols., price £2 16s.; half roan, £3 3s.

CATALOGUE OF NOVELS, &c.

ROUTLEDGE'S POPULAR NOVELS.

Well bound in **cloth**, full gilt back, black printing on the side, and cut edges, price 1s. 6d. each.

MARRYAT, Captain. 1s. 6d.
- Peter Simple.
- King's Own.
- Midshipman Easy.
- Rattlin the Reefer.
- Pacha of Many Tales.
- Newton Forster.
- Jacob Faithful.
- Dog Fiend.
- Japhet in Search of a Father.
- Poacher.
- Phantom Ship.
- Percival Keene.
- Valerie.
- Frank Mildmay.
- Olla Podrida.
- Monsieur Violet.
- Pirate and Three Cutters.

AINSWORTH, W. H. 1s. 6d.
- Windsor Castle.
- Tower of London.
- Miser's Daughter.
- Old St. Paul's.
- Crichton.
- Guy Fawkes.
- Spendthrift.
- James the Second.
- Star Chamber.
- Flitch of Bacon.
- Lancashire Witches.
- Mervyn Clitheroe.
- Ovingdean Grange.
- St. James's.
- Auriol.
- Rookwood.
- Jack Sheppard.

COOPER, J. Fenimore. 1s. 6d.
- Lionel Lincoln.
- Deerslayer.
- Waterwitch.
- Two Admirals.
- Red Rover.
- Afloat and Ashore.
- Wyandotte.
- Headsman.
- Homeward Bound.
- Sea Lions.
- Ned Myers.
- Last of the Mohicans.
- Spy.
- Pilot.

The Complete Set of 26 Vols., £1 19s.

HAWTHORNE, Nath.
- The Scarlet Letter.
- House of the Seven Gables.
- Mosses from an Old Manse.

DUMAS, Alexandre.

For the order of the Sequels, see page 12.
- Dr. Basilius.
- Twin Captains.
- Captain Paul.
- Memoirs of a Physician, 2 vols.
- Two Dianas.
- Conspirators.
- Queen's Necklace.
- Page of the Duke of Savoy.
- Regent's Daughter.
- Taking of the Bastile, 2 vols.
- Countess de Charny.
- Chevalier de Maison Rouge.
- Monte Cristo, 2 vols.
- Nanon.
- Chicot the Jester.
- Ascanio.
- Three Musketeers.
- Twenty Years After.
- Black Tulip.
- Forty-five Guardsmen.
- Isabel of Bavaria.
- Beau Tancrede.
- Pauline.
- Catherine Blum.
- Ingenue.
- Russian Gipsy.
- Watchmaker.

GEORGE ROUTLEDGE & SONS'

Popular Novels at 1/6, *continued*.

CARLETON, *William*.
Emigrants.
Jane Sinclair.
Clarionet.
Fardarougha.
Tithe Proctor.

GRIFFIN, *Gerald*.
Colleen Bawn.
Munster Festivals.
The Rivals.

In limp cloth, gilt, One Shilling each.

Uncle Tom's Cabin.
Robinson Crusoe.
Vicar of Wakefield.
Essays of Elia.

Gulliver's Travels.
Irving's Sketch Book.
Lamplighter.

AMERICAN LIBRARY:

A Series of the most Popular American Works, in fancy covers, each **One Shilling**.

BURNETT, *Mrs. Hodgson*.
The Tide on the Moaning Bar.
That Lass o' Lowrie's. 1s.
Dolly. 1s.
Pretty Polly Pemberton. 1s.
Kathleen. 1s.
Our Neighbour Opposite. 1s.
Miss Crespigny. 1s.
Lindsay's Luck. 1s.

ADELER, *Max*.
Out of the Hurly-Burly. 1s.
Elbow Room. 1s.

ADAMS, *Charles F*.
Leedle Yawcob Strauss. 1s.

ANONYMOUS.
Some Other Babies, very like Helen's, only more so. 1s.
Artemus Ward: His Book, his Travels. 1s.
The Man who was Not a Colonel. 1s.
Dot and Dime, Two Characters in Ebony. 1s.
The Four Irrepressibles. 1s.
My Mother-in-Law. 1s.

That Wife of Mine. 1s.
That Husband of Mine. 1s.

TWAIN, *Mark*.
Celebrated Jumping Frog. Author's edition, with a Copyright Poem. 1s.
Roughing It (copyright). 1s.
Innocents at Home (copyright). 1s.
Mark Twain's Curious Dream (copyright). 1s.
Innocents Abroad. 1s.
New Pilgrim's Progress. 1s.
Information Wanted, and other Sketches. 1s.
Roughing It, and Innocents at Home. 2s.
Sketches. 2s.
Innocents Abroad, and New Pilgrim's Progress. 2s.
Celebrated Jumping Frog, and Curious Dream. 2s.
Gilded Age, a Novel, by Mark Twain & C. D. Warner. 2s.

"Messrs. GEORGE ROUTLEDGE and SONS are my only authorized London publishers. (Signed) MARK TWAIN."

CATALOGUE OF NOVELS, &c.

AMERICAN LIBRARY, *continued*.

HARTE, *Bret*.
Luck of Roaring Camp, with Preface by Tom Hood. 1s.
Bret Harte's Poems (complete). 1s.
Mrs. Skaggs's Husbands. 1s.
Condensed Novels. 1s.
An Episode of Fiddletown. 1s.
The Fool of Five Forks. 1s.
Wan Lee, the Pagan. 1s.
Thankful Blossom. 1s.
My Friend the Tramp. 1s.
Story of a Mine. 1s.
The Man on the Beach.
Jinny.
Prose and Poetry. 2s.
Condensed Novels, and Mrs. Skaggs's Husbands. 2s.

LOWELL, *James R*.
Biglow Papers, 1st and 2nd Series. 1s.

MILLER, *Joaquin*.
First Families of the Sierras. 1s.

ALDRICH, *T. B*.
Prudence Palfrey. 1s.
Marjorie Daw. 2s. and 5s.
The Cloth of Gold. 1s. and 3s. 6d.
Queen of Sheba. 1s.
Flower and Thorn. 3s. 6d.
Baby Bell (Illustrated). 3s. 6d.

DANBURY NEWSMAN.
Life in Danbury. 1s.
Mr. Miggs of Danbury. 2s.

HABBERTON, *John*.
Helen's Babies. 1s.
Barton Experiment. 1s.
Jericho Road. 1s.
Other People's Children. 1s.
Helen's Babies, and Other People's Children, in 1 vol. 2s.
Scripture Club of Valley Rest. 1s.
Some Folks. 2s.
Cruise of the *Sam Weller*. 1s.

EGGLESTON, *E*.
Hoosier Schoolmaster. 1s.
End of the World. 1s.
Mystery of Metropolisville. 1s.
Circuit Rider. 2s.

HOLMES, *O. Wendell*.
Poet at the Breakfast Table. 1s. and 2s.
Autocrat at the Breakfast Table. 1s.
Elsie Venner. 2s.

HOLLAND, *Dr*.
Arthur Bonnicastle. 2s.

VICTOR, *Mrs*.
Maum Guinea. 1s.

E. S. PHELPS.
The Story of Avis. 2s.

BILLINGS, *Josh*.
Wit and Humour. 2s.

NASBY, *Petroleum V*.
Eastern Fruit on Western Dishes. 1s.

GEORGE ROUTLEDGE & SONS'

EVERY BOY'S LIBRARY.

In fancy covers, 1s. each; cloth gilt, 1s. 6d.; cloth, gilt edges, 2s.

English at the North Pole, by *Jules Verne*, with 6 plates.
Wild Man of the West, by *R. M. Ballantyne*.
Louis' School Days, by *E. J. May*.
Digby Heathcote, by *W. H. G. Kingston*.
Dick Rodney, by *James Grant*.
Boy Voyagers, by *Anne Bowman*.
Field of Ice, by *Jules Verne*, with 6 page plates.
Five Weeks in a Balloon, by *Jules Verne*.
A Journey to the Centre of the Earth, by *Jules Verne*.
20,000 Leagues under the Sea, by *Jules Verne*, 2 vols.
Ernie Elton, the Lazy Boy, by *Mrs. Eiloart*.
Ernie Elton at School, by *Mrs. Eiloart*.
The Midshipman.
Robert and Harold; or, The Young Marooners.
Holiday Camp, by *St. John Corbet*.
Voyage Round the World—South America, by *Jules Verne*.
—— Australia, by *Jules Verne*.
—— New Zealand, by *Verne*.
Robinson Crusoe.
A Floating City, and the Blockade Runners, by *Jules Verne*.
Swiss Family Robinson.
From the Earth to the Moon and Round the Moon, by *Jules Verne*.
Two Years Before the Mast.
Indian Boy, by *Rev. H. C. Adams*.
Young Gold Digger, by *Gerstaecker*.
Story of a Bad Boy, by *T. B. Aldrich*.
Three Englishmen and Three Russians, by *Jules Verne*.
Being a Boy, by *C. D. Warner*.
Archie Blake, by *Mrs. Eiloart*.
John Hartley; or, How we Got on in Life.
Joshua Hawsepipe, by *C. R. Low*.
Southey's Life of Nelson.
His Own Master, by *J. T. Trowbridge*.
Round the World in 80 Days, by *Jules Verne*.
Æsop's Fables, with 50 Illustrations by Weir.
The Fur Country, by *Jules Verne*, 2 vols.
Cousin Aleck; or, Boy Life among the Indians.
The Lost Rifle, by the *Rev. H. C. Adams*.

POPULAR LAW BOOKS, 1s. each (2d. postage).

Landlord and Tenant—Useful Forms — Glossary of Law Terms—New Stamp Act.
Wills, Executors and Administrators, with Useful Forms.
Master and Servant.
The Education Act, revised to 1873.
The Ballot Act.
Bills, Cheques, and I. O. U.s.
Friendly Societies' Act, 1875.

CATALOGUE OF NOVELS, &c.

ROUTLEDGE'S 3/6 STANDARD NOVELS, *continued*.

LEVER, *Chas.* 3s. 6d. each vol.
  Harry Lorrequer.
  Jack Hinton.
  Charles O'Malley. 2 vols.
  Con Cregan.
  O'Donoghue.
  Tom Burke. 2 vols.
  One of Them.
  The Daltons. 2 vols.
  Knight of Gwynne. 2 vols.
  Arthur O'Leary.
  Roland Cashel. 2 vols.
  Barrington.
  Dodd Family. 2 vols.
  Luttrell of Arran.
  Davenport Dunn. 2 vols.
  Bramleighs of Bishop's Folly.
  Lord Kilgobbin.
  Martins of Cro' Martin. 2 vols.
  That Boy of Norcott's.
  Fortunes of Glencore.
  Sir Jasper Carew.
  Maurice Tiernay.
  A Day's Ride: A Life's Romance.
  Tony Butler.
  Sir Brooke Fosbrooke.
  Horace Templeton.
    Or the Set Complete in 34 vols., green cloth, price £5 15s. 6d.; half-roan, gilt tops, £6 10s.

SMOLLETT, *Tobias*.
  Roderick Random.
  Peregrine Pickle.
  Humphrey Clinker.

RICHARDSON, *Samuel*.
  Clarissa Harlowe.
  Pamela.
  Sir Charles Grandison.

RUSSELL, *Dora*.
  The Miner's Oath, and Underground, 6 illustrations by J. D. Watson.

STERNE, *Lawrence*.
  Tristram Shandy and Sentimental Journey.

SUE, *Eugene*.
  Wandering Jew.
  Mysteries of Paris.

DICKENS, *Charles*.
  Pickwick, Oliver Twist, and Sketches by Boz, bound together in One Volume, cloth gilt, 3s. 6d.

DUMAS' NOVELS, a New Edition, reset, with full-page Illustrations, crown 8vo, each **3s. 6d.** The Letters before the Titles show the order of the Sequels.

  a. Three Musketeers.
  b. Twenty Years After.
  c. Vicomte de Bragelonne. 2 vols. 3s. 6d. each.
  a. Marguerite de Valois.
  b. Chicot the Jester.
  c. The Forty-five Guardsmen.

  a. The Conspirators.
  b. The Regent's Daughter.
  a. Memoirs of a Physician.
  b. The Queen's Necklace.
  c. The Taking of the Bastile.
  d. The Countess de Charny.
  Count of Monte Cristo.

  ILLUSTRATED EDITION, 14 vols., crown 8vo, cloth, £2 9s.; cloth, gilt tops, £2 12s. 6d.

GEORGE ROUTLEDGE & SONS'

ROUTLEDGE'S 3/6 STANDARD NOVELS, *continued*.

AINSWORTH'S NOVELS, in crown 8vo Monthly Volumes, with Illustrations, red cloth, **3s. 6d.** each.

- Tower of London, with 6 steel plates by George Cruikshank
- Windsor Castle, with 6 steel plates by George Cruikshank
- Lancashire Witches, with plates by Sir John Gilbert, R.A.
- Guy Fawkes, with steel plates by Cruikshank.
- St. James's, with steel plates by Cruikshank.
- Old St. Paul's, with steel plates by Cruikshank.
- Rookwood, with 6 steel plates.
- Crichton, with plates by H. K. Browne.
- Mervyn Clitheroe, with plates by Phiz.
- Boscobel, with plates by Phiz.
- Ovingdean Grange, with plates by Phiz.
- The Miser's Daughter, with steel plates by Cruikshank.

More Volumes to follow.

YANKEE DROLLERIES.—THREE SERIES, in cloth, **3s. 6d.** each.

1. Artemus Ward, his Book—Major Jack Downing—The Nasby Papers—Orpheus C. Kerr—The Biglow Papers.
2. Artemus Ward, his Travels—Hans Breitmann—The Professor at the Breakfast Table—Biglow Papers, Part 2—Josh Billings.
3. Artemus Ward's Fenians—Autocrat of the Breakfast Table—Bret Harte—The Innocents Abroad—The Jumping Frog.

CROWN 8vo. NOVELS, set from new type, well Illustrated, bound in cloth, each **3s. 6d.**

- Mysteries of Paris.
- Wandering Jew.
- Valentine Vox.
- Handy Andy.
- Frank Fairlegh.
- Tom Jones.
- Harry Lorrequer.
- Les Misérables, by *Victor Hugo*.

ROUTLEDGE'S STANDARD 2/6 NOVELS
In cloth.

DICKENS, Charles.
- The Pickwick Papers.
- Sketches by Boz.
- Nicholas Nickleby.
- Oliver Twist.
- Martin Chuzzlewit.

CATALOGUE OF NOVELS, &c.

STANDARD HALF-CROWN NOVELS, *continued.*

LORD LYTTON'S NOVELS, in 27 vols., fcap. 8vo, with Frontispiece, cloth gilt, 2s. 6d. each.

The Caxtons.	Zanoni.
Night and Morning.	Last Days of Pompeii.
My Novel. 2 vols.	Devereux.
Leila, and the Pilgrims of the Rhine.	Rienzi.
Eugene Aram.	Disowned.
Harold.	Pelham.
Last of the Barons.	What will He Do with It?
Lucretia.	A Strange Story. [2 vols.
Ernest Maltravers.	The Coming Race.
Alice.	Kenelm Chillingly.
Godolphin.	The Parisians. 2 vols.
Paul Clifford.	Falkland and Zicci.

NAVAL AND MILITARY SERIES, cloth, 2s. 6d. each.

Sword and Gown, by *Author of "Guy Livingstone."*	The Girl He Left Behind Him, by *R. Mounteney Jephson.*
Hussar, by *Rev. G. R. Gleig.*	Hearths and Watchfires, by *Col. Colomb.*
Chelsea Veterans, by ditto.	Will He Marry Her? by *Lang.*
Light Dragoon, by ditto.	
Pride of the Mess, by *Captain Neale.*	Sans Merci, by the *Author of Guy Livingstone.*
Tom Bulkeley, by *R. Mounteney Jephson.*	Guy Livingstone.

MARRYAT'S (Captain) NOVELS and TALES, fcap. 8vo, cloth gilt (Standard Novels), 2s. 6d.

Jacob Faithful.	Poacher.
Japhet in Search of a Father.	Phantom Ship.
King's Own.	Dog Fiend.
Midshipman Easy.	Percival Keene.
Newton Forster.	Frank Mildmay.
Pacha of Many Tales.	Peter Simple.
Rattlin the Reefer.	

COOPER'S NOVELS, in fcap. 8vo, cloth, with Steel Plate, 2s. 6d. each.

Last of the Mohicans.	Miles Wallingford.
Lionel Lincoln.	Afloat and Ashore.
Borderers.	Bravo.
Waterwitch.	Homeward Bound.
Deerslayer.	Headsman.
Pathfinder.	Wyandotte.
Heidenmauer.	

GEORGE ROUTLEDGE & SONS'

Standard Half-Crown Novels, *continued*.

The following in double columns, 2s. 6d. *each, containing 4 Novels, in cloth.*

COOPER, J. Fenimore.
1. Spy — Pilot — Homeward Bound—Eve Effingham.
2. Pioneers — Mohicans — Prairie—Pathfinder.
3. Red Rover—Two Admirals — Miles Wallingford—Afloat and Ashore.
4. Borderers — Wyandotte—Mark's Reef—Satanstoe.
5. Lionel Lincoln — Oak Openings—Ned Myers—Precaution.
6. Deerslayer— Headsman—Waterwitch — Heidenmauer.
7. Bravo—Sea Lions—Jack Tier—Mercedes.

MARRYAT, Captain.
1. King's Own—Frank Mildmay—Newton Forster—Peter Simple.
2. Pacha of Many Tales—Jacob Faithful—Midshipman Easy—Japhet.
3. Phantom Ship—Dog Fiend—Olla Podrida—Poacher.
4. Percival Keene—Monsieur Violet—Rattlin—Valerie.

SCOTT, Sir Walter.
1. Waverley — Monastery — Kenilworth—Rob Roy.
2. Pirate — Ivanhoe — Fortunes of Nigel—Old Mortality.
3. Guy Mannering—Bride of Lammermoor—Heart of Midlothian—Antiquary.
4. Peveril of the Peak — Quentin Durward — St. Ronan's Well—Abbot.
5. The Black Dwarf—Woodstock — Anne of Geierstein—Betrothed.
6. The Fair Maid of Perth—Surgeon's Daughter — Talisman—Count Robert of Paris—Redgauntlet.

RADCLIFFE, Ann.
Romance of the Forest—The Italian—Mysteries of Udolpho (2 parts).

Cloth, 2s. 6d. each.

Handy Andy, by *Lover*.
Rory O'More, by *Lover*.
Arthur O'Leary, by *Lever*.
Con Cregan, by *Lever*.
Mysteries of Paris, by *E. Sue*.
Wandering Jew, by *E. Sue*.
Guy Livingstone.
Running the Gauntlet, by *E. Yates*.
Kissing the Rod, by *E. Yates*.
Tom Jones, by *Fielding*.
Notre Dame, by *Victor Hugo*.
Valentine Vox, by *Cockton*.
Scottish Chiefs, by *Porter*.
Tom Cringle's Log.
Rob Roy, by *Scott*.
Ivanhoe, by *Scott*.
Salathiel, by *Dr. Croly*.
The Clockmaker, by *Sam Slick*.

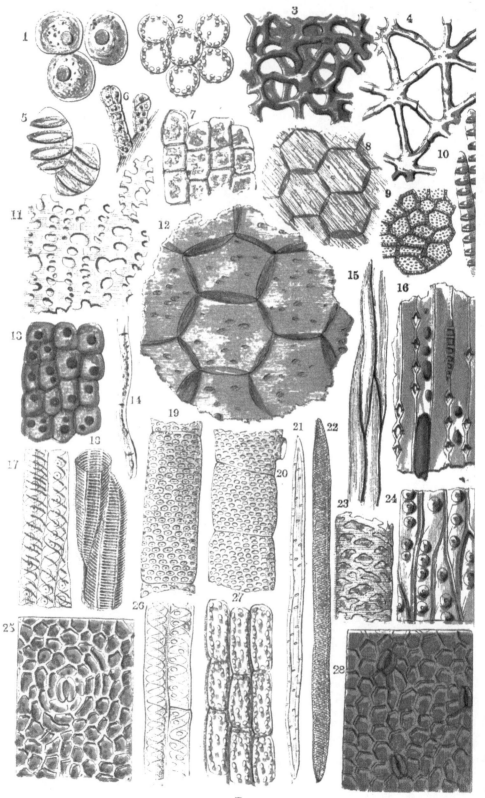

I.

COMMON OBJECTS

OF

THE MICROSCOPE.

BY THE
REV. J. G. WOOD, M.A. F.L.S. Etc.

AUTHOR OF "COMMON OBJECTS OF THE COUNTRY AND SEA-SHORE,"
"ILLUSTRATED NATURAL HISTORY," ECT. ECT.

WITH ILLUSTRATIONS BY TUFFEN WEST.
PRINTED IN COLOURS BY EVANS.

LONDON:
GEORGE ROUTLEDGE AND SONS;
THE BROADWAY, LUDGATE.

WOOD'S
COMMON OBJECTS OF THE MICROSCOPE.

A Cheap Edition of this Work, with the Illustrations, is to be obtained, price One Shilling.

PREFACE.

In my two previous handbooks, the "Common Objects" of the Sea-shore and Country, I could but slightly glance at the minute beings which swarm in every locality, or at the wonderful structures which are discovered by the Microscope within or upon the creatures therein described. Since that time a general demand has arisen for an elementary handbook upon the Microscope and its practical appliance to the study of nature; and in order to supply that want, this little volume has been produced.

I must warn the reader that he is not to expect a work that will figure and describe every object which may be found on the sea-shore or in the fields, but merely one by which he will be enabled to guide himself in microscopical research, and avoid the loss of time and patience which is almost invariably the lot of the novice in these interesting studies. Upwards of four hundred objects have been figured, including many representatives of the animal, vegetable, and mineral kingdoms, and among them the reader will find types sufficient for his early guidance.

Neither must he expect that any drawings can fully render the lovely structures which are revealed by the microscope. Their form can be given faithfully enough, and their colour can be indicated; but no pen, pencil, or brush, however skilfully wielded, can reproduce the soft, glowing radiance, the delicate pearly translucency, or the flashing effulgence of living and ever-changing light with which God wills to imbue even the smallest of his creatures, whose very existence has been hidden for countless ages from the inquisitive research of man, and whose wondrous beauty astonishes and delights the eye, and fills the heart with awe and adoration.

Owing to the many claims on my time, I left the selection of the objects to Mr. Tuffen West, who employed the greater part of a year in collecting specimens for the express purpose, and whose well-known fidelity and wide experience are the best guarantees that can be offered to the public. To him I also owe many thanks for his kind revision of the proof-sheets. My thanks are also due to Messrs. G. and H. Brady, who lent many beautiful objects, and to Messrs. Baker, the well-known opticians of Holborn, who liberally placed their whole stock of slides and instruments at my disposal.

THE MICROSCOPE.

CHAPTER I.

INTRODUCTION—USES OF THE MICROSCOPE—VALUE OF CAREFUL OBSERVATION—EARLY DISCOVERIES—EXTEMPORIZED INSTRUMENTS.

In the following pages I propose to carry out, as far as possible, with regard to the MICROSCOPE, the system which I have previously followed in the "Common Objects of the Sea-Shore and Country," and to treat in a simple manner of those wonderful structures, whether animal, vegetable, or mineral, which are found so profusely in our fields, woods, streams, shores, and gardens. Moreover, I intend to restrict my observations wholly to that class of instrument which can be readily obtained and easily handled, and to those supplementary pieces of microscopic apparatus which can be supplied by the makers at a cost of a few shillings,

B

or extemporized by the expenditure of a few pence and a little ingenuity on the part of the observer. As in the former works, ordinary and familiar English terms will in every case be used where their employment is possible; but as, on account of their extremely minute dimensions, no popular name has been given to very many objects, we must be content to accept the more difficult language of science and render it as little abstruse as possible.

Within the last few years, the microscope has become so firmly rooted among us, that little need be said in its praise. The time has long passed away when it was held in no higher estimation than an ingenious toy; but it is now acknowledged, that no one can attain even a moderate knowledge of any physical science without a considerable acquaintance with the microscope and the marvellous phenomena which it reveals. The geologist, the chemist, the mineralogist, the anatomist, or the botanist, all find the microscope a useful companion and indispensable aid in their interesting and all-absorbing researches, and, with every improvement in its construction, have discovered a corresponding enlargement and enlightenment of the field displayed by the particular science which they cultivate.

But even to those who aspire to no scientific eminence, the microscope is more than an amusing companion, revealing many of the hidden secrets of Nature, and unveiling endless beauties which were heretofore enveloped in the impenetrable obscurity of their own minuteness.

No one who possesses even a pocket-microscope of the most limited powers can fail to find amusement and instruction, even though he were in the midst of the Sahara itself. There is this great advantage in the microscope, that no one need feel in want of objects as long as he possesses his instrument and a sufficiency of light.

Many persons who are gifted with a thorough appreciation of nature in all her vivid forms are debarred by the peculiarity of their position from following out the impulses of their beings, and are equally unable to range the sea-shore in search of marine creatures or to traverse the fields and woods in the course of their investigations into the manifold forms of life and beauty which teem in every nook and corner of the country. Some are confined to their chambers by bodily ailments, some are forced to reside within the very heart of some great city, without opportunities of breathing the fresh country air more than a few times in the course of the year; and yet there is not one who may not find an endless series of Common Objects for his microscope within the limits of the tiniest city chamber. So richly does nature teem with beauty and living marvels, that even within the closest dungeon-walls a never-failing treasury of science may be found by any one who knows how and where to seek for it.

It is rather a remarkable fact, that the real value of observation is often in inverse ratio to the multitude of the objects examined; and we all know the extreme interest which attaches itself to minute and faithful records of the events which take place in some very limited sphere. For example, the annals of an obscure village in Hampshire have long risen into a standard work, merely by virtue of the close and trustworthy observations made by a resident in the place; the Tour round a Garden has enchanted thousands and proved quite as attractive as any tour round the world could be made; and many most curious and valuable original observations now committed to my note-book were made by an old lady in her daily perambulation of a little scrap of a back yard in the suburbs of London, barely twelve yards long by four wide.

The world-famous labours of Huber on the Honey-Bee, Lyonnet on the Goat Moth, and Strauss Durckheim on the Cockchaffer, are familiar to every student of zoology, and have done more towards advancing the study of animal life than hundreds of larger works

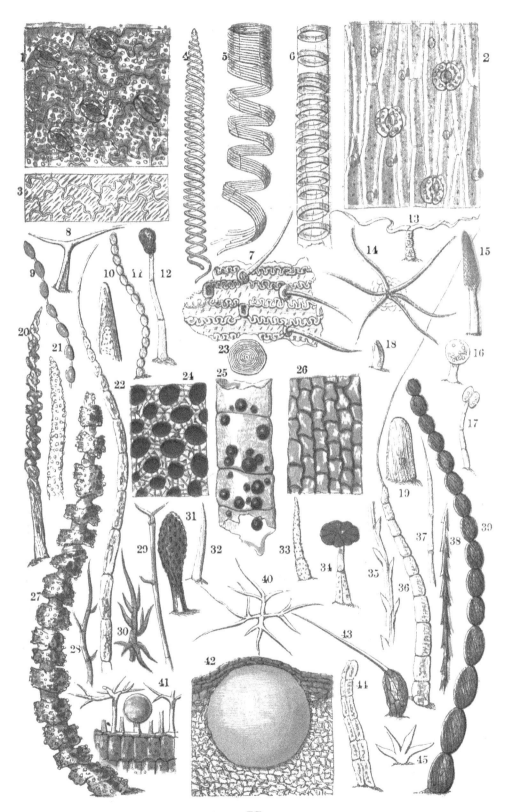

II.

which embrace thousands of species in their scope. There is little doubt but that if any one with an observant mind were to set himself to work determinately merely at the study of the commonest weed or the most familiar insect, he would, in the course of some years' patient labour, produce a work that would be most valuable to science, and enrol the name of the investigator among the most honoured sons of knowledge. There is not a mote that dances in the sunbeam, not a particle of dust that we tread heedlessly below our feet, that does not contain within its form mines of knowledge as yet unworked. For if we could only read them rightly, all the records of the animated past are written in the rocks and dust of the present.

Up to this time the powers of the microscope, as indeed is the case with all scientific inventions, are but in their infant stage; and though we have obtained instruments of very great perfection, it must be remembered, that many of the earliest and greatest discoveries were made with common magnifying glasses, such as are now sold for a few pence, and which would be despised by the generality of microscopical observers. Indeed, there are few instances where a person so minded may not possess himself of a microscope that will do a considerable amount of sound work and at an inappreciable cost. Many of my readers will doubtlessly have purchased one of those penny microscopes, composed of a pill-box and a drop of Canada balsam, which are hawked about the streets by the ingenious and deserving manufacturer; and upon a pinch, a very respectable microscope may be extemporized out of a strip of card, wood, or metal, and a little water.

There are, indeed, few branches of science which admit of such varied modes of handling as the use of the microscope. No two practical microscopists ever set about their work in the same manner; each will have his own special method of manipulation, which he thinks superior to any other, and each will arrive at most valuable results, though by different and sometimes opposite roads. The scope which it gives to ready invention is unlimited. Exigencies are continually occurring, when the observer is deprived for the time of some valuable adjunct, and is forced to invent and manufacture on the spur of the moment an efficient, though perhaps unsightly substitute. So well do some of these make-shift contrivances answer their purpose, that the inventor often prefers them to the more elegant and expensive articles which are purchased from the optician.

For example, I once patched up an extemporized dissecting microscope out of an old retort-stand, a piece of cane, and six inches of elder branch, which did its work as effectually as the shining-lackered brass instrument which it was intended to imitate. Moreover, by a very simple addition of a piece of wire, it answered as a movable stand for a camera lucida, thus performing a duty which would not have been achieved by the expensive brass microscope of the optician. All kinds of subsidiary apparatus may, in like manner, be made by any one who really cares about the beautiful pursuit in which he is engaged; and it is a matter of no light importance to those whose purses may not be overstocked, and whose hearts fail them at the price-lists of the opticians, that a great proportion of the adjuncts to the microscope may be manufactured at the cost of a very few shillings, where the regular makers charge many pounds.

The greater part of the imposing and glittering paraphernalia which decorate the dealer's counter or the table of the wealthy amateur may be replaced by apparatus that can be made at a very trifling cost from the most ordinary materials, and, for a while at least, the remainder may be dispensed with altogether. It is not the wealthiest, but the acutest and most patient observer who makes the most discoveries, for a workman is not made, nor even known by his tools, and a good observer will discover with a common pocket-magnifier many a secret of nature which has escaped the notice of a whole array of *dilettanti* microscopists, in spite of all their expensive and accurate instruments.

It is for those who desire to be of the former class that this little work is written, and in the course of the following pages many examples will be given, where a slight exertion of thought and ingenuity has been found equivalent to the purchase of costly and complicated apparatus.

CHAPTER II.

SIMPLE AND COMPOUND MICROSCOPES—MEASUREMENT OF POWER—HINTS FOR EXAMINATION—DISSECTING MICROSCOPE—KNIVES, SCISSORS, AND NEEDLES—DISSECTING TROUGHS—ARRANGEMENT OF ARTIFICIAL LIGHT—INSTRUMENT MAKING—DIPPING TUBES—CODDINGTON LENS—COMPOUND MICROSCOPE AND APPARATUS.

MICROSCOPES may be divided into two classes, Simple and Compound. The former class may contain several lenses or glasses, and generally consists of a single lens; but the Compound Microscope must consist of at least two glasses, the one near the object to be examined, and the other near the eye. We will first mention one or two forms of the Simple Microscope.

For all general purposes, the intending observer can do no better than supply himself with a common pocket-magnifier, which can be bought at any optician's for a very small sum, containing one, two, or three lenses, the last-mentioned being the most advisable. These lenses, or "powers" as they are technically called, vary in their magnifying capacities, those which increase the size of the object to the greatest degree being the smallest of size and the most decided in their convexity, and are required to be held nearest to the object.

In a work of this character it will be useless to waste time and space by mentioning the abstruse problems by which the construction of microscopes is governed, as the full account of them would more than occupy the entire book, and a compressed description would be wholly impossible. Suffice it to say that all those who desire to study the beautiful science of optics, and its application to the microscope, may find full information in the larger and more scientific works to which this little book is intended merely as an introduction.

According to this plan, I will here mention that the power of any lens is known by the distance at which it must be held from the object. Thus, the inch power of a compound microscope will magnify an object about forty times, while the quarter-inch magnifies not less than two hundred. Among microscopists the degree to which objects are magnified is always designated by "diameters," so that if an object be magnified ten diameters, we mean that it appears ten times as long and as broad as it really is. The reader must bear this in mind, for the glowing descriptions of magnifying powers that are so often seen in advertisements are not according to diameters, but superficial measure, so that a lens which magnifies ten diameters is set down as one which magnifies a hundred times, and one of two hundred and fifty diameters is advertised as magnifying five thousand times.

The pocket-magnifier has this advantage over a lens fixed in a stand, that it can be turned in every direction together with the object, so that the general details of structure can always be better made out with one of these simple instruments than with the most elaborate compound microscope ever made. The higher powers are only intended for the purpose of elucidating the minute structure of smaller points, and are rarely employed until after the observer has made good use of the pocket-lens. For example, in learning the structure of an insect, say a common gnat, it should first be thoroughly examined with the lowest power of the simple lens, and afterwards by each of the higher powers in succession, until the observer has obtained a good general idea of the form and position of the various organs, together with hints as to the portions which are best adapted for the larger instruments. After learning all those details, the observer next removes a small portion of the insect, say a wing, or a leg, and submits it to the lowest power of his compound microscope, adding successively the higher powers until he has gone over the whole of the object. By setting to work at a single subject, of whatever nature it may be, and examining it first in general and afterwards by detail, the observer will find that he has gained more than he would have learned by volumes of reading alone.

I know of no pursuit more fascinating than this, or more calculated to make him who pursues it forgetful of time, place, hunger, and cold. There is something so entrancing in the manner in which Nature gives up her wondrous secrets, that the mind seems to be entirely taken out of the world, the hours fly past as in a dream, and the day becomes too short for the pleasant labour.

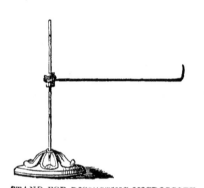
STAND FOR DISSECTING MICROSCOPE.

The simple lens already mentioned can be employed in various ways, and by a little ingenuity can be made serviceable either as a pocket-magnifier or a dissecting microscope. The latter object is thus accomplished, requiring a very trifling exercise of patience or cunning of hand. Get an iron or brass rod fixed into an iron or leaden foot, as seen in the engraving. Then bore a hole longitudinally through a rather large wine-cork, so as to

A CHEAP INSTRUMENT. 13

slide rather stiffly over the upright rod. Then take a stout brass wire, twist one end of it into a spiral, inclosing the cork in the centre, cut it off to the required length, turn up the end of it at right angles, and slightly sharpen the point. The turned-up end may then be passed through a hole drilled in the handle of the pocket-magnifier, and the microscope is complete. The sliding cork will permit the lens to be raised to a higher or lower level, while the length of horizontal wire will permit the hands to be used with freedom.

In my own instrument the upright rod is simply a common retort-stand, and the horizontal bar is a piece of hollow elder branch lashed to one of the wire rings of the stand, and carrying the lens at its extremity. Moreover, by substituting a wire ring for the upturned point, it will hold a camera lucida, and this is very useful in sketching any object that needs rapid but accurate drawing.

For those who do not possess even the small amount of mechanical skill which is required for the construction of so simple an instrument as that which has just been mentioned, Ross's dissecting microscope is one of the best. As may be seen, it is capable of motion in every direction, and permits the lenses of different powers to be fitted or secured without any screwing or waste of time. This is often a considerable

14 DISSECTING TOOLS.

advantage when much time is given to microscopic dissection. The very best dissecting microscope that I have seen was one that was employed by Dr. Acland at the Anatomical Museum at Oxford, and was formed something on the same principle as that mentioned above. The horizontal bar was so made as to be raised or lowered by turning a screw, while it was so affixed to the upright bar, that it could be pushed aside in

ROSS'S DISSECTING MICROSCOPE.

order to examine the dissection with the naked eye, and again drawn into its place without disturbing the stand.

The only practical objection to these forms of the dissecting microscope is, that they do not permit the object to be seen by means of light thrown from below, but this defect is easily remedied by cutting a hole in the dissecting table and placing a mirror beneath.

The tools employed for dissection need not be many nor complicated, and can all be purchased for a very few shillings. A very small scalpel, with a double edge, is always useful, and should be extremely flat and thin in the blade, as well as kept to the very acme of sharpness by an occasional touch of a hone and razor-strap. Three pairs of scissors are needful: one tolerably stout, for cutting hard substances, such as the wing-cases and external skeletons of beetles; another, very long in the handles, and very short and delicate in the blades, for the purpose of severing minute tissues; and the third pair bent like the beak of the avocet, to enable the dissector to snip off those little projections which are

continually getting in the way, and which cannot be reached by a straight blade without running the risk of damaging the dissection.

Two pairs of forceps will also be required, one straight and strong, and the other fine and curved, as seen in the engraving. In order to insure the accurate meeting of the points—a matter of very great importance—the blades play upon a pin which may be seen inserted near the curved extremity of the instrument. These forceps are generally made of brass, and are most useful. I generally carry a pair in my pocket whenever I go into the fields, for they serve to draw little insects out of their hiding places, to pick up objects too minute for the fingers, and to hold them while undergoing examination with the pocket-magnifier.

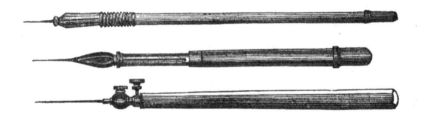

But the sheet anchor of the microscopic dissector is made of needles, which should be kept of different sizes ready to hand. Fastened into wooden handles, they are employed in "teasing" out delicate structures, so as to separate the tissues of which they are composed without tearing or cutting them. Some persons recommend that a lady's crochet-case be used, which not only contains a store of needles, but admits of changing the point whenever needed. There are also other forms of

FORMS OF NEEDLES.

dissecting needles manufactured, three of which are represented in the illustration.

For my own part, I always find that the ivory or metal handle is too heavy, and invariably employ common camel's hair brush handles, in which the needles can be readily fastened. Five forms are all that are really useful, and many of them can be made in a few minutes. In order to fix the needles firmly in their handles, the following plan is the best. Get a convenient handle, and wrap about a third of an inch with waxed thread, leaving a little of the wood projecting without any thread. Take the needle, break it off to a convenient length, push the point into the handle so as to make a hole, reverse it, and with a pair of pliers drive the needle well into the handle, the thread preventing the wood from splitting. Now trim the wood to a point, so as to make it all look neat, and a light handy instrument is at once made.

The five forms are employed for different purposes. The first, No. 1, is a short thick needle, set in a large handle, and used for boring holes in wood, mica, cork, or wax, as may be required. It is also useful for making the holes in new handles. No. 2 is a rather fine, straight needle, and is the most generally useful of the set. Several of these should be made of different degrees of fineness. No. 3 is a slightly bent needle, valuable in getting at tissues that lie hidden under other substances. Several of these should be made, bent at different curvatures, and of different strength. No. 4 is occasionally useful for pulling thready tissues aside in order to permit another instrument to be used; and No. 5 is required for lifting a delicate structure without injuring it. The reader will observe that it has no point, but that its extremity is defended by a little knob. I may also mention that it will be an improvement if the fine scissors have also one blade terminated by a little knob.

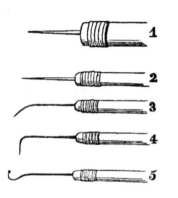

There is no need for making a great supply of these instruments before commencing work. I always used to fit up several of 2 and 3, and to make the others as they are required. I can strongly recommend these simple little instruments, as they are very light, and are little liable to injury.

The other appliances for insect dissection are equally simple.

As all delicate structures are dissected under fluid, a shallow glass dish is required. My own are simple flat round glass cells, about one inch in depth and four in diameter. For very large objects, a dish of corresponding length is of course required. This is plentifully filled with water or spirit—generally a mixture of the two—and the dissection is sunk to the bottom by being fastened to a flat cork attached to a strip of sheet lead. The simplest way of making this loaded cork is to cut a piece of flat cork to the required size, lay it on a piece of sheet lead, cut the lead rather wider than the cork, turn it up over the edge, and fasten it with a few blows of a hammer.

Some persons prefer to make a very shallow dish of lead, and to pour melted wax into it, so as to fix the object upon the wax. I, however, prefer the cork, as the pins are apt to break away from the wax. The cork should be very fine-grained, as if the holes are large and deep, the pin is sure to plunge through the dissection, carrying with it the tip of the forceps, and thereby doing irremediable damage.

Of course the whole affair must be set in a good light, or the dissection will be impracticable. Daylight is by far the best, for I always find that by artificial light the shadows are thrown so perplexingly that it is almost impossible to make out the real structure of a delicate object, and an important tissue may be broken under the impression that it is but a

20 ARRANGEMENT OF LIGHT.

shadow. If, however, artificial light **must be used,** the accompanying illustration will show the **manner** of arranging it.

The light is thrown perpendicularly upon the object by means of a common "condenser," and the hands

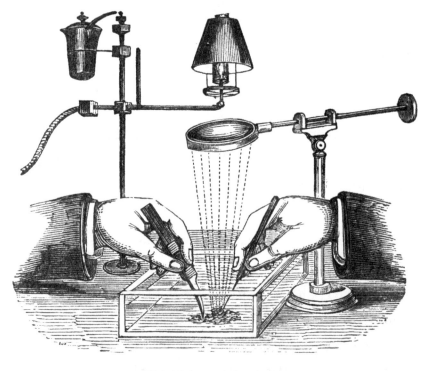

DISSECTING UNDER WATER.

are so placed as to avoid getting in the way of the light. In practice it will be found very useful to support the hands by means of a book or piece of wood on each side of the glass cell, as the handling becomes very awkward and fatiguing without some such precaution. The best support is made of a thick piece of board of the same height as the edge of the glass cell, flat for two inches or so, and then sloped away so as to form an inclined plane.

Great care must be taken with the points of the needles that they are perfectly smooth and polished, as if there is the least roughness they will hitch in the more delicate structures and tear them woefully. Also, the needles should not be too long, as the elasticity of the steel is apt to make them spring when pressed. The length given in the engraving is amply sufficient. The bending of the needles is easily accomplished by holding them in a candle until red-hot, when they can readily be bent into any form that is desired. By this mode of treatment they become soft and yielding, but can be immediately restored to their original hardness by reheating, and then plunging them into cold water. A spirit lamp will serve better than a candle, as it does not blacken the needle, and permits the dissector to work more freely.

A large supply of variously sized pins should be at hand, some of tolerable size, a great many minikins, and a box of fine entomological pins. I always have a loaded cork close to the dissecting cell, the cork being filled with pins of different sizes, graduated according to their position, so that they can be taken at **any** moment without search or disturbance. These **are** employed for fastening the object to the loaded **cork in** the cell, and also for keeping aside the various **structures** as they are dissected out. The number of pins **that is** required is really remarkable, for in the ordinary dissection of an insect some fourteen or fifteen pins **will** gradually be used.

A very fine-nosed syringe is a most useful article, **but** its place can be very well supplied by glass tubes **made** after the following fashion.

Get a glass tube or two from the chemist—the diameter is of little consequence, provided that the glass be of soft quality—light the spirit lamp, and hold one of these tubes by the ends, keeping the centre over the flame and turning it continually to prevent the glass from cracking. Lower it gradually into the flame, until it becomes of a bright red heat and quite soft. Then draw the two ends rapidly asunder, and there will be two tubes, each ending in a point of very thin glass.

Break away the extremity of the point, and you will have a tube with a very fine outlet. Of course if you want a large diameter, you have only to break away the glass a little higher.

The broken end should then be held for a moment in the margin of the flame so as to round its sharp edges.

DIPPING TUBES. 23

These tubes are extremely useful, being employed when very fine for washing aside any tissue that is too

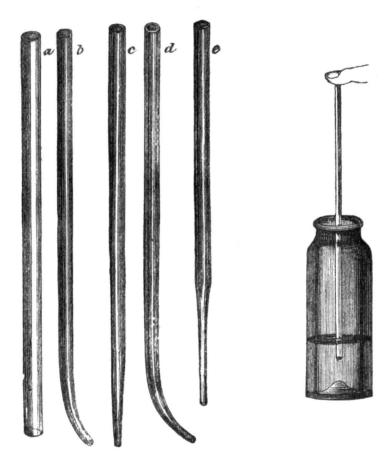

DIPPING TUBES AND MODE OF USING.

delicate to be handled with the steel, and are used by putting the large end into the mouth, drawing the liquid into them by suction, and expelling it by the breath.

These and similar tubes are also useful for "dipping" out minute organisms from the water in which they live, and are therefore termed "dipping tubes." Several forms of dipping tubes are represented in the engraving, and the way that they are used is by pressing the finger firmly on the top, plunging the other end of the tube into the water, placing it close to the object, and then suddenly removing the finger, when the water will immediately rush into the tube, carrying with it the desired object. Of all these forms, that which is marked *d* is perhaps the most generally useful. The reader will see that they can be made of any shape or size, according to the occasion.

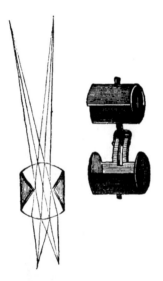

CODDINGTON LENS.

Before bidding farewell to the simple lens, we must glance at a very useful and portable form, which can be carried in the waistcoat pocket, and is quite as powerful and welldefining a magnifier as many compound microscopes. It is termed the "Coddington" Lens, and is nothing more than a polished sphere of glass, with a deep groove cut round its circumference and the hollow filled with black cement. In the illustration one of these useful little instruments is depicted, together with a section of the same, showing the manner in which the rays of light are forced to pass through the lens under similar circumstances. The "field" of this little microscope is very flat, and the definition is good in whatever way it may be held.

For the sake of convenient holding, the handle should be three or four times as long as that of the figure; and after a little practice the observer will set great value on this lens. There is another lens which bears some external resemblance to the Coddington, and is called by the name of the "Stanhope" lens, against which the reader is hereby warned. Dealers often try to induce their chance customers to purchase the Stanhope lens, and the reader may distinguish between them by the fact that both ends of the Coddington lens are alike; whereas in the Stanhope, one has twice the convexity of the other. The Coddington is extremely useful for a cursory examination of any object that may be found in the field, as its great power will enable the observer to make out the details of any minute structure, and to decide whether the object will be worth bringing home and placing under the compound microscope.

COMPOUND MICROSCOPE.

HAVING thus given some little attention to the simple microscope, we will pass to the more complicated apparatus termed the COMPOUND MICROSCOPE.

This invaluable instrument is made in various ways, the chief essential being that one glass is placed close to the object, and the other near the eye. In former days, the tube that contained these glasses was several feet in length, so that the whole affair might easily be mistaken for a great astronomical telescope. Into the details of structure I do not intend to enter, nor to describe the varieties of compound microscopes that are constantly produced by different makers, as the whole of the work would be absorbed in that one department alone. The accuracy with which these instruments are made is almost fabulous, and the number and beauty of their accessories is so great, that the first-rate compound microscope is said to be the only instrument in the world where the performance equals the theory on which it is made.

Such an instrument is beyond the reach of most persons, costing from forty or fifty pounds and upwards, and is therefore quite unsuited to the purposes of the present work. There is, however, a compound microscope which is a really admirable instrument, giving a flat though small field, great magnifying powers, clear definition, and is quite achromatic, *i.e.* without those

EDUCATIONAL MICROSCOPE.

fringes of rainbow colouring which are always seen surrounding the objects in inferior microscopes.

It is furnished with three powers, named the inch, half-inch, and quarter-inch object-glasses, has a sliding stage for the purpose of conveniently moving the object

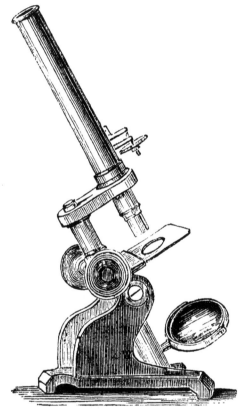

EDUCATIONAL MICROSCOPE.

horns on two pivots so as to suit the position of the head, the "body" or tube where the glasses are set is moved to and from the object by large and fine screws, called technically the "quick and slow motion," and is also supplied with dissecting forceps, a stage forceps, and a "live-box," all fitting into a neat mahogany box, so managed that supplementary appurtenances can be packed when obtained. I have subjected this instrument to careful testing, and can report in very high terms of it. In fact, for every purpose except that of scientific controversy, it is quite as good an instrument as any one could wish to see, and only costs *three guineas;* not half the price of a single object-glass belonging to the larger microscopes.

As the various objects here mentioned require some little explanation, we will treat of them separately.

The lenses constituting the three powers screw on to the lower extremity of the tube, and are so made that, in order to obtain a higher power, all that is needed is to employ all the three, which screw into each other; two giving a less power, and one the least of all. The reader will see the convenience of this arrangement. When an object has been well examined with the lowest power, a second can be added, and the large screw turned so as to bring the glass nearer to the object, which is sure of being in the exact field of the glass. This is of no small consequence, as the hunting for a little object on a large slide under a high power is one of the most provoking of chases, and often forces the

HANDLING THE MICROSCOPE.

observer to remove the high power, replace it by a lower, find the object, get it in the centre of the field, change the glass again, and then bring it down upon the object. The highest power should always be nearest the object.

On the stage, *i.e.* the flat plate of metal immediately under the object-glass, may be seen the raised ledge against which the glass slide holding the object is laid, and which, by sliding up and down the stage, carries the object with it. This movement is necessary, in order to bring every portion of a large object into the field of the microscope. The large-headed screws which form the quick movement or coarse adjustment are seen just behind the stage, and raise or depress the tube by means of a rack and pinion movement. The screw of the fine adjustment is seen just above the horizontal rim, into which the body is fixed, and acts by means of a screw working against a spring.

The mirror, which may be seen below the stage, is so fitted that it can be turned in any direction, so as to throw the rays of light straight or obliquely through the object, either method being equally useful under different circumstances. The heavy stand is made of iron, and affords a firm and solid support to the instrument, a matter of no trifling consequence when the reader reflects that motion is magnified as well as substance, so that if the instrument trembles in the least degree, the object becomes almost invisible, seeming to flutter before the eyes of the observer like the whirring wings of a hovering fly.

All these portions of the instrument are affixed to the stand, but there are other parts of the apparatus which are furnished and used separately.

The dissecting forceps have already been described and figured on page 12, so that no further mention is needful. The stage forceps are of a very different

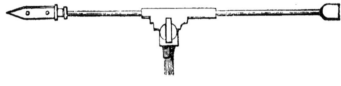

STAGE FORCEPS.

appearance, as may be seen by the accompanying illustration.

The dark-coloured pin at the bottom fits into a hole in one corner of the stage, in which it can be turned freely. A brass socket is hinged to the pin, and bears a steel bar, which passes horizontally through it, and carries at one end the forceps, and at the other end is either sharply pointed, or fitted with a brass cap, into which a piece of cork is firmly pressed. The reader will see that this instrument is capable of being turned in every direction, and as the horizontal bar revolves freely in the socket, any object held in the forceps can be turned round so as to afford a view of every side. To hold the object, one of the pins in the forceps blades is pressed, which separates one blade from the other, and when the pressure is removed, the elasticity of the blades, which are made of steel, brings the points together, and holds the object firmly between them.

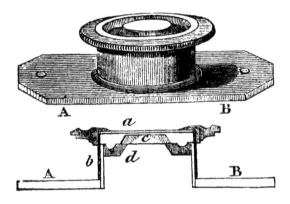

VARLEY'S ANIMALCULE CAGE, OR LIVE-BOX.

Under a low power the stage forceps are almost indispensable, but for the higher can hardly be employed at all, as the light and the focussing are both so difficult of management that the comparatively coarse forceps cannot be successfully used except by a very practised hand.

The Live-Box or Animalcule cage is also a most useful, and in fact a necessary part of the apparatus. In the illustration it is shown, together with a section exhibiting the details of its structure.

It consists of two brass tubes, sliding in each other, and each being furnished at the top with a plate of glass, so arranged that when the upper tube is pressed down to the fullest extent, the glass plates are all but in contact with each other. This instrument is used for examining animalcules, microscopic plants, and other substances, and is very simple in its operation. The upper tube or cap is removed, and a little drop of water containing the object is placed on the glass of the lower tube. The cap is then replaced, and as it is pressed down the drop of water is proportionately flattened.

The live-box represented in the engraving is of a superior description, and is so managed that the glass of the lower tube is thick, and has a groove running round it like the moat surrounding an old castle. The reason of this arrangement is, that the superabundant fluid only runs over the glass into the groove, while the objects remain in their places. A B is the flat brass plate on which the lower tube is fastened, d is the brass-grooved ledge of the tube, c is the thick glass top of that tube, and a is the glass cover of the cap, whose sides are represented by the black perpendicular lines outside the tube.

This kind of cage is especially valuable, as the cap can be fitted with glasses of different strength, so that it can either be employed in the investigation of microscopic animalcules or in flattening sundry substances which need pressure to bring out their details. There is another instrument made for this purpose,

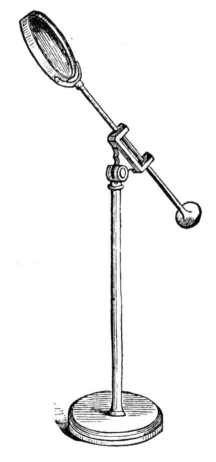

CONDENSER.

termed a "compressorium," but in careful and steady hands the live-box will answer nearly as well.

A "condenser" is generally supplied with the

microscope, and, as its name imports, is used for condensing the light upon an opaque object. It is mounted in various ways, sometimes fitting into a socket on the stage like the stage forceps, but generally on a separate stand as in the engraving. The upright rod consists of two tubes one within the other, which draw out in telescopic fashion, so that the "bull's-eye" lens can be raised to any convenient height. Some little practice is required to use this instrument properly, but when rightly managed it is quite invaluable, bringing out effects which would otherwise be totally invisible. Indeed, with the exception of those objects which are viewed by polarized light, there are none which have so splendid an effect as those which are illuminated by the condenser. So large an amount of light is concentrated in so small a space that when it is refracted from hairs, feathers, or scales, especially the wing-scales of several insects, the whole field of the microscope seems to be filled with resplendent gems, flashing with a radiance that is almost painful in its intensity.

To the under surface of the stage is generally affixed a circular plate of metal, pierced with holes of various diameters, and called a "diaphragm." It is a useful instrument, and is employed for modifying the amount of light which is thrown by the mirror through the

DIAPHRAGM.

hole in the stage. By turning this plate the holes can be successively brought under the hole in the stage, and their centres are made to coincide with the centre of the object-glass by a little spring catch which is

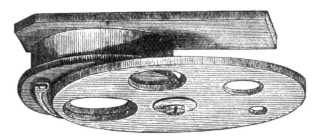

THE DIAPHRAGM.

seen on the left hand of the engraving, and which fits slightly into a notch. As a general rule, the smaller holes should be used with the higher powers, as the "pencil" of light ought to be rather smaller than the diameter of the object-glass. Should the observer wish to shut off all the light, he has only to turn the diaphragm beyond the smallest hole, when the blank portion of the metal will pass over the hole in the stage and effectually answer that purpose.

This little preliminary dissertation is rather uninteresting, but is needful in order to enable the reader to comprehend that which is to follow.

CHAPTER III.

VEGETABLE CELLS AND THEIR STRUCTURE—STELLATE TISSUES—SECONDARY DEPOSIT—DUCTS AND VESSELS—WOOD-CELLS—STOMATA, OR MOUTHS OF PLANTS—THE CAMERA LUCIDA, AND MODE OF USING—SPIRAL AND RINGED VESSELS—HAIRS OF PLANTS—RESINS, SCENTS, AND OILS—BARK CELLS.

WE will now suppose the young observer to have obtained a microscope, and learned the use of its various parts, and will proceed to work with it. As with one or two exceptions, which are only given for the purpose of further illustrating some curious structure, the whole of the objects figured in this work can be obtained without any difficulty, the best plan will be for the reader to procure the plants, insects, &c. from which the objects are taken, and follow the book with the microscope at hand. It is by far the best mode of obtaining a systematic knowledge of the matter, as the quantity of objects which can be placed under a microscope is so vast, that without some guide the tyro flounders hopelessly in the sea of unknown mysteries, and often becomes so bewildered that he gives up the

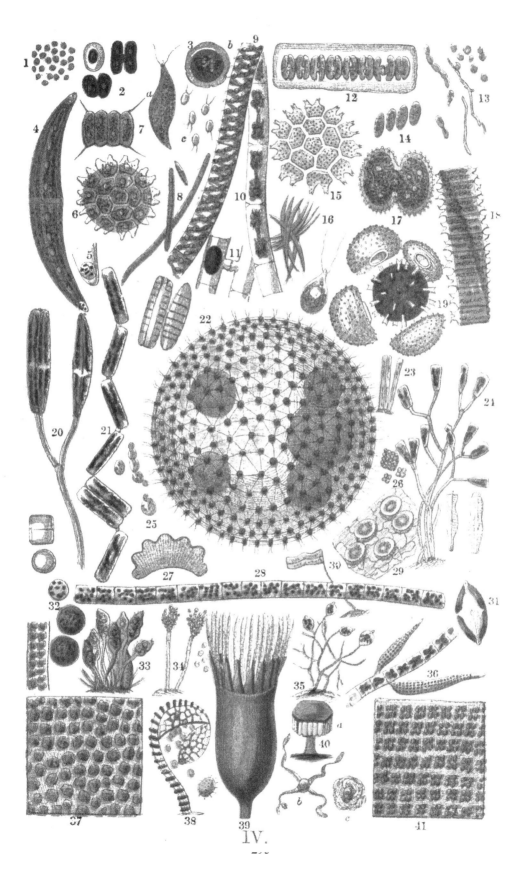

IV.

study in despair of ever gaining any true knowledge of it. I would therefore recommend the reader to work out the subjects which are here mentioned, and then to launch out for himself in the voyage of discoveries. I speak from experience, having myself known the difficulties under which a young and inexperienced observer has to labour in so wide a field, without any guide to help him to set about his work in a systematic manner.

The objects that can be easiest obtained are those of a vegetable nature, as even in London there is not a square, an old wall, a greenhouse, a florist's window, or even a greengrocer's shop, that will not afford an exhaustless supply of microscopic employment. Even the humble vegetables that make their daily appearance on the dinner-table are highly interesting; and in a crumb of potato, a morsel of greens, or a fragment of carrot, the enthusiastic observer will find occupation for many hours.

Following the best examples, we will commence at the beginning, and see how the vegetable structure is built up of tiny particles, technically called "cells."

That the various portions of every vegetable should be referred to the simple cell is a matter of some surprise to one who has had no opportunity of examining the vegetable structure, and indeed it does seem more than remarkable that the tough, coarse bark, the hard wood, the soft pith, the green leaves, the delicate flowers, the almost invisible hairs, and the pulpy fruit should all start from the same point, and owe their origin to the simple vegetable cell. This, however, is the case; and by means of a few objects chosen from different portions of the vegetable kingdom, we shall obtain some definite idea of this curious phenomenon.

On plate 1, fig. 1, may be seen three cells of a somewhat globular form, taken from the common strawberry. Any one wishing to examine these cells for himself may readily do so by cutting a very thin slice from the fruit, putting it on a slide, covering it with a piece of thin glass, which may be cheaply bought at the optician's, together with the glass slides on which the objects are laid, and placing it under a power of two hundred diameters. Should the slice be rather too thick, it may be placed in the live-box and well squeezed, when the cells will exhibit their forms very distinctly. In their primary form, the cells seem to be spherical; but as in many cases they are pressed together, and in others are formed simply by the process of subdivision, the spherical form is not very often seen. The strawberry, being a soft and pulpy fruit, permits the cells to assume a tolerably regular form, and they consequently are more or less globular.

STRUCTURE OF THE CELL.

Where the cells are of nearly equal size, and are subjected to equal pressure in every direction, they force each other into twelve-sided figures, having the appearance under the microscope of flat six-sided forms. Fig. 8, taken from the stem of a lily, is a good example of this form of cell, and many others may be found in various familiar objects.

We must here pause for a moment to define a cell before we proceed farther.

The cell is a closed sac or bag formed of a substance called from its function "cellulose," and containing certain fluid contents as long as it retains its life. In the interior of the cell may generally be found a little dark spot, termed the "núcleus," and which may be seen in fig. 1, to which we have already referred. The object of the nucleus is rather a bone of contention among the learned, but the best authorities on this subject consider it to be the vital centre of the cells, to and from which tends the circulation of the contained fluid. In point of fact, the nucleus may be considered as the heart and brain of the cell. On looking a little closer at the nucleus, we shall find it marked with several small light spots, which are termed "nucléöli."

On the same plate (fig. 2) is a pretty group of cells taken from the internal layer of the buttercup leaf, and chosen because they exhibit the series of tiny and brilliant green dots to which the colour of the leaf is due. The technical name for this substance is " chlorophyll," or "leaf-green," and it may always be found thus dotted in the leaves of different plants, the dots being very variable in size, number, and arrangement.

In the centre of the same plate (fig. 12) is a group of cells from the pith of the elder-tree. This specimen is notable for the number of little " pits " which may be seen scattered across the walls of the cells, and which resemble holes when placed under the microscope. In order to test the truth of this appearance, the specimen was coloured blue by the action of iodine, when it was found that the blue tint spread over the pits together with the cell-walls, showing that the membrane is continuous over the pits.

Fig. 7 exhibits another form of cell, taken from the Spargánium, or bur-reed. These cells are tolerably equal in size, and have assumed a squared shape. They are obtained from the lower part of the leaf. The reader who has any knowledge of entomology will not fail to observe the similarity in form between the six-sided and square cells of plants and the hexagonal and squared facets of the compound eyes belonging to insects and crustaceans. In a future page these will be separately described.

Sometimes the cells take most singular and unexpected shapes, several examples of which will be briefly noticed.

In certain loosely made tissues, such as are found in the rushes and similar plants, the walls of the cells grow very irregularly, so that they push out a number of arms which meet each other in every direction, and assume the peculiar form which is termed "stellate," or star-shaped tissue. Fig. 3 shows a specimen of stellate tissue taken from the seed-coat of the privet, and rather deeply coloured, exhibiting strongly the beautiful manner in which the various arms of the stars meet each other. A smaller group of stellated cells may be seen in fig. 4, taken from the stem of a large Rush, and exemplifying the peculiarities of the structure.

The reader will at once see that this mode of formation leaves a vast number of interstices, and gives great strength with little expenditure of material. In water-plants, such as the reeds, this property is extremely valuable, as they must be greatly lighter than the water in which they live, and at the same time must be endued with considerable strength, in order to resist its pressure.

A less marked example of stellate tissue is given in fig. 11, where the cells are extremely irregular in their form, and do not coalesce throughout. This specimen is taken from the pithy part of a Bulrush. There are very many other plants from which the stellate cells may be obtained, among which the Orange affords very good examples in the so-called "white" that lies under the yellow rind, a section of which may be made with a very sharp knife, and laid under the field of the microscope.

Looking towards the bottom of the plate, and referring to fig. 27, the reader will observe a series of nine elongated cells, placed end to end, and dotted profusely with chlorophyll. These are obtained from the stalk of the common chickweed. Another example of the elongated cell is seen in fig. 14, which is a magnified representation of the rootlets of wheat. Here the cells will be seen in their elongated state, set end to end, and each containing its nucleus. On the left hand of the rootlet (fig. 13) is a group of cells taken from the lowest part of the stem of a wheat plant which had been watered with a solution of carmine, and had taken up a considerable amount of the colouring substance. Many experiments on this subject were made by the Rev. Lord S. G. Osborne, and may be seen at full length in the pages of the *Microscopical Journal*, the subject being too

MULTIPLICATION OF CELLS. 43

large to receive proper treatment in the very limited space which can here be given to it.

One very remarkable point is, that the carmine was always found to be taken most plentifully into the nucleoli, and to give them a very deep colouring. These specimens exhibited the phenomenon which has already been casually mentioned, that the rotation of the granules in the interior of the cell takes place to and from the nuclei.

Fig. 9 on the same plate exhibits two notable peculiarities—the irregularity of the cells, and the copiously pitted deposit with which they are covered. The irregularity of the cells is mostly produced by the way in which the multiplication takes place, namely, by division of the original cell into two or more portions, so that each portion takes the shape which is assumed when a component part of the parent cell. In this case the cells are necessarily very irregular, and when they are compressed from all sides, they form solid figures of many sides, which, when cut through, present a flat surface marked with a variety of irregular outlines. This specimen is taken from the rind of a Gourd.

The "pitted" structure which is so well shown in this figure is caused by a layer of matter which is deposited in the cell and thickens its walls, and which is perforated with a number of very minute holes called

SECONDARY DEPOSIT.

"pits." This substance is called "secondary deposit." That these pits do not extend through the real cell wall has already been shown in fig. 12, p. 29.

This secondary deposit—I pray the reader's pardon for using such language, but there is no alternative—is exhibited in more modes than one. In some cases it is deposited in rings round the cell, and is clearly placed there for the purpose of strengthening the general structure. Such an example may be found in the Mistletoe, fig. 5, where the secondary deposit has formed itself into clear and bold rings, that evidently give considerable strength to the delicate walls which they support. Fig. 10 gives another good instance of a similar structure; differing from the preceding specimen in being much longer and containing a greater number of rings. This object is taken from an anther of the Narcissus. Among the many plants from which similar objects may be obtained, the Yew is, perhaps, one of the most prolific, as ringed wood-cells are abundant in its formation, and probably aid greatly in giving to the wood the strength and elasticity which have long made it so valuable in the manufacture of bows.

Before taking leave of the cells and their remarkable forms, we will just notice one example which has been drawn in fig. 6. This is a congeries of cells, containing their nuclei, starting originally end to end, but swelling

DUCTS AND VESSELS.

and dividing at the top. This is a very young group of cells from the inner part of a Lilac bud, and is here introduced for the purpose of showing the great similarity of all vegetable cells in their earliest stages of existence. No one who did not know the history of that little group could imagine what would be its perfected condition, for it might either spread itself into a leaf, or extend itself into a flower, or end its days as a hair, for all the indications that it affords of its future.

Having now examined the principal forms of cells, we arrive at the "ducts," a term which is applied to those long and delicate tubes which are formed of a number of cells set end to end, their walls of separation being absorbed. At first the young microscopist is apt to puzzle himself between ducts and vessels, but may easily set himself right by remembering that ducts are squared at their ends, and vessels or wood-cells are pointed.

In fig. 19 the reader will find a curious example of the "dotted duct," so called from the multitude of little markings that cover its walls, and which are arranged in a spiral order. Like the pits and rings already mentioned, the dots are composed of secondary deposit in the interior of the tube, and vary very greatly in number, function, and dimensions. This example is taken from the wood of the willow, and is remarkable for the extreme closeness with which the dots are packed together.

Immediately on the right hand of the preceding figure may be seen another example of a dotted duct (fig. 20), taken from a Wheat stem. In this instance the cells are not nearly so long, but are wider than in the preceding example, and are marked in much the same way with a spiral series of dots. About the middle of the topmost cell is shown the short branch by which it communicates with the neighbouring duct.

Fig. 23 exhibits a duct taken from the common Carrot, in which the secondary deposit is placed in such a manner as to resemble a net of irregular meshes wrapped tightly round the duct. For this reason it is termed a "netted duct." A very curious instance of these structures is given in fig. 26, at the bottom of the plate, where are represented two small ducts from the wood of the Elm. One of them—that on the left hand—is wholly marked with spiral deposit, the spires being complete; while in the other instance the spiral is comparatively imperfect, and the cell walls are marked with pits. If the reader would like to examine these structures more attentively, he will find plenty of them in many familiar garden vegetables, such as the common Radish, which is very prolific in these interesting portions of vegetable nature.

There is another remarkable form in which this secondary deposit is sometimes arranged, that is well

SCALARIFORM TISSUE.

worthy our notice. An example of this structure is given in fig. 18, taken from the stalk of the common Fern or Brake. It is also found in very great perfection in the Vine. On inspecting the illustration, the reader will observe that the deposit is arranged in successive bars or steps, like those of a winding staircase. In allusion to the ladder-like appearance of this formation, it is called "scalariform," or ladder-like form.

In the wood of the Yew, to which allusion has already been made, there is a very peculiar structure, found only in those trees that bear cones, and therefore termed the coniferous glandular structure. Fig. 16 is a section of a common Cedar pencil, the wood, however, not being that of the true cedar, but of a species of fragrant Juniper. This specimen shows all the peculiar formation which has just been mentioned, and in addition exhibits the situation of the oil-cells which give to the wood its well-known fragrance.

Any piece of deal or pine will exhibit the same peculiarities in a very marked manner, as is seen in fig. 24. A specimen may be readily obtained by making a very thin shaving with a sharp plane. In this example, the deposit has taken a partially spiral form, and the numerous circular pits with which it is marked are only in single rows. In several other specimens of coniferous woods, such as the Araucaria, or Norfolk Island Pine, there are two or three rows of pits.

A peculiarly elegant example of this spiral deposit may be seen in the wood of the common Yew, fig. 17. If an exceedingly thin section of this wood be made, the very remarkable appearance will be shown which is exhibited in the illustration. The deposit has not only assumed the perfectly spiral form, but there are two complete spirals, arranged at some little distance from each other, and producing a very pretty effect when seen through a good lens.

The pointed, elongated shape of the wood-cells, is very well shown in the common Elder-tree (see fig. 15). In this instance the cells are without markings, but in general they are dotted like fig. 21, an example cut from the woody part of the Chrysanthemum stalk. This affords a very good instance of the wood-cell, as its length is considerable, and both ends perfect in shape. On the right hand of the figure is a drawing of the wood-cell found in the Lime-tree (fig. 22), remarkable for the extremely delicate spiral markings with which it is adorned. In these wood-cells the secondary deposit is so plentiful, that the original membranous character of the cell-walls is entirely lost, and they become elongated and nearly solid cases, having but a very small cavity in their centre. It is to this deposit that the hardness of wood is owing, and the reader will easily see the reason why the old wood is so much harder than the young and new shoots. In order to permit the passage of the fluids which maintain the life of the part, it is needful that the cell wall be left thin and permeable in certain places, and this object is attained either by the "pits" described on page 29, or by the intervals between the spiral deposit.

At the right-hand bottom corner of Plate I. (fig. 20), may be seen a prettily marked object, which is of some interest. It is a slice stripped from the outer coat of the Holly-berry, and is given for the purpose of illustrating the method by which plants are enabled to breathe the atmospheric air on which they depend as much as ourselves, though their respiration is slower. Among the mass of net-like cells may be seen three curious objects, bearing a rather close resemblance to split kidneys. These are the mouths, or "stómata" as they are scientifically called, scientific people always liking to use a long Greek word where a short English one would do as well.

In the centre of the mouths may be seen a dark spot, which is the aperture through which the air communicates with the passages between the cells in the interior of the structure. In the flowering plants their shape is generally rounded, though they sometimes take a

E

squared form, and they regularly occur at the meeting of several surface cells. Their edges are protected by certain "pore-cells," or "guard-cells," so called from their function, which, by their change of form, cause the mouth to open or shut, as is best for the plant. In young plants these guard-cells are very little below the surface of the leaf or skin, but in others they are sunk quite beneath the layer of cells, forming the outer coat of the tissue. There are other cases, where they are slightly elevated above the surface.

Stomata are found chiefly in the green portions of plants, and are most plentiful on the under side of leaves. It is, however, worthy of notice, that when an aquatic leaf floats on the water, the mouths are only to be found on the upper surface. These curious and interesting objects are to be seen in many structures where we should hardly think of looking for them, for they may be found existing on the delicate skin which envelops the kernel of the common walnut. As might be expected, their dimensions vary with the character of the leaf on which they exist, being large upon the soft and pulpy leaves, and smaller upon those of a hard and leathery consistence. The reader will find ample amusement, and will gain great practical knowledge of the subject, by taking a plant, say a tuft of Groundsel, and stripping off portions of the external skin or "epidermis" from the leaf or stem, &c., so as to note the different sizes and shapes of the stomata.

Let me here notice that the young microscopist should always sketch every object which he views, as

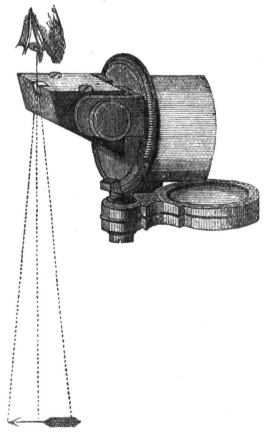

CAMERA LUCIDA.

he will thus obtain a far clearer and more lasting impression of the subject than can be gained merely by examining one object after another. By far the best mode of sketching is to use the Camera Lucida, a figure of which is given on the preceding page.

It consists of a prism of glass, affixed to a brass tube, which is slid upon the eye-piece of the microscope, after removing the perforated cap. The microscope is then laid horizontally, and the eye applied to the little oblong aperture, as seen in the figure. A piece of white paper is then laid immediately below the eye-piece, where the large arrow is seen in the illustration, and the object will apparently be thrown upon the paper. A hard and very sharply pointed pencil should then be taken, and the outline of the object traced upon the paper.

On the first two or three trials, the beginner will perhaps fail utterly, as there is some difficulty in seeing the object and the pencil simultaneously, but with a little practice success is certain. The secret of this instrument is to keep the pupil of the eye exactly upon the edge of the prism, so that the one half of the pupil looks at the object, and the other half at the paper and pencil.

The great value of the camera lucida lies in the fact that a perfectly accurate sketch of any object can be taken by one who knows nothing whatever of drawing, the process being much the same as that of drawing on a transparent slate. The saving of time is very great, as a sketch can be made with a rapidity

STEEL MIRROR AND NEUTRAL GLASS.

that seems almost marvellous to a bystander, and the drawing is sure to be in perfect proportion. The reader will see by the diagram of the rays that the size of the drawing will be precisely according to the distance of the camera from the paper, so that if he wishes to make a small sketch, he places the paper close to the prism, and if he desires a large drawing, removes it to some distance. Indeed, a very large diagram, suitable for the walls of a lecture-room, can be made by laying the paper on the floor, and drawing with a pencil fixed in a very long handle. Minute details of colour, &c. are added after the outline sketch is completed.

There are several modes of attaining the same end, among which the steel mirror and the neutral glass are the best. The former is a very tiny circular plate of steel, placed in the same position as the edge of the camera lucida, and turned at such an angle that the eye looks *in* the mirror at the object, and *around* it at the paper. The cheapest instrument for this purpose is the neutral glass plate, invented by Dr. L. Beale, and which will be found to perform admirably the functions of the camera lucida, without costing more than one-fourth of the price.

By means of a little ingenuity, the camera can be mounted on the stand of the dissecting microscope, and can be thus employed for sketching solid objects or copying drawings. By a neat adjustment of the height of the instrument the camera lucida will always produce a sketch exactly to scale, and will give a drawing of twenty, thirty, or a hundred diameters at will. It will therefore be seen that the draughtsman must be very careful to have the instrument placed at an established series of heights above the paper, and he should always append to each sketch the magnifying power with which it was drawn. Thus—

"Epidermis and stomata of Holly-berry, $\frac{4}{10} \times 150$."

By which formula will be understood that the object-glass employed is that called the "four-tenths" glass, and that the object is magnified one hundred and fifty diameters. Unless this precaution be taken, great confusion in respect to the size of the object is sure to arise.

To return to our former subject.

On the opposite bottom corner of Plate I., fig. 25 is an example of a stoma taken from the outer skin of a Gourd, and here given for the purpose of observing the curious manner in which the cells are arranged about the mouth, no less than seven cells being placed round the single mouth, and the others arranged in a partially circular form around them.

Turning to Plate II. we find several other examples of stomata, the first of which, fig. 1, is obtained from the under surface of the Buttercup leaf, by stripping off the external skin, or "epidermis," as it is scientifically termed. The reader will here notice the slightly waved outlines of the cell-walls, together with the abundant spots of chlorophyll with which the leaf is coloured. In this example, the stomata appear open. The closure or expanding of the mouth depends most on the state of the weather, and, as a general rule, they are open by day and closed at night. Some plants are provided with four guarding cells, which are arranged in different manners.

A remarkably pretty example of stomata and elongated cells is to be obtained from the leaf of the common Iris, and may be prepared for the microscope by simply tearing off a strip of the cuticle from the under side of the leaf, laying it on a slide, putting a little water on it, and covering it with a piece of thin glass. (See Plate II. fig. 2.) The peculiar elongated cells will not be seen equally spread over the whole surface of the leaf, as they are hidden by a congeries of shorter and thicker cells, covered with chlorophyll which conceals their shape. There are, however, a number of longitudinal bands running along the leaf, where these cells and mouths appear, without anything to veil their beauty. The stomata are not placed at regular intervals, for it often happens that the whole field of the microscope will be filled with cells without a single stoma, while a group of three or four are often seen clustered closely together.

Fig. 3 on the same plate exhibits a specimen of beautifully waved cells, without mouths, which are found on the upper surface of the Ivy leaf. These are difficult to arrange from the fresh leaf, but are easily shown by steeping the leaf in water for some time, and then tearing away the cuticle. The same process may be adopted with many leaves and cuticles, and in some cases the immersion must be continued for many days, and the process of decomposition aided by a very little nitric acid in the water, or by boiling.

On the same plate are three examples of spiral and ringed vessels, being types of an endless variety of these beautiful and interesting structures. Fig. 4 is a specimen of a spiral vessel taken from the Lily, and is a beautiful example of a double spire. The deposit which forms this spiral is very strong, and it is to the vast number of these vessels that the stalk owes its well-known elasticity. In many cases the spiral vessels are sufficiently strong to be visible to the naked eye, and to bear uncoiling. For example, if a leaf-stalk of geranium be broken across, and the two fragments gently drawn asunder, a great number of spiral vessels will be seen connecting the broken ends. In this case

ANALOGIES OF NATURE.

the delicate membranous walls of the vessel are torn apart, and the stronger fibre which is coiled spirally within them unrolls itself in proportion to the force employed. In many cases these coils are so strong that they will sustain the weight of an inch or so of the stalk.

In fig. 5 is seen a still more bold and complex form of this curious structure; being a coil of five threads laid closely against each other, and forming while remaining in their natural position an almost continuous tube. This specimen is taken from the root of the Water Lily, and requires some little care to exhibit its structure properly.

Every student of nature must be greatly struck with the analogies between different portions of the visible creation. These spiral structures which we have just examined are almost identical in appearance, and entirely so in functions, with the threads that are coiled within the breathing tubes of insects. Their object in both cases is twofold, namely to give strength and elasticity to a delicate membrane, and to preserve the tube in its proper form despite the flexure to which it may be subjected. When we come to the anatomy of the insect in a future page, we shall see this structure further exemplified.

In some cases the deposit, instead of forming a

spiral coil, is arranged in a series of rings, and is then termed "annulated." A very good example of this formation is given in fig. 6, being a sketch of a ringed vessel, taken from a stalk of the common Rhubarb. To see these ringed vessels properly, the simplest plan is to boil the rhubarb until it is quite soft, then to break down the pulpy mass until it is flattened, to take some of the most promising portions with the forceps, lay them on the slide and press them down with a thin glass cover. They will not be found scattered at random through the fibres, which elsewhere present only a congeries of elongated cells, but are seen grouped together in bundles, and with a little trouble may be well isolated, and the pulpy mass worked away so as to show them in their full beauty. As may be seen in the illustration, the number of the rings and their arrangement is extremely variable.

The Hairs of plants always form very interesting objects, and are instructive to the student, as they afford valuable indications of the mode in which plants grow. They are all appendages of and arising from the skin or epidermis; and although their simplest form is that of a projecting and elongated cell, the variety of shapes which are assumed by these organs is inexhaustible. On Plate II. are examples of some of the more striking forms, which will be briefly described.

The simple hair is well shown in figs. 18, 19, and 32, the first being from the flower of the Heartsease, the second from a Dock-leaf, and the third from a Cabbage. In fig. 18 the hair is seen to be but a single projecting cell, consisting only of a wall and the contents. In fig. 19 the hair has become more decided in shape, having assumed a somewhat dome-like form, and in fig. 32 it has become considerably elongated, and may at once be recognised as a true hair.

In fig. 8 is a curious example of a hair taken from the white Arabis, one of the cruciform flowers, which is remarkable for the manner in which it divides into two branches, each spreading in opposite directions. Another example of a forked hair is seen in fig. 13, but in this instance the hair is composed of a chain of cells, the three lower forming the stem of the hair, and the two upper being lengthened into the lateral branches. This hair is taken from the common Southernwood.

In most cases of long hairs, the peculiar elongation is formed by a chain of cells, varying greatly in length and development. Several examples of these hairs will be seen on the same plate.

Fig. 9 is a beaded hair from the Marvel of Peru, which is composed of a number of separate cells placed end to end, and connected by slender threads in a manner that strongly reminds the observer of a chain of beads strung loosely together, so as to show the thread by which they are connected with each other. Another good example is seen at fig. 11, in a hair taken from the leaf of the Sow-thistle. In this case the beads are strung closely together, and when placed under a rather high power of the microscope, have a beautifully white and pearly aspect. The leaf must be dry and quite fresh, and the hairs seen against the green of the leaf. Fig. 39 represents another beaded hair taken from the Virginian Spiderwort, or Tradescantia. This hair is found upon the stamens, and is remarkable for the beautifully beaded outline, the fine colouring, and the spiral markings with which each cell is adorned.

A still further modification of these many-celled hairs is found in several plants, where the hairs are formed by a row of ordinarily shaped cells, with the exception of the topmost cell, which is suddenly elongated into a whip-like form. Fig. 22 represents a hair of this kind, taken from the common Groundsel; and fig. 36 is a still more curious instance, found upon the leaf of the Thistle. The reader may have noticed the peculiar white "fluffy" appearance of the thistle leaf when it is wet after a shower of rain. This appearance is produced by the long lash-like ends of the hairs, which are bent down by the weight of the moisture, and lie almost at right angles with the thicker portions of the hair.

An interesting form of hair is seen in the "sting" of the common Nettle. This may readily be examined by holding a leaf edgeways in the stage forceps, and laying it under the field of the microscope. In order to get the proper focus throughout the hair, the finger should be kept upon the screw movement, and the hair brought gradually into focus from its top to its base. The general idea of this hair is not unlike that which characterises the sting of the bee or wasp. The acrid fluid which causes the pain is situated in the enlarged base of the hair, and is forced through the long straight tubular extremity by means of the pressure exerted when the sting enters the skin. At the very extremity of the perfect sting is a slight bulb-like swelling, which serves to confine the acrid juice, and which is broken off on the least pressure. The sting is seen in fig. 43. The extremities of many hairs present very curious forms, some being long and slender, as in the examples already mentioned, while others are tipped with knobs, bulbs, clubs, or rosettes in endless variety.

Fig. 12 is a hair of the Tobacco leaf, exhibiting the two-celled gland at the tip, containing the peculiar principle of the plant, which goes scientifically by the

62   BRANCHED HAIRS.

name of "nicotine." The reader will see how easy it is to detect adulteration of tobacco by means of the microscope. The leaves most generally used for this purpose are the Dock and the Cabbage, so that if a very little portion of leaf be examined, the character of the hairs will at once inform the observer whether he is looking at the real article or its substitute.

Fig. 15 is a hair from the flower of the common yellow Snapdragon, which is remarkable for the peculiar shape of the enlarged extremity, and for the spiral markings with which it is decorated. Fig. 16 is a curious little knobbed hair found upon the Moneywort, and fig. 17 is an example of a double-knobbed hair, taken from the Geum. Fig. 34 affords a very curious instance of a glandular hair, the stem being built up of cells disposed in a very peculiar fashion, and the extremity being developed into a beautiful rosette-shaped head. This hair came from the garden Verbena.

Curiously branched hairs are not at all uncommon, and some very good and easily obtained examples are given on Plate II.

Fig. 28 is one of the multitude of branched hairs that surround the well-known fruit of the Plane-tree, the branches being formed by some of the cells pointing outward. These hairs do not assume precisely the same shape, for fig. 30 exhibits another hair from the

PERFUME CELLS. 63

same locality, on which the spikes are differently arranged, and fig. 29 is a sketch of another such hair, where the branches have become so numerous and so well developed that they are quite as conspicuous as the parent stem.

One of the most curious and interesting forms of hair is that which is found upon the Lavender leaf, and which gives it the peculiar bloom-like appearance on the surface.

This hair is represented in figs. 40 and 41. On fig. 40 the hair is shown as it appears when looking directly upon the leaf, and in fig. 41 a section of the leaf is given, showing the mode in which the hairs grow into an upright stem, and then throw out horizontal branches in every direction. Between the two upright hairs, and sheltered under their branches, may be seen a glandular appendage, not unlike that which is shown in fig. 16. This is the reservoir containing the perfume, and it is evidently placed under the spreading branches for the benefit of their shelter. On looking upon the leaf by reflected light, the hairs are beautifully shown, extending their arms on all sides; and the globular perfume-cells may be seen scattered plentifully about; gleaming like pearls through the hair branches under which they repose. They will be found more numerous on the under-side of the leaf.

This object will serve to answer a question which the reader has probably put to himself ere this, namely, Where are the fragrant resins, scents, and oils stored? On Plate I. fig. 16, will be seen the reply to the first question, fig. 41 of the present plate has answered the second question, and fig. 42 will answer the third. This figure represents a section of the rind of an Orange, the flattened cells above constituting the delicate yellow skin, and the great spherical object in the centre being the reservoir in which the fragrant essential oil is stored. The covering is so delicate that it is easily broken, so that even by handling an orange some of the scent is sure to come off on the hands, and when the peel is stripped off and bent double, the reservoirs burst in myriads, and fling their contents to a wonderful distance. This may be easily seen by squeezing a piece of orange peel opposite a lighted candle, and noting the distance over which the oil will pass before reaching the flame, and bursting into little flashes of light. Other examples are given on the same Plate.

Returning to the barbed hairs, we may see in fig. 35 a highly magnified view of the "pappus" hair of a Dandelion, *i.e.* the hairs which fringe the arms of the parachute-like appendage which is attached to the seed. The whole apparatus will be seen more fully on Plate III. figs. 44, 45, 46. This hair is composed of a

KNOBBED HAIRS.

double layer of elongated cells lying closely against each other, and having the ends of each cell jutting out from the original line. A simpler form of a double-celled, or more properly a "duplex" hair, will be seen in fig. 44. This is one of the hairs from the flower of the Marigold, and has none of the projecting ends to the cells.

In some instances the cell-walls of the hairs become exceedingly hardened by secondary deposit, and the hairs are then better known by the name of spines. Two examples of these strengthened hairs are seen in figs. 37 and 38, the former being picked from the Indian Fig-cactus, and well known to those persons who have been foolish enough to handle the fig roughly before feeling it. The wounds which these spines will inflict are said to be very painful, and have been compared to those produced by the sting of the wasp. The latter hair is taken from the Opuntia.

The mode in which hairs increase in length is well shown in fig. 10, which represents the extreme tip of a hair from the Hollyhock leaf, subjected to a lens of very high power.

Many hairs assume a star-like appearance, an aspect which may be produced in different ways. Sometimes a number of simple hairs start from the same base, and by radiating in different directions produce the stellate effect. An example of this kind of hair may be seen in fig. 14, which is a group of hairs from the Hollyhock leaf. There is another mode of producing the star-shape, which may be seen in fig. 45, a hair taken from the leaf of the Ivy.

Several hairs are covered with curious little branches or protuberances, and present many other peculiarities of form, which throw a considerable light upon certain problems in scientific microscopy.

Fig. 33 represents a hair of two cells taken from the flower of the well-known Dead-nettle, which is remarkable for the number of knobs scattered over its surface. A similar mode of marking is seen in fig. 31, a club-shaped hair covered with external projections, found in the flower of the Lobelia. In order to exhibit these markings well, a power of two hundred diameters is needed. Fig. 21 shows this dotting in another hair from the Dead-nettle, where the cell is drawn out to a great length, but is still covered with these markings.

Fig. 20 is an example of a very curious hair taken from the throat of the Pansy. This hair may readily be obtained by pulling out one of the petals, when the hairs will be seen at its base. Under the microscope it has a particularly beautiful appearance, looking just like a glass walking-stick covered with knobs, not unlike those huge, knobby, club-like sticks in which some farmers delight, where the projections have been formed by the pressure of a honeysuckle or other climbing plant.

A hair of a similar character, but even more curious is found in the same part of the flower of the garden Verbena (see fig. 27), and is not only beautifully translucent, but is coloured according to the tint of the flower from which it is taken. Its whole length is covered with large projections, the joints much resembling antennæ belonging to certain insects, and each projection is profusely spotted with little dots, formed by elevation of the outer skin or cuticle. These are of some value in determining the structure of certain appearances upon petals and other portions of flowers, and may be compared with figs. 33 to 35 on Plate III.

Fig. 26 offers an example of the squared cells which usually form the bark of trees. This is a transverse section of Cork, and perfectly exhibits the form of bark cells. The reader is very strongly advised to cut a delicate section of the bark of various trees, a matter very easily accomplished with the aid of a sharp razor and a steady hand.

Fig. 24 is a transverse section through one of the scales of a Pine-cone, and is here given for the purpose of showing the numerous resin-filled cells which it displays. This may be compared with fig. 16 of Plate I.

Fig. 25 is a part of one of the "vittæ," or oil reservoirs from the fruit of the Caraway, showing the cells containing the globules of caraway oil. This is rather a curious object, because the specimen from which it was taken was boiled in nitric acid, and yet retained some of the oil globules. Immediately above it may be seen (fig. 23) a transverse section of the Beechnut, showing a cell with its layers of secondary deposit. These are but short and meagre accounts of a very few objects, but space will not permit of further elucidation, and the purpose of this little work is not to exhaust the subjects of which it treats, but to incite the reader to investigation on his own account, and to make his task easier than if he had undertaken it unaided.

In the cuticles of the Grasses and the Mare's-tails is deposited a large amount of pure flint. So plentiful is this substance, and so equally is it distributed, that it can be separated by heat or acids from the vegetable parts of the plant, and will still preserve the form of the original cuticle, with its cell-walls, stomata, and hairs perfectly well defined.

Fig. 7, Plate II., represents a piece of wheat chaff or "bran," that has been kept at a white heat for some time, and then mounted in Canada balsam. I prepared the specimen from which the drawing was made by laying the chaff on a piece of platinum, and holding it

VI.

over the spirit-lamp. A good example of the silex or flint in wheat is often given by the remains of a straw fire, where the stems may be seen still retaining their tubular form, but fused together into a hard glassy mass. It is this substance that cuts the fingers of those who handle the wild grasses too roughly, the edges of the blades being serrated with flinty teeth, just like the obsidian swords of the ancient Mexicans, or the shark's-tooth falchion of the New Zealander.

CHAPTER IV.

STARCH, ITS GROWTH AND PROPERTIES—SURFACE CELLS OF PETALS—POLLEN AND ITS FUNCTIONS—SEEDS—MOUNTING OBJECTS IN CANADA BALSAM—MOUNTING OBJECTS DRY AND IN CELLS—HOW TO MAKE CELLS—TURN-TABLE—PRESERVATIVE FLUIDS.

THE white substance so dear to the laundress under the name of Starch is found in a vast variety of plants, being distributed more widely than most of the products which are found in the interior of vegetable cells.

The starch grains are of very variable size even in the same plant, and their form is as variable as their size. Sometimes the grains are found loosely packed in the interior of the cells, and are then easily recognised by their peculiar form and the delicate lines with which they are marked; but in many places they are pressed so closely together, that they assume an hexagonal shape under the microscope, and bear a close resemblance to ordinary twelve-sided cells. In other plants again, the grains never advance beyond the very minute form in which they seem to commence their existence; and in some, such as the common Oat, a great number of very little granules are compacted together so as to resemble one large grain.

There are several methods of detecting starch in those cases where its presence is doubtful; and the two modes that are usually employed are polarized light and the iodide of potassium. When polarized light is employed—a subject on which we shall have something to say presently—the starch grains assume the characteristic "black-cross," and when a plate of selenite is placed immediately beneath the slide containing the starch grains, they glow with all the colours of the rainbow. The second plan is to treat them with a very weak solution of iodide of potassium, and in this case the iodine has the effect on the starch granules of staining them blue. They are so susceptible of this colour, that when the liquid is too strong, the grains actually become black from the amount of iodine which they imbibe.

Nothing is easier than to procure the starch granules in the highest excellence. Take a raw Potato, and with a razor cut a very thin slice from its interior, the direction of the cut not being of the slightest importance. Put this delicate slice upon a slide, drop a little water upon it, cover it with a piece of thin glass, give it a good squeeze, and place it under a power of a hundred or a hundred and fifty diameters. Any part of the slice, provided that it be very thin, will then present the appearance shown in Plate III. fig. 9, where an ordinary cell of potato is seen filled loosely with starch-grains of different sizes. Around the edges of the slice a vast number of starch granules will be seen, which have been squeezed out of their cells by pressure, and which are now floating freely in the water. As cold water has no perceptible effect upon starch, the grains are not altered in form by the moisture, and can be examined at leisure.

On focussing with great care, the surface of each granule will be seen to be covered with very minute dark lines, arranged in a manner which can be readily comprehended from fig. 4, which represents two granules of potato-starch, as they appear when removed from the cell in which they took their origin. All the lines evidently refer to the little dark spots at the end of the granule, called technically the "hilum."

The mode by which the starch granules increase in size has long been a problem to microscopists, and seems likely to remain so for the present, one party asserting that the outermost layers are the youngest, and others that they are the oldest. All, however, are agreed upon the one point, that the delicate lines are the boundaries of successive layers of the substance of which the granule is composed.

In the earliest stages of their growth, the starch granules appear to be destitute of these markings, or at all events they are so few and so delicate as not to be visible even with the most perfect instruments, and it is not until they assume a comparatively large size that the external markings become distinctly perceptible.

We will now glance at the examples of starch which are given in the plate, and which are a very few out of the many that might be figured.

Fig. 2 represents the starch of Wheat, the upper grain being seen in front, the one immediately below it being in profile, and the two others being examples of smaller grains. Fig. 6 is a specimen of a very minute form of starch, where the granules do not seem to advance beyond their earliest stage. This specimen is obtained from the Parsnip; and although the magnifying power is very great, their dimensions are exceedingly small, and, except to a very practised eye, they could not be recognisable as starch grains.

Fig. 3 is a good example of a starch-grain of Wheat, exemplifying the change that takes place by the effects of combined heat and moisture. It has already been observed that cold water exercises little, if any, per-

COMPOUND GRANULES.

ceptible influence upon the starch; but it will be seen from the illustration, that hot water has a very powerful effect. When subjected to this treatment, the granule swells rapidly, and at last bursts, the contents escaping in a flocculent cloud, and the external membrane collapsing into the form which is shown in fig. 3, which was taken out of a piece of hot pudding. A similar form of wheat starch may also be detected in bread, and accompanied unfortunately by several other substances not generally presumed to be component parts of the "staff of life."

In fig. 7 are represented some grains of starch from West Indian Arrowroot, and fig. 8 exhibits the largest kind of starch-grain known, obtained from the tuber of a species of Canna, supposed to be *C. edúlis*, a plant similar in characteristics to the arrowroot. The popular name of this starch is "Tous les Mois," and under that title it may be obtained from the opticians.

Fig. 10 shows the starch-granules from Indian Corn, as they appear before they are compressed into the honeycomb-like structure which has already been mentioned. Even in that state, however, if they are treated with iodine, they exhibit the characteristics of starch in a very perfect manner. Fig. 11 is starch from Sago, and fig. 12 from Tapioca, and in both these instances the several grains have been injured by the heat employed in preparing the respective substances for the market.

Fig. 13 exhibits the granules obtained from the root of the Water-Lily, and fig. 14 is a good example of the manner in which the starch-granules of Rice are pressed together so as to alter the shape and puzzle a novice. Fig. 16 is the compound granule of the Oat, which has already been mentioned, together with some of the simple granules separated from the mass; and fig. 15 is an example of the starch-grains obtained from the underground stem of the Horse-bean. It is worthy of mention that the close adhesion of the Rice starch into those masses is the cause of the peculiar grittiness which distinguishes rice flour to the touch.

In Plate III. fig. 1, may be seen a curious little drawing, which is a sketch of the Laurel-leaf cut transversely, and showing the entire thickness of the leaf. Along the top may be seen the delicate layer of "varnish" with which the surface of the leaf is covered, and which serves to give to the foliage its peculiar polish. This varnish is nothing more than the translucent matter which binds all the cells together, and which is poured out very liberally upon the surface of the leaf. The lower part of this section exhibits the cells of which the leaf is built, and towards the left hand may be seen a cut end of one of the veins of the leaf, more rightly called a wood-cell.

We will now examine a few examples of surface cells.

Fig. 5 is a portion of epidermis stripped from a Capsicum pod, exhibiting the remains of the nuclei in the centre of each cell, together with the great thickening ot the wall-cells and the numerous pores for the transmission of fluid. This is a very pretty specimen for the microscope, as it retains its bright red colour, and even in old and dried pods exhibits the characteristic markings.

In the centre of the plate may be seen a wheel-like arrangement of the peculiar cells found on the petals of six different flowers, all easily obtainable, and mounted without difficulty.

Fig. 30 is the petal of a Geranium (Pelargonium), a very common object on purchased slides. It is a most lovely subject for the microscope, whether it be examined with a low or a high power, in the former instance exhibiting a most beautiful "stippling" of pink, white, and black, and in the latter showing the six-sided cells with their curious markings.

In the centre of each cell is seen a radiating arrangement of dark lines with a light spot in the middle, looking very like the mountains on a map. These lines were long thought to be hairs; but Mr. Tuffen

West, in an interesting and elaborate paper on the subject, has shown their true nature. From his observations it seems that the beautiful velvety aspect of flower petals is owing to these arrangements of the surface cells, and that their rich brilliancy of colour is due to the same cause. The centre of each cell-wall is elevated as if pushed up by a pointed instrument from the under side of the wall, and in different flowers this elevation assumes different forms. Sometimes it is merely a slight wart on the surface, sometimes it becomes a dome, while in other instances it is so developed as to resemble a hair. Indeed, Mr. West has concluded that these elevations are nothing more than rudimentary hairs.

The dark radiating lines are shown by the same authority to be formed by wrinkling of the membrane forming the walls of the elevated centre, and not to be composed of "secondary deposit," as has generally been supposed.

Fig. 31 represents the petal of the common Periwinkle, differing from that of the geranium by the straight sides of the cell-walls, which do not present the toothed appearance so conspicuous in the former flower. A number of little tooth-like projections may be seen on the interior of the cells, their bars affixed to the walls and their points tending towards the centre, and these teeth are, according to **Mr. West,** formed of secondary deposit.

In fig. 32 the petal of the common garden **Balsam** is shown, where the cells are observed to be elegantly waved on their outlines, and with plain walls. The petal of the Primrose is seen in fig. 34, and that of the yellow Snapdragon in fig. 33, in which latter instance the surface cells assume a most remarkable shape, running out into a variety of zigzag outlines that quite bewilder the eye when the object is first placed under the microscope. Fig. 35 is the petal of the common Scarlet Geranium.

In several instances these petals are too thick to be examined without some preparation, and glycerine will be found well adapted for that purpose. The young microscopist must, however, beware of forming his ideas upon preparations of dried leaves, petals, or hairs, and should always procure them in their fresh state whenever he desires to make out their structure. Even a fading petal should not be used, and if the flowers are gathered for the occasion, their stalks should be placed in water, so as to give a series of leaves and petals as fresh as possible.

We now pass from the petal of the flower to the Pollen, that coloured dust, generally yellow or white,

POLLEN. 79

which is found upon the stamens, and which is very plentiful in many flowers, such as the Lily and the Hollyhock.

This substance is found only upon the stamens or anthers of full-blown flowers which represent the male sex, and is intended for the purpose of enabling the female portion of the flower to produce fertile seeds. In form the pollen grains are wonderfully diverse, affording an endless variety of beautiful shapes. In some cases the exterior is smooth and only marked with minute dots, but in many instances the outer wall of the pollen grain is covered with spikes or decorated with stripes or belts. A few examples of the commonest forms of pollen will be found on Plate III.

Fig. 17 is the pollen of the Snowdrop, and, as will be seen, is covered with dots and marked with a definite slit along its length. The dots are simply tubercles in the outer coat of the grain, and are presumed to be formed for the purpose of strengthening the otherwise too delicate membrane, after the same principle that gives to "corrugated" iron such strength in proportion to the amount of material. Fig. 18 is the pollen of the Wall-flower, shown in two views, and having many of the same characteristics as that of the snowdrop. Fig. 19 is the pollen of the Willow-herb, and is here given as an illustration of the manner in which the pollen aids in the germination of plants.

In order to understand its action, we must first examine its structure.

All pollen-grains are furnished with some means by which their contents when thoroughly ripened can be expelled. In some cases this end is accomplished by sundry little holes called pores; in others, certain tiny lids are pushed up by the contained matter; and in some, as in the present instance, the walls are thinned in certain places so as to yield to the internal pressure.

When a ripe pollen-grain falls upon the stigma or a nectary of a flower, it immediately begins to swell, and seems to "sprout" like a potato in a damp cellar, sending out a slender "pollen-tube" from one or other of the apertures already mentioned. In fig. 19, a pollen-tube is seen issuing from one of the projections, and illustrates the process better than can be achieved by mere verbal description. The pollen-tubes then insinuate themselves among the cells of the stigmas, and continually elongating, worm their way down the "style" until they come in contact with the "ovules." By very careful dissection of a fertilizing stigma, the beautiful sight of the pollen-tubes winding along the tissues of the style may be seen under a high power of the microscope.

CONTENTS OF POLLEN.

The pollen-tube is nothing more than the interior coat of the grain, very much developed, and filled with a substance technically named "fovilla," composed of a "protoplasm," or that liquid substance which is found in the interior of cells, very minute starch-grains, and some apparently oily globules.

In order to examine the structure of the pollen-grains properly, they should be examined under various circumstances—some dry, others placed in water to which a little sugar has been added, others in oil, and it will often be found useful to try the effect of different acids upon them.

Fig. 20 is the pollen of the common Violet, and is easily recognisable by its peculiar shape and markings. Fig. 21 is the pollen of the Musk-plant, and is notable for the curious mode in which its surface is belted with wide and deep bands, running spirally round the circumference. Fig. 22 exhibits the pollen of the Apple, and fig. 23 affords a very curious example of the raised markings upon the surface of Dandelion pollen. In fig. 24 there are also some very wonderful markings, but they are disposed after a different fashion, forming a sort of network upon the surface, and leaving several large free spaces between the meshes. The pollen of the Lily is shown in fig. 25, and is a good example of a pollen-grain

G

covered with the minute dottings which have already been described.

Figs. 26 and 27 show two varieties of compound pollen, found in two species of Heath. These compound pollen-grains are not of unfrequent occurrence, and are accounted for in the following manner.

The pollen is formed in certain cavities within the anthers, by means of the continual subdivision of the "parent-cells" in which it is developed. In many cases the form of the grain is clearly owing to the direction in which these cells are divided, but there is no great certainty on this subject. It will be seen, therefore, that if the process of subdivision be suddenly arrested, the grains will be found adhering to each other in groups of greater or smaller size, according to the character of the species and the amount of subdivision that has taken place. The reader must, however, bear in mind that the whole subject is as yet rather obscure, and that further discovery may throw a different light on many theories which at present are accepted as established rules.

Fig. 28 shows the pollen of the Furze, in which are seen the longitudinal slits and the numerous dots on the surface; and fig. 29 is the curiously shaped pollen of the Tulip. The two large yellow globular figures at each side of the plate represent the pollen of two common flowers; fig. 36 being that of the Crocus, and fig. 37 a pollen-grain of the Hollyhock. As may be seen from the illustration, the latter is of considerable size, and is covered with very numerous projections. These serve to raise the grain from a level surface, over which it rolls with a surprising ease of motion, so much so indeed that if a little of this substance be placed on a slide and a piece of thin glass laid over it, the glass slips off as soon as it is in the least inclined, and forces the observer to fix it with paper or cement before he can place it on the inclined stage of the microscope. The little projections have a very curious effect under a high power, and require careful focussing to observe them properly; for the diameter of the grain is so large, that the focus must be altered to suit each individual projection. Their office is, probably, to aid in fertilizing.

The Seeds of plants are easier of examination even than the pollen, and in most cases require nothing but a pocket lens and a needle for making out their general structure. The smaller seeds, however, must be placed under the microscope, many of them exhibiting very curious forms. The external coat of seeds is often of great interest, and needs to be dissected off before it can be rightly examined. The simplest plan in such a case is to boil the seed well, press it while still warm into a plate of wax, and then dissect with a pair of needles, forceps, and scissors under water. A few examples of the seeds of common plants are given at the bottom of Plate III.

Fig. 38 exhibits the fruit, popularly called the seed, of the common Goosegrass, or Galium, which is remarkable for the array of hooklets with which it is covered. Immediately above the figure may be seen a drawing of one of the hooks much magnified, showing its sharp curve, fig. 39. It is worthy of remark that the hook is not a simple curved hair, but a structure composed of a number of cells terminating in a hook.

Fig. 40 is the seed, or rather the fruit, of the common red Valerian, and is introduced for the purpose of showing the plumed extremity, which acts as a parachute and causes it to be carried about by the wind until it meets with a proper resting-place. It is also notable for the series of strong longitudinal ribs which support its external structure. On fig. 41 is shown a portion of one of the parachute hairs much more magnified.

The seed of the common Dandelion, so dear to children in their play-hours, when they amuse themselves by puffing at the white plumy globes which tip the ripe dandelion flower-stalks, is a very interesting object even to their parents, on account of its beauti-

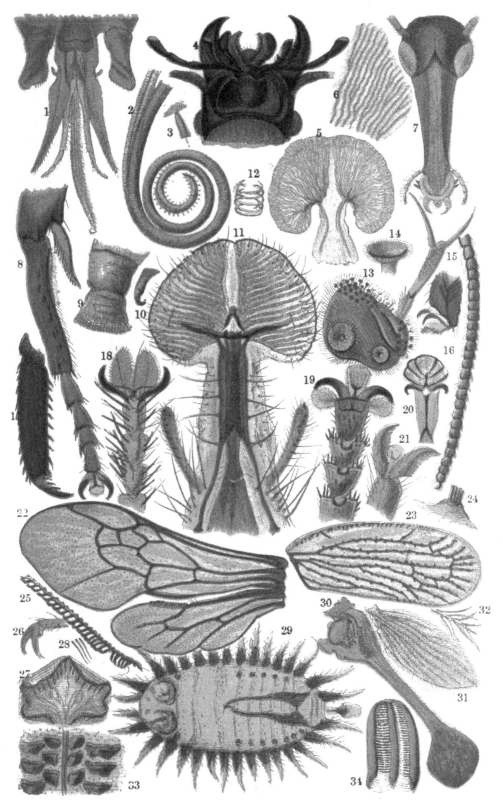

VII.

WINGED SEED.

ful structure, and the wonderful way in which it is adapted to the place which it fills. Fig. 45 represents the seed portion of one of these objects, together with a part of the parachute stem, the remainder of that appendage being shown lying across the broken stem.

The shape of the seed is not unlike that of the valerian, but it is easily distinguished from that object by the series of sharp spikes which fringe its upper end, and which serve to anchor the seed firmly as soon as it touches the ground. From this end of the seed proceeds a long slender shaft, crowned at its summit by a radiating plume of delicate hairs, each of which is plentifully jagged on its surface, as may be seen in fig. 46, which shows a small portion of one of these hairs greatly magnified. These jagged points are evidently intended to serve the same purpose as the spikes below, and to arrest the progress of the seed as soon as it has found a convenient spot.

Fig. 42 is the seed of the Foxglove, and fig. 43 the seed of the Sunspurge or Milkwort. Fig. 47 shows the seed of the yellow Snapdragon; remarkable for the membranous wing with which the seed is surrounded, and which is composed of cells with partially spiral markings. When viewed edgewise, it looks something like Saturn with his ring, or to use a more homely, but perhaps a more intelligible simile, like a marble set in the middle of a penny. Fig. 48 is a seed of Mullein, covered with net-like markings on its external surface. These are probably to increase the strength of the external coat, and are generally found in the more minute seeds.

On fig. 50 is shown a seed of the Burr-reed; a structure which is remarkable for the extraordinary projection of the four outer ribs, and their powerful armature of reverted barbs. Fig. 51 shows another form of parachute seed, found in the Willow-herb, where the parachute is not expanded nearly so widely as that of the valerian; neither is it set upon a long slender stem like that of the dandelion, but proceeds at once from the top of the seed, widening towards the extremity, and having a very comet-like appearance. Two more seeds only remain, fig. 49 being the seed of Robin Hood, and the other, fig. 52, that of the Musk-mallow, being given in consequence of the thick coat of hairs with which it is covered.

Many seeds can be well examined when mounted in Canada balsam, the manner of performing which task is simple enough, and yet is often very perplexing to a beginner.

VERY little apparatus is required. A sixpenny bottle of the best balsam, a spirit-lamp, a metal-plate standing on four legs, so as to form a little table about four

PUTTING ON THE COVER.

inches in height, some ether, spirits of turpentine, and liquor potassæ in bottles, are all the essentials, besides a supply of slides and thin glass covers. The great difficulty in mounting objects in Canada balsam is to keep them free from air-bubbles; but by proceeding in the following manner, very little difficulty will be found.

Take one of the curved dipping tubes, put some balsam into it, cork up the wide end and let it stand on its head until wanted. Lay the glass slide on the metal table, light the spirit-lamp, and place it under the table so as to warm the slide throughout, but not to overheat it. Then take the seed, put it into the spirit of turpentine, and let it wait there while the slide is being warmed.

The next process is to remove the lamp, and to hold

PUTTING UP AN OBJECT IN A CELL OR CANADA BALSAM.

the glass tube containing the balsam over the flame, when the balsam will immediately run towards the orifice, and a drop will ooze out. This drop should be placed on the centre of the slide, and the seed taken out of the turpentine, laid on the warm balsam, and gently pressed into it with one of the dissecting needles. With the needle turn the seed about a little, so as to make sure that no air-bubbles are clinging to its surface, and then take a piece of perfectly clean thin glass, warm it over the spirit-lamp, and lay it on the balsam in the manner here shown. Lower the glass very carefully, and slowly cover the balsam; or when you come to press it down, the object will shoot out at the side, and all your trouble have to be taken over again. When you have laid it nicely level, press it down with the separated points of your curved forceps, and see that no bubbles have made their appearance. Having satisfied yourself about this matter, lay a small circular piece of thick pasteboard or a slice of a small cork on the glass cover, put it within the jaws of one of those American paper clips which you can get in almost any stationer's shop for a penny, let the clip close gradually, and lay it aside to harden. Another simple mode of holding the thin glass cover firmly on the slide is by tying two pieces of whalebone together as in the engraving, and placing the slide between them, a piece of cork or pasteboard being previously laid on the cover as already recommended.

During the pressing process a large amount of balsam

will be squeezed around the edges of the thin glass, and may easily be removed by scraping it with the heated blade of an old knife kept for the purpose, and then rubbing the edges clean with a rag moistened with ether or spirits of turpentine.

Some structures require to be soaked for a considerable time in turpentine, and others in liquor potassæ, before they can be made sufficiently transparent to be mounted. The ether will be found very useful for cleaning the balsam from the fingers and points of the needles, and is, moreover, of great service in removing the unpleasant smell that results when turpentine is

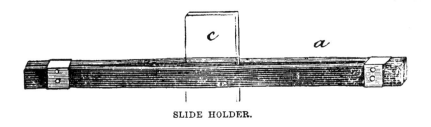

SLIDE HOLDER.

spilt on the hands. If, in spite of all precaution, air-bubbles will make their appearance in the balsam, try to stir them to the top with one of the needles, and then break them by heating the needle and touching them with the point.

After waiting until the balsam is quite hard-set, which will not safely take place for six or eight hours, and is most certain when suffered to remain quiet for a night or two, the slide may be cleaned as directed above, and either kept as it is or covered neatly with paper, perforated on the spot where the object appears. The microscopist should be careful to label every preparation as soon as it is made, and it is best to write with ink on one end of the slide before proceeding to put up the object upon it, and to wash off the writing just before the label is affixed. A little want of such precautions will cause great confusion and loss of time, and often renders a valuable collection quite useless.

The dry mode of preparing permanent objects for the microscope is much more simple, and is managed as follows.

After taking care that the object itself, the slide, and the thin glass are quite clean, put a little dot of ink in the very centre of the slide, on the opposite side to the object, and immediately below the place which it is intended to occupy, so as to act as a guide during the process. Lay the object carefully over the ink dot, place the thin glass very lightly upon it, and fasten it down with two or more strips of thin paper pasted or gummed round its edges. Some persons prefer to fasten it with gold size; but I have always found the paper to answer quite as well, and not to be nearly so troublesome.

Now lay it aside to dry, and get ready a piece of

small-patterned ornamental paper, with a hole punched through it just where it comes upon the object. The easiest plan is to cut a large supply of paper covers, and make the holes with a common gun-wad punch. In order to save materials and space, I frequently mount two dry objects in the same slide, one at each end, and ticket them on a rather large label pasted in the intermediate space between them. When the object is firmly fixed, take some thin coloured paper, blue or red is the best, and cover all the edges of the slide with it, pressing it very closely upon the glass, and then apply two of the ornamental paper covers, one above and the other below, so that the two holes are opposite each other, smooth them carefully down, and lay them aside. Covers, stamped and punched for the purpose, are sold at most of the opticians', but I recommend every one to depend as little as possible on his purse, and as much as possible on his fingers.

As we are speaking of the mode of preparing permanent objects for the microscope, we may as well glance at another method which is extremely useful in many cases which require much more care and time than with the dry mode or the Canada balsam. It is termed mounting in cells, and is principally employed for those objects which require to be immersed in fluid.

The cells in which the fluid is contained are made in

various ways, some being hollows sunk in the glass slide, others built up like glass boxes, by the aid of cement at their edges; others being made of glass tubes cut into segments, and cemented firmly on the slide; and others, of a ring of varnish in which the fluid is contained, thus forming very shallow cells holding a mere film of fluid. The first kind of cell can easily be obtained by purchase, as the slides are made expressly by the glass manufacturers. As, however, they are not one whit more useful than the mere varnish cells, they need no further mention.

The "built-up" cells are made of slips of glass cut to the required size, and laid flat upon the slide if the cell is to be a shallow one, and set on their edges if a deep cell is required. The cements used for this purpose are various, and seem to answer according to the hand which uses them, each person preferring his own mode. Marine glue is an admirable cement, especially for deep cells, but it requires a very high temperature to get it to work freely, and the risk of scorching the fingers is rather great, as I can personally testify.

Whatever cement may be used with deep cells, it is always better to fasten narrow strips of glass over the junctions, and a triangular slip at each interior angle adds greatly to the strength of the fabric. The great fault of these deep cells is their unpleasant habit of leaking after a while, and the consequent admission of air-bubbles. As, however, they are very seldom required for the microscope, the amateur may as well purchase the few that he will want, and only make those of easy manufacture.

The tube-cells are easily made by cutting a glass tube into pieces of the required length, grinding down each surface on a level stone with water, and cementing one end to the glass slide. The tubes are easily cut by notching them with a file, and then running a hot iron round them, when they are tolerably sure to break off level. At first, they are apt to crack upwards or downwards, but a little practice soon sets matters right. Tubes fit for this purpose can be obtained at an easy rate at any glass manufacturer's, circular, oval, or squared, and almost of any needful diameter. Some very large cells made by a friend of mine were cut out of bottles, but I never could master the art of their manufacture myself, the glass always splitting in a wrong direction.

The most useful cell, and that which is in general use among microscopists, is the cement cell, which is nothing more than a ring of cement drawn on the slide, which, when hard, holds the fluid. Some persons who do not care for appearances make their cells square, by drawing the figure on the glass with a pen, or laying the slide upon a square ready drawn on paper, and then painting the varnish upon it. The circular cell is, however, much neater, and can be made by substituting a circle for the square, and painting the cement neatly upon the line, taking care not to let the brush trespass within the circle.

A very neat little apparatus, invented by Mr. Shadbolt, is used for making circular cells, and can be purchased for five shillings, or constructed by a little ingenuity. It is called a turn-table, and consists merely of a horizontal revolving plate of metal, on which the slide is laid, and fastened with two clips while the table revolves. The centre of the slide must of course be brought over the centre of the table, and then, if the brush, charged with varnish or cement, is held to the slide, it immediately strikes a perfect circle, which may be made smaller or larger, according to the distance of the brush from the centre. Generally, the turn-table spins by its own weight, when propelled by the hand, but the addition of a multiplying wheel is very simple, and makes a much more convenient form of apparatus. The ring of cement must be made very wide, and nicely flattened on the upper surface.

Cells of thin glass are sometimes made, but are of

MOUNTING OBJECTS IN CELLS.

no very great value, the varnish cells answering every purpose quite as well.

The practical microscopist will find it useful to devote a spare hour or so to the manufacture of a few dozen cells of different sizes and depths, the depth of course being in proportion to the number of layers of cement. They should be put away in some place where dust will not reach them, and they will then be quite hard and ready for the reception of the fluid when needed.

The method of mounting an object in a cell is as follows:—

Pour into the cell a little of the fluid in which the object is to be mounted, and then lay the object carefully within it. Sometimes, if it is of a very delicate structure, the best plan is to immerse the whole slide in the fluid, float the object into the cell, and then lift it all out together. This precaution, however, is very seldom needed. Having laid the object in the cell, pour some more of the fluid upon it, and fill it up like a "bumper" of wine, letting the fluid stand well above the level of the cell. Lay it aside for a time in order to let all air-bubbles rise to the surface, and be sure to cover it with a shade, or put it in some place where the dust will not get into the cell.

A very expeditious mode of getting rid of the bubbles is to place the cell in the receiver of an air-

AIR-PUMP.

pump, when, after a very few strokes, all the bubbles will come to the surface, break, and disappear. An air-pump, such as that represented in the engraving, is made by Mr. Baker, of Holborn, for a small sum.

Having been quite satisfied that the bubbles have been expelled, take a circular thin glass cover, not quite large enough to reach to the outer edges of the ring of cement, and lay it carefully on the cell, in the manner employed in mounting an object in Canada balsam, page 87. When it lies quite level, take some blotting paper, and carefully dry up the superfluous fluid which will have run out of the cell, and which must be totally removed before the cover can be fastened down.

When the edges of the cover are perfectly dry, hold it down with the forceps with the left hand, and paint a thin layer of gold-size all round the edge of the cover, so as to fasten it to the cell. Do no more to it for at least six hours, but lay a little weight of lead—a bullet with a flattened side answers admirably—on the cover, and leave it to harden. After a sufficient lapse of time, another layer of varnish may be added, until the cover is hermetically sealed on the cell, and neither air can enter nor fluid escape. Unless the first layers of varnish be extremely thin, and very little material used, it is sure to run into the cell, and mar the beauty of the preparation: asphalte varnish is best for the few last layers. In all cases, another coat of varnish after the lapse of a few months can do no harm, and may save a really valuable object.

Sealing-wax varnish is often useful for the double purpose of cementing and giving a neat outside to preparations, and is made by breaking the best black or red sealing-wax into little pieces, pouring spirits of wine over them, and letting them dissolve. The bottle should be frequently shaken, as the sealing-wax is apt to settle at the bottom. It is very useful when time is a matter of consideration, as the spirit evaporates very quickly, and the varnish will become quite hard in a very few minutes.

The fluids employed for mounting many soft objects are very various; some of the best and most easily made are here given.

For vegetable tissues, algæ, &c.: 1. Distilled water with a little camphor. 2. Distilled water with a little corrosive sublimate. This substance is very useful in preventing the growth of fungi, which are apt to develop themselves in preparations, and totally disfigure the object therein. The microscopist can, however, take an appropriate revenge by magnifying and drawing them. 3. Distilled water 1 ounce, salt 5 grains, a very little corrosive sublimate. 4. Glycerine, either pure or dissolved in water, in various proportions. Deane's gelatine is very handy for vegetables, as they can be placed in it while wet, and only need to be laid on a warm slide, covered with a drop of gelatine, and then covered with thin glass as if they were set in Canada balsam. There are many other fluids employed by different microscopists, but these are amply sufficient for all ordinary purposes.

For animal tissues: 1. Chloride of zinc 20 grains, distilled water 1 ounce. This is one of the best substances known for this purpose. 2. Goadby's solution,

No. 1. Bay-salt 4 ounces, alum 2 ounces, corrosive sublimate 2 grains, boiling distilled water 1 quart. This is the strongest and most astringent of his three solutions, and is not very often employed. 3. Goadby's solution, No. 2. Same as the preceding, except that there are 4 grains of corrosive sublimate, and the quantity of water is doubled. Various modifications of this fluid can be made, so as to suit particular objects. Spirit should be avoided in cells as much as possible, as in process of time it is tolerably sure to make its way through the cement. Canada balsam is useful both for the harder vegetable and animal substances, and has already been mentioned. For mounting many crystals, castor oil is a very good preservative, but Canada balsam answers admirably in most instances.

CHAPTER V.

ALGÆ AND THEIR GROWTH — DESMIDIACEÆ, WHERE FOUND — DIATOMS, THEIR FLINTY DEPOSIT — VOLVOX — MOULD, BLIGHT, AND MILDEW — MOSSES AND FERNS — MARE'S-TAIL AND THE SPORES — COMMON SEA-WEEDS AND THEIR GROWTH.

On Plate IV. will be seen many examples of the curious vegetables called Algæ, which exhibit some of the lowest forms of vegetable life, and are remarkable for their almost universal presence in all parts of this globe, and also almost all conditions of cold, heat, or climate. Many of them are well known under the popular name of Sea-weeds, others are equally familiar under the titles of "mould," "blight," or "mildew," while many of the minuter kinds exhibit such capability of motion, and such apparent symptoms of volition, that they have been long described as microscopic animalcules, and thought to belong to the animal rather than to the vegetable kingdoms.

Fig. 1 represents one of the very lowest forms of vegetable life, being known to the man of science as the Palmella, and to the general public as "Gory dew."

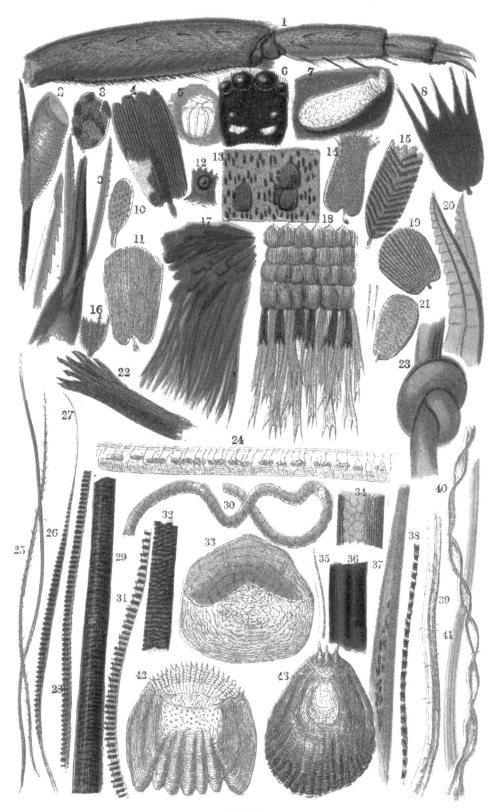

VIII.

CONJUGATION. 101

It may be seen on almost any damp wall, extending in red patches of various sizes, looking just as if some blood had been dashed on the wall, and allowed to dry there. With a tolerably powerful lens, this substance can be resolved into the exceedingly minute cells depicted in the figure. Generally, these cells are single, but in many instances they are double, owing to the process of subdivision by which the plant grows, if such a term may be used.

Fig. 2 affords an example of another very low form of vegetable, the Palmoglæa, or that green slimy substance which is so common on damp stones. When placed under the microscope, this plant is resolvable into a multitude of green cells, each being surrounded with a kind of gelatinous substance. The mode of growth of this plant is very simple. A line appears across one of the cells, and after a while it assumes a kind of hour-glass aspect, as if a string had been tied tightly round its middle. By degrees the cell fairly divides into two parts, and then each part becomes surrounded with its own layer of gelatine, so as to form two separate cells joined end to end.

One of the figures, that on the right hand, represents the various processes of "conjugation," *i.e.* the reunion and fusion together of the cells. Each cell throws out a little projection, which meet together, and then uniting form a sort of isthmus, cementing the two main bodies. By rapid degrees this neck widens, until the two cells become fused into one large body. The whole subject of conjugation is very interesting, and may be seen treated at great length in the Micrographic Dictionary of Messrs. Griffith and Henfrey, a work to which the reader is referred for further information on many of the subjects that, in this small work, can receive but a very hasty treatment.

Few persons would suppose that the slug-like object on fig. 3, the little rounded globules, with a pair of hair-like appendages, and the round disc with a dark centre, are only different forms of the same being. Such, however, is the case, and these are three of the modifications which the Protococcus undergoes. This vegetable may be seen floating like green froth on the surface of rain-water.

On collecting some of this froth and putting it under the microscope, it is seen to consist of a vast number of little green bodies, moving briskly about in all directions, and guiding their course with such apparent exercise of volition, that they might very readily be taken for animals. It may be noticed that the colour of the plant is sometimes red, and in that state it has been called the Hæmatococcus.

The "still" state of this plant is shown in the

round disc. After a while the interior substance splits into two portions; these again subdivide, and the process is repeated until sixteen or thirty-two cells become developed out of the single parent cell. These little ones then escape, and being furnished with two long "cilia" or thread-like appendages, whirl themselves merrily through the water. When they have spent some time in this state, they lose the cilia, become clothed with a strong envelope, and pass into the still stage from which they had previously emerged. This curious process is repeated in endless succession, and causes a very rapid growth of the plant. The moving bodies are technically called zoospores, or living spores, and are found in many other plants besides those of the lowest order. On fig. 13 is delineated a very minute plant, called from its colour Chlorococcus. It may be found upon tree-trunks, walls, &c. in the form of green dust, and has recently been found to be the first stage of lichens.

A large and interesting family of the "confervoid algæ," as these low forms of vegetable life are termed, is called the Desmidiaceæ, or in more common parlance Desmids. A few examples of this family are given in Plate IV.

They may be found in water, always preferring the cleanest and the brightest pools, mostly congregating in masses of green film at the bottom of the water, or investing the stems of plants. Their removal is not very easy, but is best accomplished by very carefully taking up this green slippery substance in a spoon, and straining the water away through fine muslin. For preservation, glycerine and gelatine seem to be the best fluids. A very full and accurate description of these plants may be found in Ralfs's "British Desmidieæ."

Fig. 4 represents one of the species of Closterium, more than twenty of which are known. These beautiful objects can be obtained from the bottom of almost every clear pool, and are of some interest on account of the circulating currents that may be seen within the living plants. A high power is required to see this phenomenon clearly. The Closteria are reproduced in various ways. Mostly they divide across the centre, being joined for a while by two half-cells. Sometimes they reproduce by means of conjugation, the process being almost entirely conducted on the convex sides. Fig. 5 represents the end of a Closterium, much magnified in order to show the active moving bodies contained within it.

Fig. 16 is a supposed Desmid, called by the long name of Ankistrodesmus, and presumed to be an earlier stage of Closterium.

Fig. 6 is a very pretty Desmid called the Pediastrum.

CURIOUS MODE OF GROWTH.   105

and is valuable to the microscopist as exhibiting a curious mode of reproduction. The figure shows a perfect plant composed of a number of cells arranged systematically in a star-like shape; fig. 15 is the same species without the colouring matter, in order to show the shape of the cells. The Pediastrum reproduces by the continual subdivision of the contents of each cell into a number of smaller cells, termed "gonidia" on account of their angular shape. When a sufficient number has been formed, they burst through the envelope of the original cell, taking with them a portion of its internal layer so as to form a vesicle, in which they move actively. In a few minutes they arrange themselves in a circle, and after a while they gradually assume the perfect form, the whole process occupying about two days. Fig. 18 exhibits an example of the genus Desmidium. In this genus the cells are either square or triangular in their form, having two teeth at their angles, and twisted regularly throughout their length, causing the wavy or oblique lines which distinguish them. The plants of this genus are common, and may be found almost in any water. I may as well mention that I have obtained nearly all the preceding species, together with many others, from a little pond on Blackheath.

Fig. 7 is another Desmid called Scenedesmus, in

which the cells are arranged in rows of from two to ten in number, the cell at each extremity being often furnished with a pair of bristle-like appendages. Fig. 14 is another species of the same plant, and both may be found in the water supplied for drinking in London, as well as in any pond.

A common species of Desmid is seen at fig. 12, called Sphærozosma, looking much like a row of stomata set chainwise together. It grows by self-division.

Fig. 17 is a specimen of Desmid named Cosmarium, plentifully found in ponds on heaths and commons, and having a very pretty appearance in the microscope with its glittering green centre and beautifully transparent envelope. The manner in which the Cosmarium conjugates is very remarkable, and is shown at fig. 19.

The two conjugating cells become very deeply cleft, and by degrees separate, suffering the contents to pour out freely, and, as at present appears, without any envelope to protect them. The mass, however, soon acquires an envelope of its own, and by degrees assumes a dark reddish brown tint. It is now termed a sporangium, and is covered with a vast number of projections, which in this genus are forked at their tip, but in others, which also form sporangia, are simply pointed. The Closteria conjugate after a somewhat similar manner, and it is not unfrequent to find a pair in this condition, but in their case the sporangium is quite smooth on its surface.

Another very remarkable family of confervoid algæ is that which is known under the name of Oscillatoriæ, from the oscillating movement of the plant. They are always long and filamentous in character, and may be seen moving up and down with a curious irregularity of motion. Their growth is extremely rapid, and may be watched under a tolerably powerful lens, thus giving many valuable hints as to the mode by which these plants are reproduced. One of the commonest species is represented at fig. 8. Dr. Carpenter is of opinion that the Oscillatoriæ may be the earlier or "motile" forms of some more perfect plants.

Figs. 9, 10, and 11 are examples of another family, called technically the Zygnemaceæ, because they are so constantly yoked together by conjugation. They all consist of a series of cylindrical cells set end to end, and having their green contents arranged in equal patterns. Two of the most common and typical species are here given.

Fig. 9 is the Spirogyra, so called from the spiral arrangement of the pattern; and fig. 10 is the Tyndaridea, or Zygnema as it is called by some writers. A casual inspection will show how easy it is to separate the one from the other. Fig. 11 represents a portion of the Tyndaridea during the process of conjugation, showing the tube of connexion between the cells and one of the spores.

WE now arrive at the Diatoms, so called because of their extreme brittleness and the ease with which they may be cut or broken into their component cells. The commonest of those plants is the Diátoma vulgáre, seen in fig. 21 as it appears while growing. The reproduction of this plant is by splitting down the centre, each half increasing to the full size of the original cell; and in almost every specimen of water taken from a pond, examples of this diatom undergoing the process of division will generally be distinguished. It also grows by conjugation. The diatoms are remarkable for the delicate shell of flinty matter which incrusts the cell-membranes, and which will retain its shape even after intense heat and the action of nitric acid. While the diatoms are alive, swimming through the water, their beautiful markings are clearly distinct, glittering as if the form were spun from crystalline glass. Just above the figure, and to the right hand, are two outlines of single cells of this diatom, the one showing the front view and the other the profile.

Fig. 20 is an example of a diatom—Cocconéma

lanceolátum—furnished with a stalk. The left-hand branch sustains a "frustule" exhibiting the front view, while the other is seen sideways.

Another common diatom is shown in fig. 23, and is known by the name of Synedra. This constitutes a very large genus, containing about seventy known species. In this genus the frustules are at first arranged upon a sort of cushion, but in course of time they mostly break away from their attachment. In some species they radiate in every direction from the cushion, like the spikes of the ancient cavalier's mace.

Fig. 24 is another stalked diatom called Gomphonéma acuminátum, found commonly in ponds and ditches. There are nearly forty species belonging to this genus. A pair of frustules are also shown which have been treated with nitric acid and heat, and exhibit the beautiful flinty outline without the coloured contents, technically called endochrome.

Fig. 27 is a side view of a beautiful diatom, called Eunótia diadéma from its diadem-like form. There are many species of this genus. When seen upon the upper surface, it looks at first sight like a mere row of cells with a band running along them; but by careful arrangement of the light, its true form may easily be made out. This specimen has been boiled in nitric acid.

Fig. 28 represents a very common fresh-water diatom, named Melosíra várians. The plants of this genus look like a cylindrical rod composed of a variable number of segments, mostly cylindrical, but sometimes disc-shaped or rounded. An end view of one of the frustules is seen at the left hand, still coloured with its dots of "endochrome," and showing the cylindrical shape. Immediately above is a figure of another frustule seen under both aspects, as it appears after having been subjected to the action of heat or nitric acid.

A rather curious species of diatom, called Cocconeïs pedículus, is seen at fig. 29 as it appears on the surface of common water-cress. Sometimes the frustules, which in all cases are single, are crowded very closely upon each other and almost wholly hide the substance on which they repose. Fig. 30 is another diatom of a flag-like shape, named Achnanthes, having a long slender filament attached to one end of the lower frustule and standing in place of the flag-staff. There are many wonderful species of such diatoms, some running almost end to end like a bundle of sticks, and therefore called Bacillaria; others spreading out like a number of fans, such as the genus Licmophora; while some assume a beautiful wheel-like aspect, of which the genus Meridion affords an excellent example.

The last of the diatoms which we shall be able to mention in this work is that represented on fig. 31. The members of this genus go by the name of Navícula, on account of their boat-like shape, and their habit of swimming through the water in a canoe-like fashion. There are many species of this genus, all of which are notable for the graceful and varied courses formed by their outlines, and the extreme delicacy of their markings. In many species the markings are so extremely minute that they can only be made out with the highest powers of the microscope and the most careful illumination, so that they serve as test objects whereby the performance of a microscope can be judged by a practical man.

THE large spherical figure in the centre of Plate IV. represents an example of a family belonging to the confervoid algæ, and known by the name of Volvox globator. There seems to be but one species known.

This singular plant has been greatly bandied about between the vegetable and animal kingdoms, but seems now to be satisfactorily settled among the vegetables. In the summer it may be found in pools of water, sufficiently large to be visible to the naked eye like a little green speck proceeding slowly through the water. When a moderate power is used, it appears as shown in the figure, and always retains within its body a number of smaller individuals, which after a while burst through the envelope of the parent, and start into independent existence. On a closer examination, a further generation may be discovered even within the bodies of the children. The whole surface is profusely covered with little green bodies, each being furnished with a pair of movable cilia, by means of which the whole affair is moved through the water. These bodies are analogous to the zoospores already mentioned, and are connected with each other by a network of filaments. A more magnified representation of one of the green bodies is shown immediately above the larger figure. The Volvox is apt to die soon, when confined in a bottle.

Fig. 25 is the common Yeast-plant, consisting simply of a chain of spores, and supposed by some authors to be a state of the ordinary blue mould. Fig. 26 is a curious object, presumed to be one of the confervoid algæ, and found in the human stomach, where it probably gets by means of the water used for drinking. It may possibly be a blanched form of some fresh-water alga. Its scientific name is Sárcina ventrículi.

We now come upon a few of the Blights and Mildews. Fig. 32 is the Urédo, or red-rust of wheat. Another species is very common on the Bramble-leaf, where it appears in spots which at first are red, then orange, and at last become reddish black. Another species of Uredo, together with a Phragmidium, once thought to be another kind of fungus, is seen on a Rose-leaf on Plate V. fig. 1. On fig. 10, however, of the same plate, the Phragmidium may be seen proceeding from Uredo, thus proving them to be but two states of the same plant. Fig. 33 is the mildew of corn, called Puccinia by scientific writers. Another species of Puccinia, found on the Thistle, is shown in Plate V. fig. 7. Fig. 34 is the mould found upon decaying grapes, and called therefrom, or from the clustered spores, Botrýtis. Some of the detached spores are seen by its side. Fig. 35 is another species of the same genus, termed Botrýtis parasítica, and is the cause of the well-known " potato disease."

The Mosses and Ferns afford an endless variety of interesting objects to the microscopist; but as their numbers are so vast, and the details of their structure so elaborate, they can only be casually noticed in the present work. Fig. 38 represents a spore-case of the Polypodium, one of the ferns, as it appears while in the act of bursting and scattering the contents around. One of the spores is seen more magnified below. The spore-cases of many ferns may be seen bursting under the microscope, and have a very curious appearance,

I

writhing and twisting like worms, and then suddenly filling the field with a cloud of spores. Fig. 9, Plate V., is a piece of the brown, chaff-like, scaly structure found at the base of the stalk of male fern cells, showing the manner in which a flat membrane is formed. Fig. 39 is a capsule of the Hypnum, one of the mosses, showing the beautiful double fringe with which its edge is crowned. Fig. 2, Plate V., is the capsule of another moss, Polytrichum, to show the toothed rim; on the right hand is one of the teeth much more magnified.

Fig. 3, Plate V., is the capsule of the Jungermannia, another moss, showing the "elaters" bursting out on every side, and scattering the spores. Fig. 4 is a single elater much magnified, showing it to be a spirally coiled filament, that, by sudden expansion, shoots out the spores just as a child's toy-gun discharges the arrow. Fig. 5 is a part of the leaf of the Sphagnum moss, showing the curious spiral arrangement of secondary fibre which is found in the cells, as well as the circular pores which are found in each cell at a certain stage of growth. Just below, and to the left hand, is a single cell greatly magnified, in order to show these peculiarities more strongly. Fig. 8 is part of a leaf of Jungermannia, showing the dotted cells.

Fig. 6, Plate V., is a part of a rootlet of moss,

showing how it is formed of cells elongated, and joined end to end.

On the common Mare's-tail, or Equisétum, may be seen a very remarkable arrangement for scattering the spores. On the last joint of the stem is a process called a fruit-spike, being a pointed head, around which are set a number of little bodies just like garden-tables, with their tops outward. One of these bodies, which are called the sporangia, is seen in fig. 40. From the top of the table depend a number of tiny pouches, lying closely against each other, and containing the spores. At the proper moment these pouches burst from the inside, and fling out the spores, which then look like round balls with irregular surfaces, as shown in fig. 40, c. This irregularity is caused by four elastic filaments knobbed at the end, which are originally coiled tightly round the body of the spore, but by rapidly untwisting themselves, cause the spore to leap about so as to aid in the distribution. A spore with uncoiled filaments is seen at fig. 40, b. By breathing on them they may be made to repeat this process at will.

Fig. 36 is a common little sea-weed, called Ectocarpus siliculósus, that is found parasitically adhering to large plants, and is given, in order to show the manner in which the extremities of the branches are developed into sporanges. Fig. 37 is a piece of the common Green laver, Ulva latíssima, showing the green masses that are ultimately converted into zoospores, and by their extraordinary fertility cause the plant to grow with such rapid luxuriance wherever the conditions are favourable. Every possessor of a marine aquarium knows how rapidly the glass sides become covered with growing masses of this plant. The smaller figure above is a section of the same plant, showing that it is composed of a double plate of cellular tissue.

Fig. 41 is a piece of Purple laver or "Sloke," Porphýra laciniáta, to show the manner in which the cells are arranged in groups of four, technically named "tetraspores." This plant has only one layer of cells.

On Plate V. may be seen a number of curious details of the higher Algæ.

Fig. 11 is the Sphaceláría, so called from the curious capsule cells found at the end of the branches, and termed sphacelæ. This portion of the plant is shown more magnified in fig. 12. Another sea-weed is represented on fig. 13, in order to show the manner in which the fruit is arranged; and a portion of the same plant is given on a larger scale at fig. 14.

A very pretty little sea-weed called Cerámium is shown at fig. 15; and a portion showing the fruit much more magnified is drawn at fig. 22. Fig. 23 is a little

IX.

alga called Myrionéma, growing parasitically on the preceding plant.

Fig. 16 is a section of a capsule belonging to the Hálydris siliquósa, showing the manner in which the fruit is arranged; and fig. 17 shows one of the spores more magnified.

Fig. 18 shows the Polysiphónia parasítica, a rather common species of a very extensive genus of sea-weeds, containing nearly three hundred species. Fig. 19 is a portion of the stem of the same plant, cut across in order to show the curious mode in which it is built up of a number of longitudinal cells, surrounding a central cell of large dimensions, so that a section of this plant has the aspect of a rosette when placed under the microscope. A capsule or "ceramídium" of the same plant is shown at fig. 20, for the purpose of exhibiting the pear-shaped spores, and the mode of their escape from the parent-cell preparatory to their own development into fresh plants. The same plant has another form of reproduction, shown in fig. 21, where the "tetraspores" are seen imbedded in the substance of the branches. There is yet a third mode of reproduction by means of "antheridia," or elongated white sacs at the extremities of the branches.

Fig. 25 is the Cladóphora, a green alga, given to illustrate its mode of growth; and fig. 26 represents one of the red sea-weeds, Ptilóta élegans, beautifully feathered, and with a small portion given also on a larger scale, in order to show its structure more fully. A good contrast with this species is seen on fig. 27, and the mode in which the long, slender, filamentary fronds are built up of many-sided cells is seen just to the left hand of the upper frond. Fig. 24 is a portion of the lovely Delesséria sanguínea, given in order to show the formation of the cells, as also the arrangement by which the indistinct nervures are formed.

The figure on the bottom left-hand corner of Plate V. is a portion of the pretty Nitophyllum lacerátum, a plant belonging to the same family as the preceding. The specimen here represented has a gathering of spores upon the frond, in which state the frond is said to be "in fruit."

Fig. 27 represents a portion of the common Sea grass (*Enteromorpha*), so common on rocks and stones between the range of high and low water. On the left hand of the figure, and near the top, is a small piece of the same plant much more magnified, in order to show the form of its cells.

CHAPTER VI.

ANTENNÆ, THEIR STRUCTURE AND USE—EYES, COMPOUND AND SIMPLE—BREATHING ORGANS—JAWS AND THEIR APPENDAGES—LEGS, FEET, AND SUCKERS—DIGESTIVE ORGANS—WINGS, SCALES, AND HAIRS—EGGS OF INSECTS—HAIR, WOOL, LINEN, SILK, AND COTTON—SCALES OF FISH—FEATHERS—SKIN AND ITS STRUCTURE—EPITHELIUM—NAILS, BONE, AND TEETH—BLOOD CORPUSCLES AND CIRCULATION—ELASTIC TISSUES—MUSCLE AND NERVE.

WE now take leave of the vegetables for a time, and turn our attention to the animal kingdom.

On Plate VI. may be seen many beautiful examples of animal structures, most of them being taken from the insect tribes. We will begin with the antennæ, or horns, as they are popularly termed, of the insect.

The forms of these organs are as varied as those of the insects to which they belong, and in most cases they are so well defined that a single antenna will, in almost every instance, enable a good entomologist to designate the genus to which the insect belonged. The functions of the antennæ are not satisfactorily ascertained. They are certainly often used as organs of speech, as may be seen when two ants meet each other, cross their antennæ, and then start off simultaneously to some task which is too much for a single ant. This pretty scene may be witnessed on any fine day in a wood, and a very animated series of conversations may readily be elicited by laying a stick across their paths, or putting a dead mouse or large insect in their way.

I once saw a very curious scene of this kind take place at an ant's nest near Hastings. A great Daddy Long-legs had unfortunately settled on the nest, and was immediately "pinned" by an ant or two at each leg so effectually, that all its struggles availed it nothing. Help was, however, needed, and away ran four or five ants in different directions, intercepting every comrade they met, and by a touch of the antennæ sending them off in the proper direction. A large number of the wise insects soon crowded round the poor victim, whose fate was rapidly sealed. Every ant took its proper place, just like a gang of labourers under the orders of their foreman; and by dint of pushing and pulling, the long-legged insect was dragged to one of the entrances of the nest, and speedily disappeared.

Many of the ichneumon-flies may also be seen quivering their antennæ with eager zeal, and evidently using them as feelers, clearly to ascertain the presence of the insect in which they intend to lay their eggs, and many other similar instances will be familiar to any one who has been in the habit of watching insects and their ways.

It is, however, most likely that the antennæ serve other purposes than that which has just been mentioned, and many entomologists are of opinion that they serve as organs of hearing.

Fig. 15, Plate VI., represents a part of one of the joints belonging to the antennæ of the common House-fly, and is seen to be covered with a multitude of little depressions, some being small, and others very much larger. A section of the same antenna, but on a larger scale, is shown by fig. 16, in order to exhibit the real form of these depressions. Nerves have been traced to these curious cavities, which evidently serve some very useful purpose, some authors thinking them to belong to the sense of smell, and others to that of hearing. Perhaps they may be avenues of sensation which are not possessed by the human race, and of which we are therefore ignorant. Fig. 17 represents a section of the antennæ of an Ichneumon-fly, to show the structure of these organs of sense.

We will now glance casually at the forms of antennæ which are depicted in the plate.

Fig. 1 is the antenna of the common Cricket, and consists of a vast number of little joints, each a trifle smaller than the preceding, so as to form a long, thread-like organ. Fig. 2 is taken from the Grasshopper, and shows the joints larger in the middle than at each end.

Figs. 3 and 5 are from two minute species of Cock-tailed Beetles (*Staphylínidæ*), which swarm throughout the summer months, and even in the winter may be found in profusion under stones and moss. The insect from which fig. 5 was taken is so small that it is almost invisible to the naked eye, and was captured on the wing by waving a sheet of gummed paper under the shade of a tree. These are the tiresome little insects that so often get into the eye in the summer, and cause such pain and inconvenience before they are removed.

Fig. 4 shows the antenna of the Tortoise Beetle (*Cássida*), so common on many leaves, and remarkable for its likeness to the reptile from which it derives its popular name. Fig. 3 is from one of the Weevils, and shows the extremely long basal joint of these beetles, as well as the clubbed extremity. Fig. 7 is the beautifully notched antenna of the Cardinal Beetle (*Pyrochróa*), and fig. 11 is the fan-like antenna of the common Cock-chaffer. This specimen is taken from a male insect, and the reader will find his trouble repaid by mounting one of these antennæ as a permanent object.

It may here be noticed that all these antennæ must

be mounted in Canada balsam, as otherwise they will be too opaque for the transmission of light through their substance.

In many cases they are all the better for being soaked for some time in liquor potassæ, then dried between two slips of glass, then soaked in turpentine, and lastly put up in the balsam. Otherwise, their characteristics will be totally invisible under the microscope, and the observer will be as bewildered as a gentleman of whom I heard, who lately purchased a good microscope, and returned it next day as useless. The maker who had guaranteed it naturally thought that it had been injured by rough treatment, but finding that it performed well in his own hands, he inquired as to the details, and especially as to the object which it would not show. The answer was, that it would not exhibit the crystals of sugar. "How large a crystal did you try?" asked the optician. "A lump out of the sugar-basin," was the answer.

Fig. 12 is an antenna from one of the common Ground Beetles (*Cárabus*), the joints looking like a string of elongated pears. The reader will find that in beetles he is sure to find eleven joints in the antennæ.

Fig. 10 is the entire antenna of a fly (*Syrphus*), one of those pretty flies that may be seen hovering over one spot for a minute, and then darting off like lightning to hang over another. The large joint is the one on which are found those curious depressions that have already been mentioned. Fig. 8 is one of the antennæ of a Tortoiseshell Butterfly (*Vanessa*), showing the slender knobbed form which butterfly antennæ assume; and figs. 13 and 14 are specimens of moth antennæ, showing how they always terminate in a point. Fig. 13 is the beautiful feathery antennæ of the Ermine Moth (*Spilosóma*); and fig. 14 is the toothed antenna of the Tiger Moth (*Arctia caja*). In all these feathered and toothed antennæ of moths, the male insects have them much more developed than the female, probably for the purpose of enabling them to detect the presence of their mates, a property which some possess in wonderful perfection. The male Oak-egger Moth, for example, can be obtained in any number by putting a female into a box with a perforated lid, placing the box in a room, and opening the window. In the course of the evening seven or eight males are seen to make their appearance, and they are so anxious to get at their intended mate, that they will suffer themselves to be taken by hand.

Fig. 9 is an antenna of the male Gnat, a most beautiful object, remarkable for the delicate transparency of the joints, and the exquisitely fine feathering with which they are adorned.

SIMPLE AND COMPOUND EYES. 125

We now arrive at the Eyes of the insects, all of which are very beautiful, and many are singularly full of interest.

In the centre of Plate VI. may be seen the front view of the head of a Bee, showing both kinds of eyes, three simple eyes arranged triangularly in the centre, and two large masses of compound eyes at the sides.

The simple eyes, termed "ocelli," are from one to three in number, and usually arranged in a triangular form between the two compound eyes. Externally they look merely like shining rounded projections, and can be seen to great advantage in the Dragon-flies. The compound eyes may be considered as aggregations of simple eyes, set closely together, and assuming a more or less perfect six-sided form. Their numbers vary very greatly; in some insects, such as the common Fly, there are about four thousand of these eyes, in the Ant only fifty, in the Dragon-fly about twelve thousand, and in one of the Beetles more than twenty-five thousand.

Fig. 18 shows a portion of the compound eye of the Atalanta Butterfly, and fig. 20 the same organ of the Death's-head Moth. A number of the protecting hairs may be seen still adhering to the eye of the butterfly. **Fig. 22** is a remarkably good specimen of the eye of a fly (*Heliophilus*), showing the nearly-squared facets, the tubes to which they are attached, and portions of the optic nerves. Fig. 23 is part of the compound eye of a lobster, showing the facets quite square. All these drawings were taken by the camera lucida from my own preparations, so that I can answer for their authenticity.

On Plate VIII. figs. 6 and 12, the reader will find two more examples of eyes, being in these cases taken from the Spiders. Fig. 6 is an example of the eight eyes of the well-known Zebra Spider, so common on our garden walls and similar situations, hunting incessantly after flies and other prey, and capturing them by a sudden pounce. The eyes are like those of the ocelli of insects, and are simple in their construction. The number, arrangement, and situation of the eyes is extremely varied in spiders, and serves as one of the readiest modes of distinguishing the species. Fig. 12, Plate VIII., represents one of the curious eyes of the common Harvest Spider, as it appears perched on a prominence or "watch-tower," as it has been aptly named, for the purpose of enabling the creature to take a more comprehensive view of surrounding objects.

RETURNING to Plate VI., on fig. 12 we see a curiously branched appearance, something like the hollow root of

SPIRACLES AND BREATHING TUBES. 127

a tree, and covered with delicate spiral markings. This is part of the breathing apparatus of the Silkworm, extracted and prepared by myself for the purpose of showing the manner in which the tubes branch off from the "spiracle" or external breathing-hole, a row of which may be seen along the sides of insects, together with the beautiful spiral filament which is wound round each tube for the purpose of strengthening it. One of these spiracles may be seen in the neck of the Gnat (fig. 27). Another spiracle, more enlarged, may be seen on Plate VII. fig. 34, taken from the Wireworm, *i.e.* the larva of the Skipjack Beetle (*Eláter*), to show the apparatus for excluding dust and admitting air. The object of the spiral coil is very evident, for as these breathing-tubes extend throughout the whole body and limbs, they would fail to perform their office when the limbs were bent, unless for some especial provision. This is achieved by the winding of a very strong but slender filament between the membranes of which the tube is composed, so that it always remains open for the passage of air throughout all the bendings to which it may be subjected. Flexible tubes for gas and similar purposes are made after the same fashion, spiral metal wire being coiled within the leather case. A little piece of this thread is seen unwound at the end of a small branch towards the top, and this thread is so strong that it retains its elasticity when pulled away from the tube, and springs back into its spiral form. I have succeeded in unwinding a considerable length of this filament from the breathing-tube of a Humble Bee.

Fig. 28 represents the two curious tubercles upon the hinder quarters of the common Green-blight, or Aphis, so very common on our garden plants, as well as on many trees and other vegetation. From the tips of these tubercles exudes a sweet colourless fluid, which, after it has fallen upon the leaves, is popularly known by the name of honey-dew. Ants are very fond of this substance, and are in the habit of haunting the trees upon which the aphides live, for the purpose of sucking the honey-dew as it exudes from their bodies. A drop of this liquid may be seen on the extremity of the lower tubercle.

The head of the same insect may be seen on fig. 24, where the reader may observe the bright scarlet eye, and the long beak with which it punctures the leaves and sucks the sap. Fig. 29 is the head of the Sheep-tick, exhibiting the organ by which it pierces the skin of the creature on which it lives. Fig. 25 is the head of another curious parasite found upon the Tortoise,

JAWS AND THEIR APPENDAGES.

and remarkable for the powerful hooked apparatus which projects in front of the head.

Turning to Plate VII. fig. 4, we find the head of a Ground Beetle (*Cárabus*), valuable as possessing the whole of the organs of the head and mouth.

Immediately above the compound eyes are seen the roots of the antennæ, those organs themselves being cut away in order to save room. Above these are two pairs of similarly constructed organs termed the "maxillary palpi," because they belong to the lesser teeth, or maxillæ, which are seen just within the pair of great curved jaws, called the mandibles, which are extended in so threatening a manner. The "labial palpi," so called because they belong to the "labium," or under lip, are seen just within the others; the tongue is seen between the maxillæ, and the chin or "mentum" forms a defence for the base of the maxillæ and the palpi. A careful examination of a beetle's mouth with the aid of a pocket lens is very instructive as well as interesting.

Fig. 1 on the same plate shows the jaws of the Hive Bee, where the same organs are seen modified into many curious shapes. In the centre may be seen the tongue, elongated into a flexible and hair-covered instrument, used for licking the honey from the interior of flowers. At each side of the tongue are the labial

K

palpi, having their outermost joints very small, and the others extremely large, and acting as a kind of sheath for the tongue. Outside the labial palpi are the maxillæ, separated in the specimen, but capable of being laid closely upon each other, and the mandibles outside all.

The curiously elongated head of the Scorpion-fly (*Panorpa*), seen at fig. 7, affords another example of the remarkable manner in which these organs are developed in different insects. Another elongated head, belonging to the Daddy Long-legs, is seen in Plate VI. fig. 27, and well shows the compound eyes, the antennæ, and the palpi. Fig. 2 represents the coiled tongue of the Atalanta Butterfly, being composed of the maxillæ very greatly developed, and having a form as if each had originally been flat, and then rolled up so as to make about three-fourths of a tube. A number of projections are seen towards the tip, and one of these little bodies is shown on a larger scale at fig. 3. These curious organs have probably some connexion with the sense of taste. Along the edges of the semi-tubes are arranged a number of very tiny hooks, by means of which the insect can unite the edges at will.

Fig. 11, in the centre of the plate, shows one of the most curious examples of insect structure, the proboscis or trunk of the common Bluebottle-fly. The maxillary palpi covered with bristles are seen projecting at each side, and upon the centre are three lancet-like appendages, two small and one large, which are used for perforating various substances on which the insect feeds. The great double disc at the end is composed of the lower lip greatly developed, and is filled with a most complex arrangement of sucking-tubes, in order to enable it to fulfil its proper functions. The numerous tubes which radiate towards the circumference are strengthened by a vast number of partial rings of strong filamentary substance, like that which we have already seen in the breathing-tube of the Silkworm. Some of these partial rings are seen on fig. 12, a little above. The mode by which the horny matter composing the rings is arranged upon the tubes is most wonderful, and requires a tolerably high power to show it.

Fig. 5 shows the tongue of the common Cricket, a most elegantly formed organ, having a number of radiating bands covered with zig-zag lines, resulting from the triangular plates of strengthening substance with which they are furnished, instead of the rings. A portion more highly magnified is shown at fig. 6, exhibiting the manner in which the branches are arranged.

THE Legs of insects now claim our attention.

Fig. 9, Plate VII., shows the "pro-leg" of a Caterpillar. The pro-legs are situated on the hinder parts of the caterpillar, and, being set in pairs, take a wonderfully firm hold of a branch or twig, by pressure against each other. Around the pro-legs are arranged a series of sharp hooks, set with their points inwards, for the greater convenience of holding. Fig. 10 represents one of the hooks more magnified.

Fig. 15 is the lower portion of the many-jointed legs of the Long-legged Spider (*Phalángium*), the whole structure looking very like the antenna of the cricket. Fig. 17 is the leg of the Glow-worm, showing the single claw with which it is armed. Fig. 26 shows the foot of the Flea, furnished with two simple claws. Fig. 16 is the foot of the Trombídium, a genus of parasitic creatures, to which the well-known Harvest-bug belongs. Fig. 26, Plate VI., shows the leg of the green Aphis of the geranium, exhibiting the double claw, and the pad or cushion, which probably serves the same purpose as the pad found upon the feet of many other insects. Fig. 8 is the lower portion of the leg of the Ant, showing the two claws and the curious pad in the centre, by means of which the insect is able to walk upon slippery surfaces. The Típula has a foot also furnished with a single pad (see Plate VI. fig. 30). This organ is seen under a very high power to be covered with long hair-like appendages, each having a little disc at the end, and probably secreting some glutinous fluid which will enable the creature to hold on to perpendicular and smooth surfaces. Many of my readers will doubtlessly have noticed the common Fly towards the end of autumn, walking stiffly upon the walls, and evidently detaching each foot with great difficulty, age and infirmity having made the insect unable to lift its feet with the requisite force.

Fig. 21 is the foot of one of the Ichneumon-flies (*Ophion*), the hairy fringe being apparently for the purpose of enabling it to hold firmly to the caterpillar in which it is depositing its eggs, and which wriggles so violently under the infliction, that it would soon throw its tormentor, had not some special means been provided for the purpose of keeping its hold. Fig. 20 is a beautiful example of a padded foot, taken from the parasitic creature which is so plentifully found upon the Dor Beetle (*Geotrupes*), and of which the afflicted insect is said to rid itself by lying on its back near an ant's nest, and waiting until the ants carry off its tormentors.

Fig. 18 is the foot of the common yellow Dung-fly, so plentiful in pasture lands, and having two claws and two pads; and fig. 19 shows the three pads and two claws found in the foot of the Hornet-fly (*Ásilus*).

Few microscopic objects call forth such general and deserved admiration as the fore-foot of the male Water-Beetle (*Acilius*), when properly prepared and mounted, for which see fig. 13.

On examining this preparation under the microscope, it is found that three of the joints are greatly expanded, and that the whole of their under surface is covered profusely with certain wonderful projections, which are known to act as suckers. One of them is exceedingly large, and occupies a very considerable space, its hairs radiating like the rays of the heraldic sun. Another is also large, but scarcely half the diameter of the former, and the remainder are small, and mounted on the extremities of delicate footstalks, looking something like wide-mouthed trumpets. In the specimen from which the drawing was taken, the smaller suckers are well shown, as they protrude from the margin of the foot.

The preparation of these feet is a very tiresome business, as the suckers hold so much air, that bubbles are constantly showing themselves, and cannot be easily extirpated without the expenditure of time and much patience. Two specimens of these feet which I prepared cost an infinity of trouble, having to be soaked

STOMACH AND GIZZARD OF INSECTS.

in spirits of turpentine, boiled several times in Canada balsam, poked about with needles, and subjected to various treatments before they showed themselves clean and translucent, as they ought to do.

One of the larger suckers is seen more magnified on fig. 14.

Plate VIII. fig. 1 well exemplifies the manner in which the muscles of insects do their work, being well attached in the limbs to the central tendon, and pulling "with a will" in one direction, thus giving very great strength. This leg is taken from the Water Boatman (*Notonecta*), and has been mounted in Canada balsam.

On Plate VII. fig. 29, may be seen a curiously formed creature. This is the larva of the Tortoise Beetle (*Cássida*), the skin having being flattened and mounted in Canada balsam. The spiracles are visible along the sides, and at the end is seen a dark fork-like structure. This is one of the peculiarities of this creature, and is employed for the purpose of carrying the refuse of its food, which is always piled upon its back, and retained in its place by the forked spines, aided probably by the numerous smaller spines that project from the side.

Fig. 33 shows part of the stomach and gastric teeth of the Grasshopper. This structure may be seen to perfection in the "gizzard," as it is called, of the great green Locust of England (*Ácrida viridíssima*). The organ looks like a sudden swelling of the œsophagus, and when slit longitudinally under water, the teeth may be seen in rows set side by side, and evidently having a great grinding power. Just above, fig. 27, is the corresponding structure in the Hive Bee, three of the teeth being shown separately at fig. 28.

We now cast a rapid glance at the Wings of insects.
They have no analogy, except in their use, with the wings of birds, as they are not modifications of existing limbs, but entirely separate organs. They consist of two membranes united at their edges, and traversed and supported by sundry hollow branches or "nervures," which admit air, and serve as useful guides to entomologists for separating the insects into their genera. Indeed, the general character of the wings has long been employed as the means of dividing the insect race into their different orders, as may be seen in any work on entomology. The primary number of wings is four, but it often happens that two are almost wholly absent, or that the uppermost pair are thickened into a shelly kind of substance which renders them useless for flight, while in many insects, such as the Ground Beetles and others, the upper wings become hardened into firm coverings for the body, and the lower pair are shrivelled and useless.

HOOKLETS OF WINGS.

Fig. 22 shows two of the wings of a Humble Bee, together with their nervures, and the peculiar system by which the upper and lower pair are united together at the will of the insect. At the upper edge of the lower wing, and nearly at its extremity, may be seen a row of very tiny hooks, shown on a larger scale at fig. 25. These hooklets hitch into the strengthened membrane of the upper wing, which is seen immediately above them, and so conjoin the two together. The curious wing-hooks of the Aphis may be seen on fig. 24, very highly magnified.

Fig. 31 is the wing of the Midge (*Psychóda*), that odd little insect which is seen hopping and popping about on the windows of outhouses and similar localities, and is so hard to catch. The whole wing is plentifully covered with elongated scales, and is a most lovely object under any power of the microscope. These scales run along the nervures and edges of the wings, and part of a nervure is shown more highly magnified at fig. 32.

At fig. 23 is shown the wing of one of the hemipterous insects, common along the banks of ditches and in shady lanes, and known by the name of Cíxius. It is remarkable for the numerous spots which stud the nervures, one being always found at each forking, and the others being very irregularly disposed.

Fig. 30 is one of the balancers or "haltéres" of the House-fly. These organs are found in all the two-winged insects, and are evidently modifications of the second pair of wings. They are covered with little vesicles, and protected at their base by scales. Some writers suppose that the sense of smell resides in these organs. Whatever other purpose they may serve, they clearly aid in the flight, as, if the insect be deprived of one or both of the balancers, it has the greatest difficulty in steering itself through the air.

The wings of insects are mostly covered with hairs or scales, several examples of which are given in Plate VIII. Fig. 4 shows one of the scales of the Adippe or Fritillary Butterfly, exhibiting the double membrane — part of which has been torn away—and the beautiful lines of dots with which it is marked. The structure of the scales is further shown by a torn specimen of Tiger Moth scale seen on fig. 16. On many scales these dots assume a "watered" aspect when the focus or illumination changes an example of which may be seen on fig. 15, a scale of the Peacock Butterfly.

Fig. 11 is one of the ordinary scales of the Azure Blue Butterfly, and fig. 10 shows one of the curious "battledore" scales of the same insect, with its rows of distinct dottings. Fig. 14 is one of the prettily tufted scales of the Orange-tip Butterfly, and fig. 8 is the splen-

SCALES OF INSECTS. 139

did branched scale of the Death's-head Moth. Fig. 19 shows a scale of the Sugar-runner (*Lepisma saccharina*), a little silvery creature with glistening skin, and long bristles at the head and tail, that is found running about cupboards, window-sills, and similar places. It is not easy to catch with the fingers, as it slips through them like oil, but if the finger be afterwards examined, some of the beautiful scales will be found adhering, and may be placed under the microscope. The Gnats also possess very pretty scales, with the ribs projecting beyond the membrane.

Fig. 21 is a scale from the common Spring-tail (*Podúra plúmbea*), a little creature which is found plentifully in cellars and other damp places, skipping about with great activity. Some flour scattered on a piece of paper is a sure trap for these little beings. Fig. 3 is one of the scales taken from the back of the celebrated Diamond Beetle, showing the cause of the magnificent gem-like aspect of that insect. We have in England many beetles of the same family—the Weevils—which, although much smaller, are quite as splendid when exhibited under a microscope by reflected light. The wing-case or "elytron" of a little green weevil, very common in the hedges, may be seen on Plate XII. fig. 10.

The reader will observe that all these scales are furnished with little root-like appendages, by means of which they are affixed to the insect. Fig. 13 shows a portion of the wing of the Azure Blue Butterfly, from which nearly all the scales have been removed, for the purpose of exhibiting the pits or depressions in which they had formerly been fastened, and one or two of the scales are left still adherent to their places. The scales are arranged in equal rows like the slates of a housetop, as may be seen on fig. 18, which represents part of the same wing, to show the scales overlapping each other, and the elegant form which they take near the edges of the wing, so as to form a delicate fringe. The long hair-like down which covers the legs and bodies of the moths and butterflies (which are called Lepidóptera or scale-winged insects in consequence of this peculiarity), is seen under the microscope to be composed of scales very much elongated, as is shown in fig. 17, a portion taken from the leg of a Tiger Moth.

The Eggs of insects are all very beautiful, and three of the most curious forms are given on Plate VIII.

Fig. 2 is the empty egg of the Gad-fly, as it appears fastened to a hair of a horse. Fig. 5 represents the pretty ribbed egg of the common Tortoiseshell Butterfly; and fig. 7 is the very beautiful egg of the very horrid Bed-bug, worthy of notice on account of the curious lid with which its extremity is closed, and by

STINGS AND SAWS. 141

means of which the young larva creeps out as soon as it is hatched.

Fig. 9 shows the penetrating portions of the Sting of the Wasp. The two barbed stings, which seem to be the minute prototypes of the many-barbed spears of the South Sea Islanders, are seen lying one at each side of their sheath, and a single barb is drawn a little to the left on a very much larger scale. It is by reason of these barbs that the sting is always left adhering to the wound, and is generally drawn wholly out of the insect, causing its death in a short while.

The sting is only found in female insects, and is supposed to be analogous to the "ovipositor" of other insects, *i.e.* the instrument by which the eggs are deposited in their places. Fig. 20 shows the curious egg-placing apparatus of one of the Saw-flies. The backs of these "saws" work in grooves, and they work alternately, so that the fly takes but a very short time in cutting a slit in the young bark of a tender shoot, and laying her eggs in the slit. When she has completed one of these channels, she sets to work upon another, and in the early spring the young branches of the gooseberry bushes may be seen plentifully covered with these grooves and the eggs. When hatched, from the eggs issue black caterpillar-like grubs, which devastate the bushes sadly, and in process of time turn into blackish flies, which are seen hovering in numbers over the gooseberries, and may be killed by thousands.

THE scales and hairs of other animals deserve great attention. Fig. 23 is a single hair of the human beard, as it often appears when tied in a knot—by Queen Mab and her fairies, according to Mercutio. Fig. 22 is a portion of the same hair as it appears when splitting at its extremity. The structure of the hair is not, however, so well seen in this object as in that represented on fig. 24, which is a beautiful example of white human hair, that once adorned the head of the victor of Waterloo. It formed one of a tiny lock given to me by a friend, and is so admirable an example of human hair, that I forthwith mounted it for the microscope. In this hair the marrow-cells may be seen extending down its centre, and the peculiar roughened surface produced by the flattened cells which are arranged around its circumference are also seen. By steeping in caustic potash, these scales can be separated, but generally they lie along the hair in such a manner that if the hair be drawn through the fingers from base to point, their projecting ends permit it to pass freely; but if it be drawn in the reverse direction, they cause it to feel very harsh to the touch.

In the Sheep's Wool, fig. 30, this structure is much

HAIRS AND FIBRES. 143

more developed, and gives to the fibres the "felting" power that causes them to interlace so firmly with each other, and enables cloth—when really made of wool—to be cut without unravelling. Fig. 37 is the smooth hair of the Badger; and fig. 34 is the curious hair of the Red Deer, which looks as if it had been covered with a delicate net.

Fig. 28 is the soft, grey, wool-like hair of the Rat; and fig. 29 is one of the larger hairs that protrude so plentifully, and form the glistening brown coat of that animal. Fig. 38 is the curiously knobbed hair of the Long-eared Bat, the knobs being formed of protuberant scales that can easily be scraped off. Fig. 31 shows a hair of the common Mole; and fig. 32 is one of the long hairs of the Rabbit. Fig. 27 is a flat hair of the Dormouse, slightly twisted, the difference in the breadth showing where the twist has taken place. Fig. 26 is one of the very long hairs that so thickly clothe the Tiger Moth caterpillar; and fig. 25 is a beautifully branched hair taken from the common Humble Bee.

The four fibres mostly used in the manufacture of apparel are: Wool, fig. 30, which has already been described; Linen, fig. 39; Cotton, fig. 40; and Silk, fig. 41. The structure of each is very well marked and easily made out with the microscope; so that an adulterated article can readily be detected by a practised eye. Cotton is mostly used in adulterations of silk and linen fabrics, and may at once be detected by its flat twisted fibre. Silk is always composed of two parallel threads, each proceeding from one of the spinnerets of the caterpillar, and it may be here remarked that if these threads are not quite parallel the silk is of bad quality. Silken fibre is always when new covered with a kind of varnish, usually of a bright orange colour, which gives the undressed "floss" silk its peculiar hue, but which is soluble and easily washed away in the course of manufacture.

Figs. 35 and 36 are the small and large hairs of that magnificent creature, the Sea Mouse (*Aphrodíte aculeáta*), whose covering, although it lies in the mud, glows with every hue of the rainbow, and in a brilliant light is almost painfully dazzling to the eye.

THE scales of some of the Fish are shown on Plate VIII. in order to exhibit their mode of growth by successive layers. The scales are always enveloped in membranous sacs, and in some cases, as in the Eel, they do not project beyond the surface, and require some little observation to detect them. A scale of an Eel is shown on Plate XI. fig. 15, and is a magnificent subject under polarized light. Fig. 33 is a scale of the Greenbone Pike; and figs. 42 and 43 are scales of

FEATHERS.

the Perch, showing the roots by which they are held in their places. The Roach, Dace, Bleak, and many other similar fish have some beautiful silvery crystals on the under surface of the scales, which were greatly used in the manufacture of artificial pearls, glass beads being thinly coated in the interior with the glittering substance, and then filled in with wax. A piece of Sole-skin, when preserved in Canada balsam and placed under the microscope, is a very beautiful object.

More examples of hairs, and other processes from the skin, together with the structure of the Skin itself, of Bone, of Blood, and the mode in which it circulates, are given on Plate X.

In all important points of their structure, the Feathers of birds are similar to the hairs of animals, and are developed in a similar manner. They are all composed of a quill portion, in which the pith is contained, and of a shaft, which carries the vane, together with its barbs. The form of each of these portions is greatly modified even in different parts of the same bird, and the same feather has almost always two kinds of barbs; one close and firm, and the other loose, floating, and downy. If a small feather be plucked from the breast or back of a sparrow or any other small bird, the upper part of the feather is seen to be close and firm, while the lower

L

is loose and downy, the upper part being evidently intended to lie closely on the body and keep out the wet, while the lower portion affords a soft and warm protection to the skin.

Fig. 12, Plate X., shows the feather of a Peacock, wherein the barbs are very slightly fringed and lie quite loosely by each other's side. Fig. 18 is part of the same structure, in a Duck's feather, wherein are seen the curious hooks which enable each vane to take a firm hold of its neighbour, and so to render the whole feather firm, compact, and capable of repelling water. The reader will not fail to notice the remarkable analogy between these hooks and those which connect the wings of the bee.

Fig. 17 is a part of the shaft of a young feather taken from the Canary, and given for the purpose of showing the form of the cells of which the pith is composed. Fig. 20 is part of the down from a Sparrow's feather, showing its peculiar structure; and fig. 21 is a portion of one of the long drooping feathers of the Cock's tail.

Fig. 13 exhibits a transverse section of one of the large hairs or spines from the Hedgehog, and shows the disposition of the firm, horn-like exterior, and the arrangement of the cells. Sections of various kinds of hair are interesting objects, and are easily made by tying a bundle of them together, soaking them in glue, letting them harden, and then cutting thin slices with a razor. A little water will dissolve the glue, and the sections of hair will be well shown. Unless some such precaution be taken, the elasticity of the hair will cause the tiny sections to fly in all directions, and there will be no hope of recovering them.

Several examples of the Skin are also given. Fig. 27 is a section through the skin of the human finger, including one of the little ridges which are seen upon the extremity of every finger, and half of two others. The cuticle, epidermis, or scarf-skin, as it is indifferently termed, is formed by flattened cells or scales, is consequently very thin, and is shown by the dark outline of the top. The true skin or "cutis" is fibrous in structure, and lies immediately beneath, the two together constituting the skin, properly so called. Beneath lies a layer of tissue filled with fatty globules, and containing the glands by which the perspiration is secreted.

One of the tubes or channels by which these glands are enabled to pour their contents to the outside of the body, and if they be kept perfectly clean, to disperse them into the air, is seen running up the centre of the figure, and terminating in a cup-shaped orifice on the surface of the cuticle. On the palm of the hand very nearly three thousand of these ducts lie within the compass of a square inch, and more than a thousand in every square inch of the arm and other portions of the body, so that the multitude of these valuable organs may be well estimated, together with the absolute necessity for keeping the skin perfectly clean in order to enjoy full health.

Fig. 1 shows a specimen of epidermis taken from the skin of a Frog, exhibiting the flattened cells which constitute that structure, and the oval or slightly elongated nuclei, of which each cell has one. In fig. 32, being a portion of a Bat's wing, the arrangement of the pigment is remarkably pretty. Immediately above, at fig. 31, is some of the pigment taken from the back of the human eyeball, which gives to the pupil that deep black aspect which it presents. The shape of the pigment particles is well shown. Fig. 33 shows the pigment in the shell of the Prawn.

On various parts of animal structures, such as the lining of internal cavities, the interior of the mouth, and other similar portions of the body, the cells are developed into a peculiar form which is called "Epithélium," and which supplies the place of the epidermis of the exterior surface of the body. The cells which form this substance are of different shapes, according to their locality. On the tongue, for example (for which

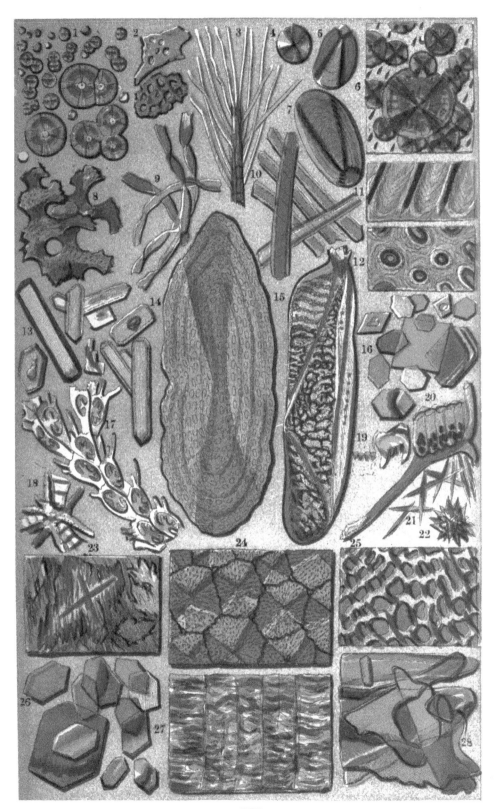

XI.

see fig. 11), they are flattened, and exhibit their nucleus, in which the nucleölus, or something which goes by that name, may be discovered with a little care. Cells of this kind are sometimes rounded, as in the case just mentioned, or angular, and in either case they are termed "pavement" or "tesselated" epithelium. Sometimes they are like a number of cylinders, cones, or pyramids, ranged closely together, and are then called cylinder epithelium. Sometimes the free ends of cylinder epithelium are furnished with a number of vibrating filaments or cilia, and in this case the structure is called "ciliated" epithelium. Cylinder epithelium may be found in the ducts of the glands which open into the intestines, as well as in the glands that secrete tears; and ciliated epithelium is seen largely in the windpipe, the interior of the nose, &c. A specimen taken from the nose is seen at fig. 15.

BONE in its various stages is figured on Plate X.

Fig. 9 is a good example of human bone, and is a thin transverse section taken from the thigh. When cut asunder, bone exhibits a whitish structure filled with little dottings that become more numerous towards the centre, and are almost invisible towards the circumference. In the centre of the bone there is a cavity, which contains marrow in the mammalia, and air in the birds. When placed under a microscope, the bone presents the appearance shown in the illustration.

The large aperture in the centre is one of the innumerable tubes that run through the bone, and that serve to allow a passage to the vessels which convey blood from one part of the bone to another. They are technically called Haversian canals, and if a longitudinal section be made, they will be found running tolerably parallel, and communicating freely with each other. Around each Haversian canal may be seen a number of little black spots with lines radiating in all directions, and looking something like flattened insects. These are termed bone-cells or "lacúnæ," and the little black lines are called "canalículi." When viewed by transmitted light, the lacunæ, together with the canaliculi, are black; but when the mirror is turned aside, and light thrown on the object by the condenser, the Haversian canals become black, and the lacunæ are white.

As these canaliculi exist equally in every direction, it is impossible to make a section of bone without cutting myriads of these across; and when a high power is employed, they look like little dots scattered over the surface. A very pretty object can be made of the bone taken from a young animal which has been fed with madder, as the colour gets into the bone and settles chiefly round the Haversian canal. A young pig is a very good subject, so is a rabbit.

Fig. 16 is a similar section cut from the leg-bone of an Ostrich.

The development of bone is beautifully shown in fig. 30, a delicate slice taken from a Pig's rib. Above may be seen the gristle or cartilage, with the numerous rows of cells; below is the formed bone, with one of the Haversian canals and its contents; while between the two may be seen the cartilage-cells gathering together and arranging themselves into form. The cartilage-cells are well shown in fig. 28, which is a portion of the cup which had contained the eye of a Haddock.

The horn-like substances at the end of our fingers, which we popularly call our Nails, are composed of innumerable flattened cells. These cells are generally so fused together as to be quite indistinguishable even with a microscope, but can be rendered visible by soaking a section of nail in liquor potassæ, which causes the cells to swell up and resume to a degree their original rounded form.

It is worthy of remark that the animal form is built up of cells, as is the case with the vegetables, although the cells are not so variable in shape. They generally may be found to contain nuclei well marked, two or more being often found within a single cell, and in many cases the tiny nucleöli are also visible. Good examples of these cells may be obtained from the yolk of an egg, and by careful management they may be traced throughout every part of the animal form. The aid of chemistry is often needed in order to force the cells to exhibit themselves in their true forms.

The Teeth have many of the constituents of bone, and in some of their parts are made precisely after the same fashion. When cut, the teeth are seen to consist of a hard substance, called Enamel, which coats their upper surfaces, of ivory within the enamel, and of "cement," which surrounds the fangs. In fig. 26, Plate X., which is a longitudinal section of the human "eye" tooth, the enamel is seen above, the ivory occupying the greater part of the tooth, and the cement at the bottom. In the centre of each tooth there is a cavity, which is plentifully filled with a pulpy substance from which the tooth is formed. A transverse section of the same tooth is seen at fig. 25.

The enamel is made of little elongated prisms, all pointing to the centre of the tooth. When viewed transversely, their ends are of a somewhat hexagonal shape, something like an irregular honeycomb. The Ivory is composed of a substance pierced with myriads of minute tubes, which give out branches that communicate freely with each other. They require a rather high power—say 300 diameters—to show them properly. The cement is found at the root of the fangs, and is best shown in the tooth of an aged individual, when it assumes very clearly the character of bone.

Sections may be made by sawing a slice in the required direction, polishing one side, and cementing it with old Canada balsam to a slide. It may then be filed down to nearly the required thinness, finished by carefully rubbing with a hone, and polished with buff leather. Canada balsam may then be dropped upon it, and a glass cover pressed firmly down. Sections of young bone form magnificent objects for the polariser.

Fig. 29 is a section cut from one of the palate teeth of the Ray (*Myliobates*).

A rather important element in the structure of animals is the " elastic ligament " which is found in the back of the neck, and other parts of the body, especially about the spine. It is made of a vast number of fibres of variable shape and length, arranged generally in bundles, and remarkable for containing very few vessels, and no nerves at all. At fig. 14 may be seen an example of elastic ligament, popularly called " paxwax," taken from the neck of a sheep.

The white fibrous tissue by which all the parts of the body are bound together is seen at fig. 10; and at fig. 11 is a beautiful example of "ultimate" fibrous tissue taken from the crystalline lens of a Sturgeon's eye.

The muscles of animals are composed of two kinds of fibre, the one termed the striped, and the other the unstriped. Of these, the latter belongs to organs which work without the will, such as the stomach, &c., while the former belongs to those portions of the body which are subject to voluntary motion, such as the arm and the leg. The unstriped muscle is very simple, consisting merely of long simple fibres; but the striped or voluntary muscle is of more complex construction. Every voluntary muscle consists of myriads of tiny fibres, bound together in little bundles, enveloped in a kind of sheath. Fig. 24 is an example of this muscular fibre, taken from beef. When soaked in spirits, it often splits into a number of discs, the edges of which are marked by the transverse lines.

A fibre of Nerve is drawn at fig. 23, and is given for the purpose of showing the manner in which the nerve is contained in and protected by its sheath, just like a telegraph-wire in its coverings. Just above is a transverse section of the same fibre, showing the same arrangement from another point of view, and also illustrating the curious phenomenon, that when nerve-fibres are treated with carmine, the centre takes up the

BLOOD AND ITS STRUCTURE. 155

colouring matter, while the sheath remains white as before. Dissection of nerves is a tedious and difficult subject, and requires the aid of good books and instruments for its successful achievement.

The Blood of animals is analogous in its office to the sap of plants, but differs greatly from it under the microscope. In sap there seem to be no microscopic characters, except that when a branch is cut, as in the vine, the flowing sap may contain certain substances formed in the wounded cells, such as chlorophyll, starch, and raphides; but the blood is known to be an exceedingly complex substance both in a microscopic and chemical point of view. When a little recent blood is placed under the microscope, it is seen to consist of a colourless fluid filled with numerous little bodies, commonly called "blood-globules," varying very greatly in size and shape, according to the animal from which they were taken. Those of the reptiles are very large, as may be seen at fig. 4, Plate X., which represents a blood corpuscle of the Proteus. In this curious reptile the globules are so large that they may be distinguished during its life by means of a common pocket-lens.

In the vertebrated animals these corpuscles are red, and give to the blood its peculiar tint. They are accompanied by certain colourless corpuscles, spherical in form, which are sometimes, as in man, larger than the red globules, and in others, as in the Siren and the Newt, considerably smaller. The general view of these red corpuscules has sufficient character to enable the practised observer to name the class of animal from which it was taken, and in some cases they are so well marked that even the genus can be ascertained with tolerable certainty. In point of size, the reptiles have the largest, and the mammalia the smallest, those of the Siren and the Goat being, perhaps, the most decidedly opposed to each other in this respect.

In shape, those of the Mammalia are circular discs, mostly with a hollowed centre; those of the Birds are more or less oval and convex; those of the Reptiles are decidedly oval, very thin, and mostly have the nucleus projecting; and those of the Fishes are oval and mostly convex. During the process of coagulation, the blood corpuscules run together into a series of rows, just as if a heap of pence had been piled on each other, and then pushed down, so that each penny overlaps its next neighbour.

These objects are illustrated by six examples on Plate X. Fig. 2 is Human blood, showing one of the white corpuscules. Fig. 3 is the blood of the Pigeon; fig. 4, of the Proteus anguinus; fig. 5, of the Tortoise; fig. 6, of the Frog, showing the projecting nucleus; and fig. 7, of the Roach. The blood possesses many curious properties, which cannot be described in these few and simple pages.

In the centre of Plate X. is a large circular figure representing the membrane of a Frog's foot as seen through the microscope, and exhibiting the circulation of the blood. The mode of arranging the foot so as to exhibit the object without hurting the frog is simple enough.

Take an oblong slip of wood—my own was made in five minutes out of the top of a cigar-box—bore a hole about an inch in diameter near one end, and cut a number of little slits all round the edge of the wooden slip. Then get a small linen bag, put the frog into it, and dip him into water to keep him comfortable. When he is wanted, pull one of his hind feet out of the bag, draw the neck tight enough to prevent him from pulling his foot back again, but not sufficiently tight to stop the circulation. Have a tape fastened to the end of the bag, and tie it down to the wooden slide.

Then fasten a thread to each of his toes, bring the foot well over the centre of the hole, stretch the toes well apart, and keep them in their places by hitching the threads into the notches on the edge of the wooden strip. Push a glass slide carefully between the foot and the wood, so as to let the membrane rest upon the glass, and be careful to keep it well wetted. If the frog kicks, as he will most likely do, pass a thin tape over the middle of the leg, and tie it down gently to the slide.

Bring the glass into focus, and the foot will present the appearance so well depicted in the engraving. The veins and arteries are seen spreading over the whole of the membrane, the larger arteries being often accompanied by a nerve, as seen in the illustration. Through all these channels the blood continually pours with a rather irregular motion, caused most probably by the peculiar position of the reptile. It is a most wonderful sight, of which the observer is never tired, and which seems almost more interesting every time that it is beheld.

The corpuscules go pushing and jostling one another in the oddest fashion, just like a British crowd entering an exhibition, each one seeming to be elbowing its way to the best place. To see them turning the corners is very amusing, for they always seem as if they never could get round the smaller vessels, and yet invariably accomplish the task with perfect ease, turning about and steering themselves as if possessed of volition, and insinuating their ends when they could not pass crosswise.

By putting various substances, such as spirits or salt, upon the foot, the rapidity of the circulation can be greatly increased or reduced at will, or even stopped altogether for a while, and the phenomenon of inflammation and its gradual natural cure be beautifully illustrated. The numerous black spots upon the surface are caused by pigment.

The tails of young fish also afford excellent objects under the microscope, as the circulation can be seen nearly as well as in the Frog's foot. The gills of Tadpoles can also be arranged upon the stage with a little care, and the same organs in the young of the common Newt will also exhibit the circulation in a favourable manner. The Frog, however, is perhaps the best, as it can be arranged on the "frog-plate" without difficulty, and the creature may be kept for months by placing it in a cool, damp spot, and feeding it with flies, little slugs, and similar creatures.

CHAPTER VII.

INFUSORIA—ROTIFERS—POND HUNTING—SPONGES—NOCTILUCA—SEA ANEMONES—JELLY FISH—FORAMINIFERA—ZOOPHYTES—ENTOMOSTRACA—LARVA OF CRAB—STRUCTURE OF SHELL—CILIA OF MUSSEL—STAR-FISHES AND ECHINI—PEDICILLARIA—CORALLINE—CAPILLARY VESSELS—INJECTIONS—COAL—GOLD-DUST AND COPPER—POLARIZED LIGHT.

TURNING back for a while to Plate IX. we come upon a series of objects which have long been termed Infusoria, because they are found in water in which vegetable or animal substances have been steeped. They get into almost every drop of water in which such substances have lain, and may be found even in the mud on the road or the roof-pipes of houses. Many of these curious beings are tolerably familiar to the public, through the medium of the oxy-hydrogen microscope so popular at exhibitions, and are generally supposed to be inhabitants of every drop of water which we drink. This, however, is not the case, as the water is always prepared for the purpose; hay, leaves, or similar substances being steeped in it for some weeks, and the turbid scrapings placed under the microscope to discompose the public mind. The whole history of these creatures is very obscure, and seems unlikely to be satisfactorily settled for the present. Suffice it to say that they may be found in almost every localityin every climate, being capable of withstanding cold far below zero and heat far above the boiling point, and may be dried over and over again, without seeming to care anything about the matter. Their increase is wonderfully rapid, taking place by subdivision, and thereby spreading the minute organisms throughout a large mass of water in an incredibly short space of time. So rapid is the process, that it may even be noticed under the microscope, and is remarkably interesting.

If a little hay or leaves be put into water, and suffered to remain in the open air for a week or two, a kind of deposit will collect round the decaying vegetation, and when submitted to the microscope is seen to contain a vast number of these minute but interesting beings. Many persons are fond of making skeleton plants by steeping them in water for a long time. The vessels in which this operation is performed are found to be extremely fertile in these curious little infusoria.

When some of the muddier portion of the water is placed under the microscope, it will be seen to be absolutely crowded with moving creatures, running restlessly in all directions, like Vathek and his companions in the Hall of Eblis, and in a similar manner avoiding each other, as if repelled by some innate force. On a closer examination, this moving crowd gradually resolves itself into variously shaped forms, among which one of the largest, and strongest, and swiftest is that which is represented in fig. 5, Plate IX., and is known by the name of Paramœcium.

It is one of a large family of Infusoria, the genera of which are reckoned at twelve in number. In size it is so large, that when the vessel of water is held between the eye and the light, it may be seen as a very minute white speck moving through the water. Its body is covered with vibrating cilia, by means of which it whirls itself through the water, and drives its food within reach of its mouth. The mode of action may be readily comprehended by putting a little carmine into the water, when the crimson particles will be seen hurled in regular currents by the cilia, and after awhile many will be distinguished within the transparent body. At each end of the Paramœcium there is a curious star-shaped, contractile vesicle. The cilia are arranged in regular rows. Three of the dart-like weapons with which it is armed are seen just below the figure.

Perhaps the lowest form of animal life is to be found in another of these Infusoria, the Amœba, which is represented at fig. 1.

This wonderful creature is remarkable for having no particular shape, altering its form momentarily, and moving by means of this curious mode of progression. At first it mostly looks like a little rounded semi-transparent mass, but in a short time it begins to push out one part of its body into a projection of some length, which gradually fixes itself to some convenient object, and by contraction draws the body after it. In fig. 1 three forms of the same creature are shown, one of which is remarkable for having included a large diatom in its structure.

The mode by which this creature feeds is the simplest imaginable. Any object which may serve as food, such as a diatom, a desmid, or another infusorial, comes in contact with the surface of its body, and is there held. Presently, that portion of the body whereon the captured organism lies begins to recede and forms a cavity, into which it is pressed. This cavity answers the purpose of a stomach, and the indigestible portions are either returned by the same way, or squeezed through some other portion of the animal. Such inclosed organisms may be seen in the figure.

At fig. 2 may be seen another curious creature named

the Arcella, which is in fact little more than an Amœba with a shell, the soft body altering its shape in precisely the same manner.

A very curious Infusorial is shown at fig. 3. This is the Sun animalcule (*Actinophrys Sol*), remarkable for the long tentacles with which its body is surrounded. The nourishment of this creature is managed after the same manner as has already been mentioned when treating of the Amœba. The organisms which serve for its food are captured by one of the tentacles, which immediately begins to contract. Those surrounding it give their aid by bending towards it, and by their united force the victim is gradually pressed into the body, where it is digested.

Fig. 6 is given in order to show the process of subdivision as it takes place in the Infusoria. The species figured is the Chilodon, another flattened creature covered with cilia, with teeth arranged in the form of a tube, and with the fore part of the so-called head having a kind of membranous lip.

These creatures, together with many minute inhabitants of the waters, are of incalculable use in devouring the decaying substances that would otherwise breed pestilence, and converting them into their own living persons.

Figs. 7 and 10 represent two examples of the singular

XII.

beings called Rotifers, on account of the wheel-like apparatus which they bear. Upon the front of the body or head is a retractile disc, whose edges are covered with cilia, which, by bending in regular succession, look exactly as if they were wheels running round at a great rate. A similar phenomenon may be seen in the cornfields, when the wind forces them to bend, and produces a succession of waves that seem to roll over the field. These cilia are their chief modes of progression ; but in many cases they get along rather fast by attaching their tails and heads alternately to the substances on which they move, after the well-known fashion of the leech, to which they then bear no small resemblance Fig. 10 represents the commonest species of rotifer, *Rótifer vulgáris*, in which this mode of progression may be seen.

They are furnished with a well-defined mouth and digestive organs, and their teeth, a pair of which may be seen in fig. 11, are very powerful.

Some species of Rotifer, such as Melicerta, fig. 7, are inclosed in a tube. In some cases, as in the present, the tube is opaque ; but in many it is beautifully transparent, and permits the inclosed creature to be seen through its substance.

I would that rapidly narrowing space did not compel me to give such brief notice of these most interesting creatures. They may, however, be readily obtained in almost any pond, provided the water be not putrid, and are so large that their movements may be watched with a rather low power. There are very many species, but they may be distinguished from the other inhabitants of the waters by the wheel-like processes from which they derive their name. Sometimes the wheel-disc is withdrawn within the body; but if the observer wait for a little while, he is sure to see the disc protruded and the wheels run their merry course.

The mode of obtaining these tiny creatures is sufficiently simple.

Get a small, rather wide-mouthed phial, and with the piece of string which every sensible man always has in his pocket, lash the bottle by the neck across the end of a walking-stick. Look out for the best hunting-grounds in ponds, rivulets, &c., an accomplishment in which practice soon makes perfect; push the inverted bottle among the flocculent greenage or the decaying leaves, and after poking it well about, turn the bottle suddenly over, when the water will rush rapidly in, carrying with it myriads of minute organisms. After a little experience in this kind of fishing, it soon becomes easy to capture any creature that may be seen in the water, by placing the inverted bottle delicately near the intended victim, and then quietly turning it over without alarming the easily frightened creature.

When the bottle is filled, the contents should be poured into another wide-mouthed bottle, and the process repeated until a sufficient amount of living organisms has been obtained. The bottle should always be labelled with the particular place, pond, or stream whence the water was obtained; and it is always well to add the date. It will be found advisable to have a number of wide-mouthed bottles always ready, which can be carried in a basket or box fitted up for the purpose by means of wooden partitions, or even by strings crossing each other at proper intervals. One reason for this precaution is the power of  identifying the exact locality in which any rare or curious creatures may be found; and another reason is, that the inhabitants of different ponds or puddles are apt to wage deadly war if put into the same bottle. A drawing of a case fitted up with bottles is here introduced, more for the purpose of giving the reader a model on which to make a hunting-case for himself, than to recommend him to purchase or order it from a carpenter.

The air must always be admitted freely into the bottles, or the creatures will soon die. But as the conveyance of uncorked wide-mouthed bottles half full of water is exceedingly inconvenient, it is needful to fill up each bottle in such a manner that the air can be freely admitted, while the water will not run out.

A very simple contrivance is to close the mouths of the bottles with corks, to bore a hole through the centre of each cork, and to pass a quill through it, projecting about half an inch through the cork within the bottle, and cut off level above. Even if the bottle should be turned upside down by any mischance, scarcely a drop of water will escape; while the air is admitted nearly as freely as if the mouth of the bottle were open.

One bottle should be supplied with some of the mud taken from the bottom of the pool or puddle, as it is sure to contain many interesting objects, and is generally a rich preserve of the flinty skeletons of these little inhabitants of the waters.

They are easily separated from the other constituents of the mud, by putting a little of the mud into the bottom of a tall test-tube, pouring some nitric acid upon it, and boiling it gently over the flame of the spirit-lamp. Very great care is needed in this operation, as the liquid is apt to rise suddenly and boil over; and

PREPARING SKELETONS OF DIATOMS.

the fumes which arise are always copious and very deleterious, so that the boiling should always be done in the open air, or at all events in some place where the fumes can be carried away as fast as they are generated.

After it has boiled for some little time, and got cool, half fill the tube with distilled water, shake it up well, and set the tube upright so as to let the solid particles sink to the bottom. When it has thoroughly settled, which will not be for some hours, remove the clear liquid by means of a syphon—a wet skein of cotton thread hung over the edge of the tube will do very well—pour in some fresh nitric acid, and boil it again. Repeat this process three or four times, and then wash the residue very thoroughly in distilled water, always allowing the solid matter to settle, and removing the liquid with a siphon; and when the acid has been entirely washed away, spread some of the residue upon a clean slide, and examine it under the microscope. The field of the instrument will then be filled with the lovely flint scaffolding upon which the living organisms of these minute creations are supported; and when some peculiarly good specimens are found, they should be preserved as permanent objects by dropping a little Canada balsam upon them and mounting them after the manner already described.

There is a considerable amount of amusement to be got out of this kind of fishing, as it is always a sort of lottery, in which the blanks are none, and the prizes many.

For the capture of the larger creatures, such as are readily visible to the naked eye, and swim with much velocity, a net is needful. This is readily made by twisting a piece of brass or copper wire into a ring, and sewing a piece of very fine net over it so as to give it a hollow about as deep in proportion as that of a watch-glass. This little net can easily be carried in the pocket, and when wanted can be attached to a stick at a moment's notice.

A very useful little net, which, however, requires the aid of the tinman, is here depicted. The reader will see that it is made of a strip of tin bent into a spoon-like shape, and with a net fastened at the bottom. The great advantage of this net is that the high walls are very effectual in inclosing any quickly moving creature, and prevent it from being washed out of the net on its being raised to the surface. The meshes of the net need not be very close, as a mesh will always secure an insect of only half its diameter.

It is convenient for many reasons to have the nets

and other apparatus made in such a manner that they can be carried without attracting observation, for at the best of times the microscopic angler is sure to be beset with inquisitive boys of all sizes, who cannot believe that any one can use a net in a pond except for the purpose of catching fish, and is therefore liable to have his sanity called in question, and his proceedings greatly disturbed. However, by a little judicious administration of "soft sawder" and a few pence, the enemies may generally be converted into allies, and rendered extremely useful.

SPONGES and their structures are very interesting. They consist chiefly of a very thin gelatinous substance not unlike that of the Amœba, which is spread over a horny skeleton, which skeleton is sustained by an internal arrangement of spiculæ, mostly of a flinty nature, but sometimes being chalky in their substance. The gelatinous envelope is covered with cilia, by means of which the water is forced to circulate throughout the entire sponge, entering through the little apertures and being expelled through the larger holes. A portion of this substance with its cilium is shown at fig. 12. Fig. 14 shows this process in a Sponge (*Grántia*).

The little granules which afterwards become mature sponges are also thrown from the parent in the same manner. One of these bodies covered with cilia is shown at fig. 18. When ejected from the parent, they swim merrily about for some time, but at last settle down and become fixed for the rest of their life. During the life of a sponge it is coloured, and often vividly, with various tints; but after the death of the living portion, the bare horny skeleton is left.

Two forms of sponge spiculæ are shown at figs. 8 and 20. The shapes, however, which these curious objects assume are almost innumerable. In them may be seen accurate likenesses of pins, needles, marlinspikes, cucumbers, grappling-hooks, fish-hooks, porters'-hooks, calthrops, knife-rests, fish-spears, barbed arrows, spiked globes, war-clubs, boomerangs, life-preservers, and many other indescribable forms. They may be obtained by cutting sponge into thin slices, and soaking it in liquor potassæ or any other substance that will dissolve the horny skeleton and leave the flinty spiculæ uninjured.

Every one who has been by or on the sea on a fine summer night must have noticed the bright flashes of light that appear whenever its surface is disturbed; the wake of a boat, for example, leaving a luminous track as far as the eye can reach. This phosphorescence is caused by many animals resident in the sea, but chiefly by the little creature represented at fig. 9, the

Noctilúca, myriads of which may be found in a pail of water dipped at random from the glowing waves. A tooth of this creature more magnified is shown immediately above.

In my "Common Objects of the Sea Shore" the Actíniæ or Sea-Anemones are treated of at some length. At fig. 16 is shown part of a tentacle flinging out the poison-darts by which it secures its prey; and fig. 17 is a more magnified view of one of these darts and its case.

The Jelly-Fishes, or Medúsæ, are partially represented at fig. 28, &c. This represents a very small and very pretty Medusa, called Thaumántias. When touched or startled, each of the purple globules round the edge flashes into light, producing a most beautiful and singular appearance. Fig. 29 exhibits the so-called compound eye of another species of Medusa. The reproduction of these creatures is too complicated a subject for the small space allowed in these pages, but is partially illustrated by figs. 26 and 27. Fig. 26 is the Hydra tuba, a creature long thought to be a distinct animal, but now known to be the young of a Medusa, which does not itself attain maturity, but throws off its joints, so to speak, each of which becomes a perfect Medusa. One of these joints is shown at fig. 27.

A large group of microscopic organisms is known to

zoologists under the name of Foraminifera, on account of the numerous holes in their beautiful shells. Their real position in the animal kingdom is somewhat doubtful. The holes are intended to permit the passage of certain thread-like tentacles, and are variously arranged upon the shell. Chalk is largely mixed with these minute shells, and whole tracts of country are composed almost wholly of these little creatures in a fossilized state. They may often be found in sand, and separated by spreading the sand on black paper and examining it with a glass. Examples of these creatures are given in Plate IX. fig. 4 (*Miliolína*), and Plate XII. fig. 7, which is a portion of the shell to show the holes, fig. 13 (*Polystomella*), fig. 14 (*Truncatulína*), fig. 15 (*Polymorphína*), fig. 16 (*Miliolína*, partly fossilized), fig. 18 (*Lagéna*), and fig. 20 (*Biloculína*).

The Zoophytes or Polypi are represented by several examples. These creatures are soft and almost gelatinous, and are furnished with tentacles or lobes by which they can catch and retain their prey. In order to support their tender structure they are endowed with a horny skeleton, sometimes outside and sometimes inside them, which is called the polýpidon. They are very common on our coasts, where they may be found thrown on the shore or may be dredged up from the deeper portions of the sea.

Fig. 13 is a portion of one of the commonest genera, Sertularia, showing one of the inhabitants projecting its tentacles from its domicile. Fig. 15 is the same species, given to show the egg-cells. This, as well as other zoophytes, is generally classed among the sea-weeds in the shops that throng all watering-places. Fig. 19 is a very curious zoophyte called Anguinaria, or Snake-head, on account of its shape, the end of the polypidon resembling the head of the snake, and the tentacles looking like its tongue as they are thrust forward and rapidly withdrawn. Fig. 21 is the same creature on an enlarged scale, and just below is one of its tentacles still more magnified. Fig. 23 is the Ladies'-slipper zoophyte; and fig. 24 is called the Tobacco-pipe zoophyte.

Fig. 22 is a portion of the Bugularia, with one of the curious "birds'-head" processes. These appendages have the most absurd likeness to a bird's head, the beak opening and shutting with a smart snap, so smart indeed that the ear instinctively tries to catch the sound, and the head nodding backward and forward just as if the bird were pecking up its food. On Plate XII. fig. 2, is a pretty zoophyte called Gemellaria, on account of the double or twin-like form of the cells; and fig. 5 represents the Antennularia, so called on account of its resemblance to the antennæ of an insect. Fig. 22 is an example of a pretty zoophyte found parasitic on many sea-weeds, and known by the name of Membranipora. Two more specimens of zoophytes may be seen on Plate XII. as they appear under polarized light. Fig. 17 is the Cellularia reptans; and fig. 20 is the Bowerbankia.

Our space is so rapidly diminishing, that we can only give one example of the curious group of animals called Entomóstraca. They belong to the great class of Crustaceans, and are found both in fresh and salt water. Their shell is often transparent, so as to permit their limbs to be seen through its substance, and when boiled it gets red like that of the lobster. Their shape is extremely various, but that of the example at Plate IX. fig. 31, the Fresh-water Flea (*Daphnia*), affords a good illustration of their general appearance. The Cyclops, another fresh-water example, is very common in our ponds, and may be known by the long body, the single eye in the head, and the egg-bags depending from the sides of the females.

Fig. 25 is the larva of the common Crab, once thought to be a separate species, and described as such under the title of Zoea. I may as well mention that many of the objects here mentioned in a cursory manner are to be found described more at length in my two previous manuals, the "Common Objects" of the Sea-shore and Country.

Parts of the so-called feet of the Serpula are shown at fig. 36, where the spears or "pushing-poles" are seen gathered into bundles as used by the creature. One of them on a larger scale is shown at fig. 32.

The structure of shell, *e. g.* oyster-shell, is well shown in three examples: Fig. 34 is a group of artificial crystals of carbonate of lime; and on figs. 38 and 39 may be seen part of an oyster-shell, showing how it is composed of similar crystals aggregated together. Their appearance under polarized light may be seen on Plate XI. figs. 1 and 6.

Before entering upon the Echinoderms, we will cast a glance at a beautiful structure found upon the gills of the common Mussel. Fig. 39 shows a portion of the gills in order to exhibit the numerous cilia with which it is covered. It is a valuable example, as the cilia attain a very large size on this organ, being about one five-hundredth of an inch in length. Their object is of course to produce circulation in the water which bathes the gills.

The old story of the goose-bearing tree is an example how truth may be stranger than fiction. For if the fable had said that the mother goose laid eggs which grew into trees, budded and flowered, and then produced new geese, it would not have been one whit a stranger tale than the truth. Plate IX., fig. 33, shows the young state of one of the common Star-fishes (*Comátula*), which in its early days is like a plant with a stalk, but afterwards breaks loose and becomes the wandering sea-star which we all know so well. In this process there is just the reverse to that which characterizes the barnacles and sponges, where the young are at first free and then become fixed for the remainder of their lives. Fig. 30 is the young of another kind of Star-fish, the long-armed Ophiúra, or Snake-Star.

Fig. 37 is a portion of the skin of the common Sun-star (*Solaster*), showing the single large spine surrounded by a circle of smaller spines, supposed to be organs of touch, together with two or three of the curious appendages called Pedicillariæ. These are found on Star-fishes and Echini, and bear a close resemblance in many respects to the bird-head appendages of the zoophytes. They are fixed on foot-stalks, some very long and others very short, and have jaws which open and shut regularly. Their object is doubtful, unless it be to act as police, and by their continual movements to prevent the spores of algæ, or the young of various marine animals, from effecting a lodgment on the skin. A group of Pedicillariæ from a Star-fish is shown on a large scale on Plate XII. fig. 6, and fig. 9 of the same plate shows the Pedicillariæ of the Echinus.

Upon the exterior of the Echini or Sea-Urchins are a vast number of spines, having a very beautiful structure, as may be seen by fig. 35, Plate IX., which is part of a transverse section of one of these species. An entire spine is shown on Plate XII. fig. 12, and shows the ball-and-socket joint on which it moves, and the membranous muscle that moves it. Fig. 8 is the disc of the Snake-Star as seen from below. Fig. 1 is a portion of skin of the Sun-Star, to show one of the curious madrepore-like tubercles which are found upon this common Star-fish. Fig. 3 is a portion of Cuttle "bone" very slightly magnified, in order to show the beautiful pillar-like form of its structure; and fig. 4 is the same object seen from above. When ground very thin, this is a magnificent object for the polarizer.

One or two miscellaneous objects now come before our notice. Fig. 11 is one of those curious marine plants, the Corallines, which are remarkable for depositing a large amount of chalky matter among their tissues, so as to leave a complete cast in white chalk when the coloured living portion of the plant dies. The species of this example is *Jania rubens*.

Fig. 19 is part of the pouch-like inflation of the skin, and the hairs found upon the Rat's tail, which is a curious object as bearing so close a similitude to fig. 22, the Sea-mat zoophyte. Fig. 23 is a portion of the skin taken from the finger. which has been injected with a coloured preparation in order to show the manner in which the minute blood-vessels or "capillaries" are distributed; and fig. 26 is a portion of a Frog's lung, also injected.

The process of injection is a rather difficult one, and needs tools of some cost. The principle is simple enough, being merely to fill the blood-vessels with a coloured substance, so as to exhibit their form as they appear while distended with blood during the life of the animal. It sometimes happens that when an animal is killed suddenly without effusion of blood, as is often seen in the case of a mouse caught in a spring trap, the minute vessels of the lungs and other organs become filled with coagulated blood, so as to form what is called a natural injection, ready for the microscope.

Before leaving the subject, I must ask the reader to refer again for a moment to the Frog's foot on Plate X. and to notice the arrangement of the dark pigment spots. It is well known that when frogs live in a clear sandy pond, well exposed to the rays of the sun, their skins are bright yellow, and that when their residence is in a shady locality, especially if sheltered by heavy overhanging banks, they are of a deep blackish brown colour. Moreover, under the influence of fear, they will often change colour instantaneously. The cause of this curious fact is explained by the microscope.

Under the effects of sunlight the pigment granules

are gathered together into small rounded spots, as seen on the left hand of the figure, leaving the skin of its own bright yellow hue. In the shade the pigment granules spread themselves so as to cover almost the entire skin and to produce the dark brown colour. In the intermediate state, they assume the bold stellate form in which they are shown on the right hand of the round spots.

Figs. 24 and 25 are two examples of Coal, the former being a longitudinal and the latter a transverse section, given in order to show its woody character. Fig. 17 is a specimen of Gold-dust intermixed with crystals of quartz sand, brought from Australia; and fig. 21 is a small piece of Copper-ore.

Every possessor of a microscope should, as soon as he can afford it, add to his instrument the beautiful apparatus for polarizing light. The optical explanation of this phenomenon is far too abstruse for these pages, but the practical appliance of the apparatus is very simple. It consists of two prisms, one of which, called the polarizer, is fastened by a catch just below the stage; and the other, called an analyser, is placed above the eye-piece. In order to aid those bodies whose polarizing powers are but weak, a thin plate of selenite is generally placed on the stage immediately below the object. The colours exhibited by this instrument are gorgeous in the extreme, as may be seen by Plate XI., which affords a most feeble representation of the glowing tints exhibited by the objects there depicted. The value of the polarizer is very great, as it often enables observers to distinguish, by means of their different polarizing powers, one class of objects from another.

Another instrument really essential to the microscopist is the micrometer, for the purpose of measuring the minute objects under examination. The cheapest and simplest is the Stage Micrometer, which may be purchased for five shillings at the opticians'. It consists simply of a glass slide on which are ruled a series of lines, some the hundredth of an inch apart, and others the thousandth. This is laid on the stage, and the object placed upon it, when with a little management the lines may be made to cut the objects so as to give their dimensions.

Another simple and even more accurate way is to slip the camera lucida on the eye-piece, and sketch the object as mentioned on page 51. Then remove the object, substitute the stage micrometer, and sketch the lines upon the drawing of the object. It will be also evident that if the "hundredth" lines coincide with an inch, the object is magnified one hundred diameters; if with two inches, two hundred diameters, and so on.

INDEX.

Air-bubbles, 95.
Algæ, 100.
Anemones, Sea, 173.
Antennæ, 119.
Ants, 119.
Balancers of Fly, 138.
Bark, 67.
Blights, 112.
Blood, 155.
Bone, 149.
Breathing-tubes, 127.
Camera Lucida, 51.
Canada balsam, 86.
Cartilage, 151.
Cells, animal, 151.
—— glass, 92.
—— mounting in, 95.
——varnish, 94.
——vegetable, 37.
Ceramidia, 117.
Chlorophyll, 40.
Cilia of Mussel, 177.
Circulation, 156.
Coal, 181.
Coddington Lens, 24.
Compound Microscope, 26.
Condenser, 33.
Confervoid Algæ, 103.
Conjugation, 101.
Copper, 181.
Corallines, 179.
Cotton, 143.
Cuttle Fish, 179.
Desmidiaceæ, 103.
Diaphragm, 35.
Diatomaceæ, 108.
Dipping-tubes, 23.
Dissecting Microscope, 12.
Dotted Ducts, 45.
Dry mounting, 90.
Ducts and Vessels, 45.
Echini, 178.
Echinodermata, 177.
Eggs of Insects, 140.
Entomostraca, 176.
Epidermis, animal, 148.
—— vegetable, 55.
Epithelium, 148.
Essential oil, 64.
Extemporized instruments, 6.
Eyes, 125.
Feathers, 145.

Ferns, 118.
Fibre, white, 153.
Flint, or Silex, 68, 108.
Footpads of Fly, &c. 132.
Foraminifera, 173.
Forceps, 16.
Frog, colour, 180.
Frog-plate, 157.
Gizzard of Insects, 135.
Glass tubes, 22.
Gold dust, 181.
Gory-dew, 100.
Gristle, 151.
Hairs of Animals, 142.
Hairs of Insects, 139.
Hairs (vegetable), 58.
Heads of Insects, 128.
Infusoria, 160.
Injections, 180.
Insects, 119.
Introduction, 1.
Jaws of Insects, 129.
Jelly Fish, 173.
Legs of Insects, 132.
Ligament, elastic, 153.
Linen, 143.
Live-box, 31.
Mare's tail, 115.
Medusæ, 173.
Micrometer, 182.
Mildew, 112.
Mosses, 113.
Mould, 112.
Mouths (vegetable), 49.
Muscle, 153.
Muscles of Insects, 154.
Nails, 151.
Needles, 16.
Nerve, 154.
Nets, 170.
Netted or reticulated Ducts, 46.
Nucleus, and Nucleolus, 39.
Oil-cells, 47.
Oscillatoriæ, 107.
Paps of Aphis, 128.
Pedicillariæ, 178.
Petals, 76.
Pigment, 150.
Pins, 21.
Pitted structure, 40.
Polarized light, 181.
Pollen, 78.
Polypi, 174.

Pond-hunting, 166.
Pouches, Rat's tail, 179.
Preservative Fluids, 98.
Proboscis of Fly, 131.
Ringed or annulated cells, 24, 58.
Rotifers, 165.
Sap, 155.
Saws of Sawfly, 141.
Scalariform deposit, 47.
Scales of Fish, 144.
Scales of Insects, 138.
Scent-glands, 64.
Seaweeds, 115.
Secondary deposit, 44.
Seeds, 83.
Serpula, 177.
Shell, 176.
Silk, 141.
Simple Microscopes, 9.
Skin of Beetle larva, 135.
—— Human, 147.
Spiculæ, 171.
Spines of Hedgehog, 146.
Spiracles, 127.
Spiral deposit, 46.
Sponges, 171.
Sporanges, 115.
Stage forceps, 30.
Stanhope Lens, 25.
Starch, 70.
Star Fishes, 177.
Stellate tissue, 41.
Sting of Insects, 141.
Stomach of Insects, 135.
Stomata, 49.
Suckers of Foot, 134.
Sweat ducts, 147.
Teeth, 152.
Tetraspores, 117.
Tongue of Cricket, 131.
Turn-table, 94.
Urchins, Sea, 178.
Uses of the Microscope, 3.
Varnish (vegetable), 75.
Vittæ, 68.
Volvox, 111.
Wings of Insects, 136.
Wool, 142.
Yeast Plant, 112.
Zoea, 176.
Zoophytes, 174.
Zygnemaceæ, 107.

DESCRIPTION OF PLATES.

I.

FIG		PAGE
1.	Strawberry, cellular tissue	38
2.	Buttercup leaf, internal layer	39
3.	Privet, Seed coat, showing star-shaped cells	41
4.	Rush, Star-shaped cells	41
5.	Mistletoe, cells with ringed fibre	44
6.	Cells from interior of Lilac bud	45
7.	Bur-reed (*Sparganium*), square cells from leaf	40
8.	Six-sided cells, from stem of Lily	39
9.	Angular dotted cells, rind of Gourd	43
10.	Elongated ringed cells, anther of Narcissus	44
11.	Irregular star-like tissue, pith of Bulrush	42
12.	Six-sided cells, pith of Elder	40
13.	Young cells from Wheat	42
14.	Do. rootlets of Wheat	42
15.	Wood-cells, Elder	48
16.	Glandular markings and resin, "Cedar" pencil	47
17.	Do. Yew	48
18.	Scalariform tissue, Stalk of Fern	47
19.	Dotted Duct, Willow	45
20.	Do. Stalk of Wheat	46
21.	Wood-cell, Chrysanthemum	48
22.	Do. Lime-tree	48
23.	Dotted Duct, Carrot	46
24.	Cone-bearing wood, Deal	47
25.	Cells, outer coat, Gourd	54
26.	Ducts, Elm	46
27.	Cellular tissue, Stalk of Chickweed	42
28.	Holly-berry, outer coat	49

FIG.		PAGE
7.	Chaff, after burning	68
8.	Bifid hair, Arabis	59
9.	Hair, Marvel of Peru	59
10.	End of hair, leaf of Hollyhock	65
11.	Hair, Sowthistle leaf	60
12.	Do. Tobacco	61
13.	Do. Southernwood	59
14.	Group of hairs, Hollyhock leaf	66
15.	Hair, Yellow Snapdragon	62
16.	Do. Moneywort	62
17.	Hair, Geum	62
18.	Do. Flower of Heartsease	59
19.	Do. Dockleaf	59
20.	Do. Throat of Pansy	66
21.	Do. Dead-nettle flower	66
22.	Do. Groundsel	60
23.	Cell, Beech-nut	68
24.	Do. Pine cone	67
25.	Vitta, Caraway seed	68
26.	Cork	67
27.	Hair, flower of Garden Verbena	67
28.	Do. fruit of Plane	62
29.	Do. do.	63
30.	Do. do.	62-3
31.	Do. Lobelia	66
32.	Do. Cabbage	59
33.	Do. Deadnettle flower	60
34.	Do. Garden Verbena flower	62
35.	Fruit-hair, Dandelion	64
36.	Hair, Thistle leaf	60
37.	Do. Cactus	65
38.	Do. do.	65
39.	Do. Virginian Spider-wort	60
40.	Do. Lavender	63
41.	Section, Lavender leaf, Hairs and perfume-gland	63
42.	Section, Orange peel	64
43.	Sting of Nettle	61
44.	Hair, Marigold flower	65
45.	Do. Ivy	66

II.

1.	Cuticle, Buttercup leaf	54-5
2.	Do. Iris	55
3.	Do. Ivy leaf	56
4.	Spiral vessel, Lily	56
5.	Do. root, (rhizome) Water Lily	57
6.	Ringed vessel, Rhubarb	58

III.

1.	Laurel leaf, transverse section	75
2.	Starch, Wheat	73
3.	Do. from Pudding	74
4.	Do. Potato	72
5.	Outer Skin, Capsicum pod	76

DESCRIPTION OF PLATES.

FIG.		PAGE
6.	Starch, Parsnip	73
7.	Do. Arrow Root, West Indian	74
8.	Do. "Tous les Mois"	74
9.	Do. in cell of Potato	72
10.	Do. Indian Corn	74
11.	Do. Sago	74
12.	Do. Tapioca	74
13.	Root, Yellow Water-Lily	75
14.	Starch, Rice	75
15.	Do. Horsebean	75
16.	Do. Oat	75
17.	Pollen, Snowdrop	79
18.	Do. Wallflower	79
19.	Do. Willow Herb, a pollen tube	79-80
20.	Do. Violet	81
21.	Do. Musk Plant	81
22.	Do. Apple	81
23.	Do. Dandelion	81
24.	Do. Sowthistle	81
25.	Do. Lily	81
26.	Do. Heath	82
27.	Do. Heath, another species	82
28.	Pollen, Furze	82
29.	Do. Tulip	82
30.	Petal, Pelargonium	76
31.	Do. Periwinkle	77
32.	Do. Golden Balsam	78
33.	Do. Snapdragon	78
34.	Do. Primrose	78
35.	Do. Scarlet Geranium	78
36.	Pollen, Crocus	83
37.	Do. Hollyhock	83
38.	Fruit, Galium, Goose-grass	84
39.	A hook of ditto more magnified	84
40.	Seed, Red Valerian	84
41.	Portion of Parachute of same, more magnified	84
42.	Seed, Foxglove	85
43.	Seed, Sunspurge	85
44.	Parachute, Dandelion seed	84
45.	Seed, Dandelion	85
46.	Do. Hair of Parachute	85
47.	Do. Yellow Snapdragon	85
48.	Do. Mullein	86
49.	Do. Robin Hood	86
50.	Do. Bur-reed	86
51.	Do. Willow Herb	86
52.	Do. Musk Mallow	86

IV.

1. Gory Dew, Palmella cruenta 100
2. Palmoglæa macrococca . . 101
3. Protococcus pluvialis, *a*, in its motile, *b*, in its fixed state, *c*, zoospores . . 102
4. Closterium 104

FIG.		PAGE
5.	Ditto, end more magnified	104
6.	Pediastrum	104
7.	Scenedesmus	105
8.	Oscillatoria	107
9.	Spirogyra	107
10.	Tyndaridea	107
11.	Do. spore	108
12.	Sphærozosma	106
13.	Chlorococcus	103
14.	Scenedesmus	106
15.	Pediastrum, to show cells	105
16.	Ankistrodesmus	104
17.	Cosmarium	106
18.	Desmidium	105
19.	Cosmarium, formation of Resting Spore	106
20.	Cocconema lanceolatum	108-9
21.	Diatoma vulgare	108
	Do. larger frustules, at the side	108
22.	Volvox globator	111
	Do. single green body, above	111
23.	Synedra	109
24.	Gomphonema acuminatum	109
	Do. larger frustules, below	109
25.	Yeast	112
26.	Sarcina ventriculi	112
27.	Eunotia diadema	109
28.	Melosira varians	110
	Do. two bleached frustules	110
29.	Cocconeis pediculus	110
30.	Achnanthis exilis	110
31.	Navicula amphisbœna	111
32.	Uredo, "Red-rust" of corn	112
33.	Puccinea, Mildew of corn	113
34.	Botrytis, mould on grapes	113
	Do. Sporules, beside it	—
35.	Do. parasitica, Potato blight	113
36.	Ectocarpus siliculosus	115
37.	Ulva latissima	116
38.	Polypodium	113
	Do. single spore, below	113
39.	Moss capsule, Hypnum	114
40.	Mare's-tail, Equisetum, *a*. Do. do. *b* and *c*.	114
41.	Porphyra laciniata	116

V.

1. Rose Leaf, with fungus . . 113
2. Moss capsule, Polytrichum . 114
3. Jungermannia, capsule . . 114
4. Do. an elater more magnified 114
5. Leaf of Moss, Sphagnum . 114
6. Rootlet, Moss 114
7. Puccinia, from Thistle . . 113
8. Jungermannia, leaf . . . 114
9. Scale from stalk of male fern 114

DESCRIPTION OF PLATES.

FIG.		PAGE
10.	Uredo	113
11.	Sphacelaria filicina	116
12.	Do. top, more magnified	116
13.	Seaweed, showing fruit	116
14.	Do. fruit, more magnified	116
15.	Ceramium	117
16.	Capsule, Halidrys	117
17.	Spore of do.	117
18.	Polysiphonia parasitica	117
19.	Do. stem, more magnified	117
20.	Do. Capsule, tetraspores escaping	117
21.	Do. fruit, another form	117
22.	Ceramium, fruit	116
23.	Myrionema, parasitic Seaweed	117
24.	Delesseria sanguinea, Frond	118
25.	Cladophora	117
26.	Ptilota elegans	118
27.	Enteromorpha clathrata	118
28.	Nitophyllum laceratum	118

VI.

1.	Antenna, Cricket	121
2.	Do. Grasshopper	122
3.	Do. Staphylinus	122
4.	Do. Cassida	122
5.	Do. Staphylinus	122
6.	Do. Weevil	122
7.	Do. Pyrochroa	122
8.	Do. Butterfly, Tortoiseshell	124
9.	Do. Gnat, male	124
10.	Do. Syrphus	123
11.	Do. Cockchaffer, male	122
12.	Do. Ground Beetle	123
13.	Do. Ermine Moth	124
14.	Do. Tiger Moth	124
15.	Antenna, Blowfly	121
16.	Do. do. section	121
17.	Do. Ichneumon	121
18.	Eye of Butterfly, Atalanta	125
19.	Eyes, Bee	125
20.	Eye, Death's Head Moth	125
21.	Breathing-tube, Silkworm	127
22.	Eye, Heliophilus	125-6
23.	Eye, Lobster	126
24.	Do. Aphis of Geranium	128
25.	Head, Parasite of Tortoise	128-9
26.	Hindleg, Aphis of Geranium	132
27.	Head, Gnat	130
28.	"Paps" of Aphis	128
29.	Head, Sheep-tick	128
30.	Foot, Tipula	132

VII.

1.	Tongue, Hive Bee	129
2.	Do. Tortoiseshell Butterfly	130
3.	Do. do. one of the barrel-shaped bodies	130
4.	Head, Violet Ground Beetle (Carabus)	129
5.	Tongue, Cricket	131
6.	Do. do.	131
7.	Head, Scorpion Fly (Panorpa)	130
8.	Leg, Ant	132
9.	Proleg, Caterpillar	132
10.	Do. do. single hook	132
11.	Proboscis, Fly	130
12.	Do. do. "modified trachea"	131
13.	Part of Foreleg of Water Beetle (Acilius)	134
14.	Do. large sucker	135
15.	Leg, long-legged Spider (Phalangium)	132
16.	Do. Harvest-bug (Trombidium)	132
17.	Do. Glow-worm	132
18.	Do. Dung-fly	133
19.	Do. Asilus	133
20.	Do. Acarus of Dor-beetle	133
21.	Claws and Pad, Ophion	133
22.	Wings, Humble Bee	137
23.	Do.	137
24.	Wing hooks, hind wing of Aphis	137
25.	Wing hooks, Humble Bee	137
26.	Foot, Flea	132
27.	Stomach and gastric teeth, Bee	136
28.	Three teeth of do.	136
29.	Cast skin, Larva of Tortoise Beetle (Cassida)	135
30.	Balancer, Blow fly	138
31.	Wing, Midge (Psychoda)	137
32.	Do. do. part of a nervure with scales	137
33.	Stomach and gastric teeth, Grasshopper	135
34.	Spiracle, Wire-worm	127

VIII.

1.	Boat-fly, leg	135
2.	Gadfly, empty egg	140
3.	Diamond Beetle, scale	139
4.	Scale, Fritillary, Adippe	138
5.	Egg, Tortoiseshell Butterfly	140
6.	Head and Eyes, Zebra Spider	126
7.	Eyes, Bed-Bug	140
8.	Scale, Death's-Head Moth	138
9.	Sting, Wasp	141
10.	Scale, battledore, Azure Blue	138
11.	Do. ordinary scale	138
12.	Eye, Harvest Spider	126
13.	Wing Membrane, Azure Blue	140

DESCRIPTION OF PLATES.

FIG.		PAGE
14.	Scale, Anthocera cardaminis	138
15.	Do. Peacock Butterfly	138
16.	Do Tiger Moth	138
17.	Do. Thigh of Tiger Moth.	140
18.	Wing and Scales, Azure Blue	140
19.	Scale, Lepisma	139
20.	Saws, Sawfly	141
21.	Scale, Podura	139
22.	Hair, Black Human	142
23.	Do. Human Beard	142
24.	Do. do. aged.	142
25.	Do. Humble Bee	143
26.	Do. Tiger Moth, Larva	143
27.	Do. Dormouse	143
28.	Do. Rat	143
29.	Do. do. long hair	143
30.	Do. Sheep	142
31.	Do. Mole	143
32.	Do. Rabbit	143
33.	Scale, Greenbone Pike	144
34.	Hair, Red Deer	143
35.	Do. fine, Sea Mouse	144
36.	Do. do. large	144
37.	Do. do. Badger	143
38.	Do. do. Long-eared Bat	143
39.	Fibre, Linen	143
40.	Do. Cotton	143
41.	Do. Silk	143
42.	Scale, Perch	144
43.	Do. do.	144

IX.

1.	Amœba diffluens	163
2.	Arcella	163
3.	Sun animalcule	164
4.	Miliolina	174
5.	Paramœcium	162
6.	Chilodon subdividing	164
7.	Melicerta ringens	164
8.	Spicula of Sponge, Grantia	172
9.	Noctiluca miliaris	172
10.	Rotifer vulgaris	165
11.	Do. jaws	165
12.	Sponge animalcule	171
13.	Sertularia operculata	175
14.	Sponge, Grantia	171
15.	Sertularia operculata, with ovicells	175
16.	Actinia, showing weapons	173
17.	Do. base of weapon more magnified	173
18.	Sponge granule, ciliated	172
19.	Anguinaria anguina	175
20.	Spicules of sponge from Oyster Shell	172
21.	Head of Snake-headed Zoophyte	17

FIG.		PAGE
22.	Bugularia circularia	175
23.	Zoophyte, Ladies' Slipper	175
24.	Zoophyte, Tobacco-pipe bearer	175
25.	Zoea, Young of Crab	176
26.	Hydra tuba	173
27.	Medusa, cast off from above	173
28.	Naked-eyed Medusa, Thaumantias	173
29.	Compound eye, Medusa	173
30.	Larva, Snake Star	178
31.	Water Flea	176
32.	Serpula, Pushing Pole	177
33.	Comatula, early stage of Starfish	177
34.	Carbonate of Lime, artificial	177
35.	Sea Urchin, transverse section of spine	179
36.	Serpula, bundle of spears	177
37.	Sunstar, part of skin	178
38.	Oyster shell in different stages	177
39.	Cilia on mussel	177

X.

1.	Skin, Frog	148
2.	Blood, Human	156
3.	Do. Pigeon	156
4.	Do. Proteus	156
5.	Do. Tortoise	156
6.	Do. Frog	156
7.	Do. Fish	156
8.	Human nail	151
9.	Bone, human	149
10.	White fibrous tissue	153
11.	Epithelial cells from tongue	149
12.	Feather, Peacock	146
13.	Spine, Hedgehog, transverse section	146
14.	Pax-wax	153
15.	Epithelial cells from nose	149
16.	Bone, Ostrich	151
17.	Feather, Shaft of Canary's	146
18.	Do. Wild Duck	146
19.	Circulation of blood, Frog's foot	157
20.	Feather, Sparrow	146
21.	Do. Cock's tail	146
22.	Fibre, crystalline lens of fish	154
23.	Nerve	154
24.	Muscle, Meat	154
25	Tooth, transverse section	152
26.	Do. Longitudinal section	152
27.	Sweat duct	147
28.	Eye of Haddock	151
29.	Myliobates, palate	153
30.	Gristle, Pig	151

DESCRIPTION OF PLATES.

FIG.		PAGE
31.	Pigment, Human eye	148
32.	Do. Wing of Bat	148
33.	Do. Shell of Prawn	148

XI.

POLARIZED LIGHT.

FIG.		PAGE
1.	Carbonate of Lime	177
2.	Starfish	179
3.	Thistle down	84
4.	Starch, Wheat	73
5.	Do. Potato	72
6.	Prawn-shell	177
7.	Starch, Tous les mois	74
8.	Bone, cancellous	149
9.	Gun-cotton	143
10.	Cow's hair	143
11.	Hoof, Donkey, longitudinal	—
12.	Do. transverse	—
13.	Nitre, Crystals	99
14.	Scale, Eel	144
15.	Wing, Water-Boatman	137
16.	Chlorate of Potash, Crystals	99
17.	Cellularia reptans	176
18.	Star-shaped hair, Stalk of Yellow Water-Lily	65
19.	Teeth, Palate of Whelk	152
20.	Zoophyte, Bowerbankia	175
21.	Raphides, *i.e.* crystalline formations in vegetable cells, Bulb of Hyacinth	—
22.	Do. Rhubarb	—
23.	Sulphate of Magnesia Crystals	99
24.	Bone, Skate	149

FIG.		PAGE
25.	Cherrystone, transverse section	—
26.	Sugar, Crystals in honey	99
27.	Tendon, Ox Calcareous plates. Tooth of Echinus	152

XII.

FIG.		PAGE
1.	Tubercle, Sun-star	179
2.	Zoophyte, Gemellaria	175
3.	Cuttle bone	179
4.	Plate of ditto from above	179
5.	Zoophyte, Antennularia	175
6.	Pedicillaria, skin of Starfish	178
7.	Shell, Foraminifer	173
8.	Snake-star, disc from below	179
9.	Pedicillaria, Echinus	178
10.	Wing-case, Weevil	139
11.	Coralline	179
12.	Spine, Echinus	179
13.	Foraminifer, Polystomella	174
14.	Do. Truncatulina	174
15.	Do. Polymorphina	174
16.	Do. Miliolina	174
17.	Gold dust, with quartz	181
18.	Foraminifer, Lagena vulgaris	174
19.	Pouches, Skin of Rat's tail	179
20.	Foraminifer, Biloculina ringens	174
21.	Ore, Copper	181
22.	Zoophyte Membranipora pilosa	175-6
23.	Human skin, injected	179
24.	Coal, Longitudinal section	181
25.	Do. Transverse section	181
26.	Lung, Frog	180

R. CLAY, SON, AND TAYLOR, PRINTERS, BREAD STREET HILL.

- 460 -

THE COMMON OBJECTS

OF

THE COUNTRY.

BY THE
REV. J. G. WOOD, M.A., F.L.S.

AUTHOR OF
"THE ILLUSTRATED NATURAL HISTORY,"
"COMMON OBJECTS OF THE SEA SHORE," "MY FEATHERED FRIENDS,"
&c., &c.

WITH ILLUSTRATIONS BY COLEMAN,

PRINTED IN COLOURS BY EVANS.

LONDON:
GEORGE ROUTLEDGE & SONS,
THE BROADWAY, LUDGATE HILL.
NEW YORK: 416, BROOME STREET.
1866.

Price 2s. 6d. each, cloth gilt, with Illustrations, printed in colours,

COMMON OBJECTS OF THE MICROSCOPE. By the Rev. J. G. WOOD.

COMMON OBJECTS OF THE SEA SHORE. By the Rev. J. G. WOOD.

COMMON OBJECTS OF THE COUNTRY. By the Rev. J. G. WOOD.

HAUNTS OF THE WILD FLOWERS. By ANNE PRATT

BRITISH BIRDS' EGGS AND NESTS. By the Rev. J. C. ATKINSON.

BRITISH FERNS AND THEIR ALLIES. By THOMAS MOORE, F.L.S., &c.

WANDERINGS AMONG THE WILD FLOWERS. By SPENCER THOMPSON.

BRITISH BUTTERFLIES. By W. S. COLEMAN.

GEORGE ROUTLEDGE AND SONS.

PREFACE.

In the following pages will be found short and simple descriptions of some of the numerous objects that are to be found in our fields, woods, and waters.

As this little work is not intended for scientific readers, but simply as a guide to those who are desirous of learning something of natural objects, scientific language has been studiously avoided, and scientific names have been only given in cases where no popular name can be found. In so small a compass but little can be done; and therefore I have been content to take certain typical objects, which will serve as guides, and to omit mention of those which can be classed under the same head.

Every object described by the pen is illustrated by the pencil, in order to aid the reader in his researches; and the subjects have been so chosen, that no one with observant eyes can walk in the fields for half an hour without finding very many of the objects described in the book.

London, *April*, 1858.

COMMON OBJECTS OF THE COUNTRY.

CHAPTER I.

EYES AND NO EYES—DIFFICULTIES OF OBSERVERS—THE BATS—LONG-EARED BAT—ITS UTILITY—SPORT AND MURDER—SONG OF THE BAT—A BRAVE PRISONER—HOW BATS FEED—HAIR OF BAT AND MOUSE—WING OF THE BAT—THE FIELD-MOUSE—ITS STEALTHY MOVEMENTS—HARVEST MOUSE—WATER RAT—AN INNOCENT VICTIM.

EVERY one has read, or at least heard of, the tale entitled "Eyes and no Eyes;" which tale is to be found in the "Evenings at Home." Now this story, or rather the moral of it, is in my opinion as often used unfairly as rightly.

Although there are those who pass through life with closed eyes and stopped ears, yet there are many more who would be glad to use their eyes and ears, but know not how to do so for want of proper teaching. To one who has not learned to read, the Bible itself is but a series of senseless black marks; and similarly, the unwritten Word that lies around, below, and above us, is unmeaning to those who cannot read it.

Many would like to read, but cannot do so ; and it is in order to help such, to bring before them the first alphabetical teaching, that the following pages are written.

It is no matter of marvel that many an observant person becomes bewildered among natural objects ; that he is lost amid the variety of animal, vegetable, and mineral life in which he lives ; and that, after vainly attempting to comprehend some simple object, he finds himself baffled, and so in despair ceases to inquire into particulars, and contents himself with admiration of and love for nature in general.

Objects change so rapidly and so constantly, that there is hardly time to note a few remarks before the season has passed away ; the object under examination has changed with it, and a year must elapse before that investigation can be continued.

From experience I know how valuable are even a few hints by which the mind can be directed in a straight course without wasting its strength and losing its time by devious wanderings. Only hints can be given, for the limits of the volume forbid any lengthened discussion of single objects ; and besides, the mind is more pleased to work out a subject according to its own individuality, than to have it laid down as completed, and to be forbidden to go any farther.

Almost every object that is described by the pen will

be figured by the pencil, in order to assist the reader in identifying the creature in an easier manner than if it were merely described in words.

Of the birds I shall not be able to treat, as they alone would occupy the entire space of this volume; and, for the same reason, only a short account can be given of each object.

As in the scale of creation the mammals fill the highest place, we will speak of them first, taking, as far as possible, each creature in its own order.

Perhaps there are few people who would not feel some surprise when they learn that the very highest of our British animals is the Bat. Usually the bat is looked upon with rather a feeling of dread, and is regarded as a creature of such ill-omen that its very presence causes a shudder, and its approach would put to flight many a human being.

There is certainly some ground for this feeling; for the night-loving propensities of the creature, its weird-like aspect, its strange devious flight, and more especially its organs of flight, are so interwoven with the popular ideas of evil and its ministers, that bats and imps appear to be synonymous terms.

Painters always represent their imps as upborne by bats' wings, furnished with several supplementary hooks; and sculptors follow the same principle.

In consequence, all bats and objects connected with

bats are viewed with great horror, with two exceptions: a cricket-bat and a bat's-wing gas-burner.

Now I cannot but think that this is very hard on the bats. It is said that the African negroes depict and describe *their* evil spirits as white; and that, in consequence, the negro children fly in consternation if perchance a white man comes into their territory.

Yet a white man is not so very horrid an object, after all, if one only dare look at him; and the same remark holds good with the bats.

A very pretty creature is a bat, more especially the long-eared species, *Plecotus communis*, as it is scienti-

COMMON LONG-EARED BAT.

fically called, and its habits are most curious. It is well worth the time to watch these little creatures on a warm summer's night, as they flit about in the air, and to note the enjoyment of their aërial hunt. They are fearless animals; and provided that the observer remains tolerably still and does not speak, bats will often flit so close to his face that he could almost catch them in his hand.

Their flight is very singular, and reminds one of the butterfly in its apparently vague flitting. Indeed, there are many large moths that fly by night who can hardly be distinguished from the bats, if the evening be rather dark, so similar are they in their mode of journeying through the air.

From this peculiarity of flight, they are accounted difficult marks for a gun; and it is unfortunately a custom with some ruthless powder-burners to practise by day at swallows and by night at bats. Now, even putting the matter in its lowest form, it is wrong to shoot swallows; for they are most useful birds, and serve to thin the host of flies and other insects that people the summer air.

As regards the swallow, this is well known, and does serve to protect it from some persons who have more compassion than the generality. Moreover, the swallows, swifts, and martins are extremely pretty birds, and their beauty is in some degree their shield.

But the bat is as useful a creature as the swallow, and in the very same way; for when the evening comes on, and the swallow retires to its nest, the bat issues from its home, and takes up the work just where the swallow leaves it, the two creatures dividing the day and night between them. Therefore, let those who refrain from swallow shooting include the bat in their free list.

Some there are whom nothing can restrain from killing, for the instinct of slaughter is strong in them. With them, nothing is valuable unless it is to be killed. If it can be eaten afterwards, so much the better; but the great enjoyment consists in the mere act of killing.

They contrive to disguise the ugliness of the thing by giving it any name but the right one; but, in spite of the name, the thing exists. And, I wonder, if they were to look very closely into themselves, whether they would not find there a decided desire to kill men, provided that they had no reason to dread the consequences. Those who have practised the sport unanimously say that nothing is so exciting as man-hunting and killing, and that all other sport is tame in comparison.

The chief name under which this profanity is disguised is that of "Sport," a word which always reminds me of the "Frog and Boys" fable. There are actually men who are audacious enough to declare that there is no cruelty in "sport;" that foxes are charmed at being hunted, and that pheasants derive a singular gratification from getting shot. Now, I never was

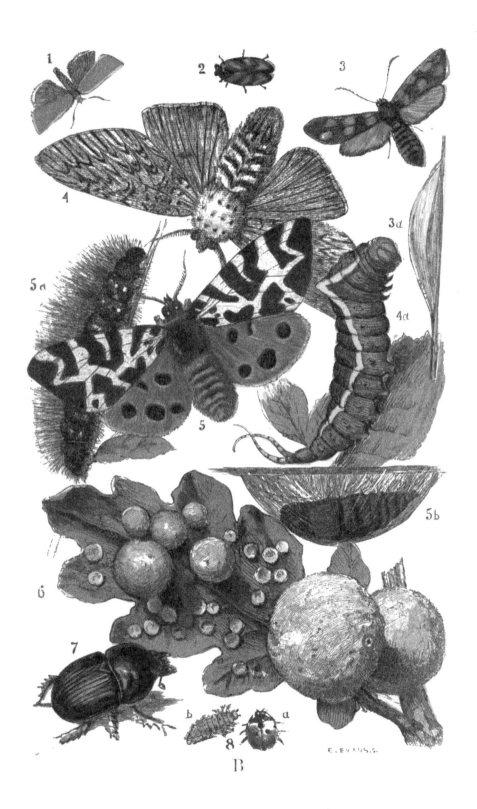

either a fox or a pheasant; but I entirely repudiate the assertion that any animal likes to be chased or to be wounded; and moreover, I disbelieve the sincerity of the man who can say such a thing. If he says openly that he finds excitement in the chase, and means to gratify himself without any reference to the feelings of the creatures which he chases, I can understand while I disapprove. But when a man justifies himself by asserting that any animal likes to be hunted, I can hardly find epithets too contemptuous for him; and I could see him run the gauntlet among the Sioux Indians with but small pangs of conscience.

Some again call themselves Naturalists, and under the shelter of that high-sounding name occupy themselves in destroying nature. The true naturalist never destroys life without good cause, and when he does so, it is with reluctance, and in the most merciful way; for the life is really the nature, and that gone, the chief interest of the creature is gone too. We should form but a poor notion of the human being, were we only to see it presented to our eyes in the mummy; and equally insufficient is the idea that can be formed of an animal from the inspection of its outward frame. Nature and life belong to each other; and if torn asunder, the one is objectless and the other gone.

Lastly, let me remind those who find such gratifica-

8   CRY OF THE BAT.

tion in destroying, that the word "Destroyer" is in the Greek language "Apollyon."

As we do not intend to treat of the dead and dried bodies of animals, but of their active life, we return to our bat flitting in the evening dusk, and, instead of shooting him, watch his proceedings.

Every creature is made for happiness, and receives happiness according to its capacity; and it is very wrong to suppose that because *we* should be miserable if we led the life of a vulture, or sloth, or a bat, therefore those creatures are miserable. In truth, the vulture is attracted to and feels its greatest gratification in those substances which would drive us away with averted eyes and stopped nostrils. The sloth is, on the authority of Waterton, quite a jovial beast, and anything but slothful when in his proper place; and as for the bat, it sings for very joy. True, the song is not very melodious, neither is that of the swift, or the peacock, nor, perhaps, that of the Cochin-China fowl, but it is nevertheless a song from the abundance of the heart.

There are many human ears that are absolutely incapable of perceiving the cry of the bat, so keen and sharp is the note; a very razor's-edge of sound.

More than once I have been standing in a field over which bats were flying in multitudes, filling the air almost oppressively with their sharp needle-like cries. Yet my companion, who was a musician, theoretically and practically, was unable to hear a sound, and could not for some time believe me when I spoke of the noisy little creatures above.

The sound bears some resemblance to that produced by a slate-pencil when held perpendicularly in writing on the slate, only the bat's cry is several octaves more acute. I never but once heard the sound correctly imitated, and that was done by a graceless urchin, during a long sermon, one Sunday morning. He had contrived to arrange two keys in such a manner, that, when grated over each other, they produced a squeaking sound that exactly resembled the cry uttered by the bat. So, by judicious management of his keys, he kept the congregation on the look-out for the bat, and beguiled the time much to his satisfaction.

Of so piercing and peculiar a nature is the cry, that it gives no clue to the position or distance of the creature that utters it, and it seems to proceed indiscriminately from any portion of the air towards which the attention happens to be directed. The note of the grasshopper lark possesses somewhat of the same quality.

Even in confinement the bat is an interesting creature, and discovers certain traits of character and peculiarities of habit which in its wild state cannot be seen. I might here refer to several stories of domesticated and tamed bats; but as they have already been given to the world, and my space is limited, I prefer to give my own experiences.

Not long ago, I received a message from a neighbouring grocer, requesting me to capture a bat which had flown into the shop, and which no one dared touch.

When I arrived, the creature had taken refuge on an upper shelf, and had crawled among a pile of sugar-loaves that were lying on their sides after the usual custom. We pulled out several loaves near the spot where the bat was last seen, and by casting a strong light from a bull's eye lantern, discovered a little black object snugly ensconced at the very back of the shelf.

I pushed my hand towards the spot, but for some time could not seize the creature, as it was so tightly packed, and squeezed into a corner. At last the bat gave a flap with one of the wings, which I caught, and so gently drew my prisoner forwards.

He was a brave little fellow, as well as discreet, and bit savagely at my fingers. However, his little tiny teeth could not do much damage, and I put him into a cage which I brought with me.

The cage was originally made for the reception of mice, and was of a rude character—the back and ends being of wood and the front of wire. In a very few minutes after his entrance into the cage, the bat climbed up the wooden back, by hitching his claws into the slight inequalities of the wood, and there hung suspended, head downwards.

When so placed, his aspect was curious enough. The claws of the hind legs being fixed into a crevice, so as to bear the weight of the body, the wings were then extended to their utmost, and suddenly wrapped round the body. At the same time the large ears were folded back under the wings and protected by them, the orifice of the ear itself being guarded in a very singular manner.

If the reader will refer to the figure of the bat on page 4, he will see that inside the great ear is a sharply-pointed membrane, somewhat resembling a second ear. This membrane is called the "*tragus*," and when the large ears are tucked away out of sight, the tragus remains exposed, and gives the creature a very strange appearance.

When the bat is living, the ears are of singular beauty. Their substance is delicate, and semi-transparent if viewed against the light; so much so, indeed, that by the aid of a microscope the circulation of the blood can be detected. As the creature moves about, the ears are continually in motion, being thrown into graceful and ever-changing curves. If people only knew what a pretty pet the Long-eared Bat can become, they would soon banish dormice and similar creatures in favour of bats.

It was rather a remarkable circumstance, that the bat of which I have just been speaking would not touch a fly, although one which I had in my possession some ten years since would eat flies and other insects readily. Whenever it took the insect, it daintily ate up the abdomen and thorax, rejecting the head, wings, and legs. But my second bat entirely refused insects of any kind, and would eat nothing but raw beef cut up into very small morsels. I never had a pet so difficult to feed or so dainty.

If the meat were not perfectly fresh, or if it were not cut small enough, the bat would hardly look at it. Now if a bit of raw meat about the size of a large pin's head be placed in the air, a few minutes will dry and harden its exterior; and when this was the case, my bat did not even notice it. So I had to make twenty or more attempts daily, before the creature would condescend to take any food.

When, however, it *did* eat, its mode of so doing was remarkable enough. It seized the meat with a sharp snap, retreated to the middle of the cage, sat upright —as in the engraving already alluded to—thrust its wings forward to form a kind of tent, and then, lowering its head under its wings, disposed of the meat unseen.

From the movement of the neck and upper portion of the head, it would be seen that the creature ate the

MODE OF FEEDING.

meat much after the manner of a cat; that is, by a series of snaps or pecks; for the teeth are all sharply pointed, and have no power of grinding the food. These teeth can be seen in the accompanying sketch of a bat's skull.

In many parts of England, the bats are called "Flitter-mice," and are thought to be simply mice plus wings. This opinion has been  formed from the resemblance between the general shape, and especially that of the fur, of the two animals. But if we look at the teeth, we find at once that those of the bat are sharp and pointed, extending tolerably equally all round the jaw-bone; while the teeth of the mice are of that chisel-shaped character found in the rabbit and other rodent animals.

Now if we turn to the fur, and examine it with a microscope, we shall there find characteristics as decided as those of the teeth.

On the next page is the magnified image of a single hair, taken from the Long-eared Bat. It will be seen that the outline of the hair is deeply cut, and the markings run in a double line. These markings and outlines are caused by the structure of the hair, which is covered with a regular series of scales adhering but loosely to its exterior. These scales can be removed by rough handling, and therefore the aspect of the hair can be much altered.

Let us now take a hair from the common mouse, and place it under the microscope. This being done, we find the result to be as shown in the accompanying cut.

The two objects here shown are two portions of the same hair; the upper one showing the middle of the hair, and the lower being taken from a portion nearer the root. Both these specimens were taken by myself from the animals, and drawn by myself by means of the Camera Lucida, so that they are to be depended on.

To return to my caged bat.

Although it did not do much in the eating way, it frequently came to the water vessel and drank therefrom; but it was so timid when drinking, that I could not see whether it lapped or drank. When disturbed, it used to scuttle away over the floor, in a most absurd manner, but with some speed. Sometimes it tried to drink by crawling to a spot just over the vessel, and lowering itself until its nose was within reach of the water; but the distance was too great for the attempt to be successful. In its wild state, the bat hunts insects, as they hover over the surface of water, and drinks as it flies, by dipping its head in the water while on the wing.

I rather think that my bat must have received some injury from the brooms and caps that were aimed at it when it entered the shop, for it only lived a fortnight or so, and one morning I found it hanging by its hind claws from the roof of the cage, quite dead.

I believe that bats generally die while thus suspended, for it is a very common thing to find plenty of suspended bats, dry and mummified, when entrance is made into an unfrequented cave, or a hollow tree cut down, or indeed when any bat-haunted spot is examined.

In speaking of the bat, I have used popular terms, and therefore have employed the word "wing." But the apparatus of the bat is not a wing at all, but only a developed hand. Let the reader spread his hand as wide as he can, and he will see that between each finger, and especially between the forefinger and the thumb, the skin forms a kind of webbing, something of the same kind as that on the feet of ducks and other aquatic birds.

Now if the bones of the fingers were drawn out like

wire until they became some seven or eight feet long, and the skin between them were extended to the nails of the elongated fingers, we should have a structure analogous to that of the bat's wing. The thumb joint is left comparatively free; and by means of this joint, and the hooked claw at its extremity, the creature walks on a level surface, or can crawl suspended from a beam or a trunk. It is very curious to see the bat stretching out its wings and feeling about for a convenient spot whereon to fix the hooks.

So tenacious are these hooks, that the baby bat is often found enjoying an airing by clinging to the body of its mother, and holding firm, while she flies in search of prey.

It is true that the little creature is suspended with its head downwards; but it appears quite comfortable, nevertheless. Bat-children do not suffer from determination of the blood to the brain. Neither do certain human children, it seems, if we are to take as a criterion those whom we see hanging half out of perambulators, fast asleep, and rolling from side to side with every movement of the vehicle.

Both my bats were very particular, not to say finicking, about their personal appearance. They bestowed much time and pains on the combing of their fur, and specially seemed to value a straight parting down the back.

It was most interesting to watch the little thing parting its hair. The claw was drawn in a line straight from the top of the head to the very tail, and the fur parted at each side with a dexterity worthy of an accomplished lady's-maid. The same habit has been observed in other bats that have been tamed.

There are more than twenty British bats, but the habits of all are very similar; and so I prefer to take the prettiest, and having described it, to leave the remaining species for a future occasion.

Pass we now from the Flittermouse to the Mouse.

In the fields, in the farm-yards, in the barns, and in the ricks are to be found myriads of certain little animals called Field-mice. Acting on the principle that I have just laid down, I shall take the most common and I think the prettiest species—the Common Short-tailed Field-mouse, represented on next page.

The fur of this creature is strongly tinged with red, and by its colour alone it is easily to be distinguished from the common grey or brown mouse. Its tail is short and stumpy, looking as if it had suffered amputation at an early period of life, and its nose is more rounded than that of the common mouse. Indeed, it has a very bluff and farmer-like aspect, and looks as if it ought to wear top-boots.

Common as these little creatures are, they are seldom seen, because they keep themselves so close to the ground, and assimilate so nearly with it in colour, that they cannot easily be descried among the grass stalks, under shelter of which they pursue their noiseless way.

Their speed is not nearly so great as that of the house-mice, but they are much more difficult to catch; for they wind among the grass so lithely, and press upon the earth so closely, that the fingers cannot readily close on them, even when they are discovered.

SHORT-TAILED FIELD-MOUSE.

From this facility of avoiding observation and capture, they seem to derive much audacity, and run about a field in fear of nothing but the kestrel.

When first I made a personal acquaintance with these creatures, it was under rather peculiar circumstances. There is a certain field, which was given up to football, cricket, hockey, and similar games, as soon as the grass was converted into hay and removed. One day I was very tired with running, and lay down to rest on a pile of coats that had been laid aside; my eyes were fixed on one spot of earth, just visible between the grass stalks, but without any particular object. Presently I thought I saw a something red glide across the spot, but was not certain. However, I leaned over the place, and a little farther on saw the same thing again. So I made a sharp pounce at the object, and found that I had caught a short-tailed field-mouse.

Now here was this impertinent little animal taking a walk close to the wicket, in spite of the bats, ball, and runners. In order to watch its proceedings, I released it, and followed it in its progress. After watching for a few minutes, I happened to look up for a moment; and when I again looked for the creature, it was gone, and I could not find it again.

Subsequently I became sufficiently expert to find them whenever I wished; and if I wanted a field-mouse, seldom had to examine more than a square yard of ground without finding one.

They are very injurious little creatures, for they are not content with eating corn, but nibble the young shoots of various plants, and sometimes strip young trees of their bark.

Fortunately we have allies in air and on earth, in the

persons of owls and kestrels, stoats and weasels, or the damage done by these red-skinned marauders would be more than serious.

Some idea of the damage that may be done by the aggregate numbers of these small quadrupeds may be formed from the fact, that in Dean Forest and the New Forest great numbers of holly plants were entirely destroyed by them, they having eaten off the bark for a distance of several inches from the ground. And other trees were favoured with the notice of the field-mice, but in a different mode. Great numbers of oak and chestnuts were found dead, and pulled up; and when pulled up, it was seen that their roots had been gnawed through, about two inches below the level of the ground.

Various modes of destroying the marauders were put in practice, such as traps, poison, &c., but the most effectual was, as effectual things generally are, the most simple.

A great number of holes were dug in the ground, about two feet long, eighteen inches wide, and eighteen inches deep. This is the measurement at the bottom of the hole; but at the top the hole was only eighteen inches long and nine wide, so that when mice fell into it, they were unable to escape.

In these holes upwards of forty thousand mice were taken in less than three months, irrespective of those that were removed from the holes by the stoats, weasels, crows, magpies, owls, and other creatures.

Like most of the mouse family, the field-mouse is easily tamed: and I have seen one that would come to the side of its cage, and take a grain of corn from its owner's fingers.

There is another kind of mouse which may be found in the autumn, together with its most curious nest. This is the Harvest-mouse, the tiniest of British quadrupeds, two harvest-mice being hardly equal in weight to a halfpenny.

HARVEST-MOUSE.

The chief point of interest in this little creature is its nest, which is not unfrequently found by mowers and haymakers when they choose to exert their eyes.

One of these nests, that was brought to me by a

mower, was about the size of a cricket ball, and almost as spherical. It was composed of dried grass-stems, interwoven with each other in a manner equally ingenious and perplexing. It was hollow, without even a vestige of an entrance; and the substance was so thin that every object would be visible through the walls. How it was made to retain its spherical form, and how the mice were to find ingress and egress, I could not even imagine. The nest was fastened to two strong and coarse stems of grass that had grown near a ditch, and had overgrown themselves in consequence of a superabundance of nourishment.

If we walk along the bank of a stream or a pond, we shall probably hear a splash, and looking in its direction, may see a creature diving or swimming, which creature we call a Water-rat; to the title of Rat, however, it has but little right, and ought properly to be called the "Water-vole."

WATER RAT.

On examining the banks we shall find the entrance to its domicile, being a hole in the earth, just above the

water, and generally, where possible, made just under a root or a large stone. Sometimes the hole is made at some height above the water, and then it often happens that the kingfisher takes possession, and there makes its home. Whether it ejects the rat or not I cannot say, but I should think that it is quite capable of doing so. Many a time I have seen the entrance to a rat-hole decorated with a few stray fish-bones, which the rustics told me were the relics of fish brought there and eaten by the water-rat. But I soon found out that fish-bones were a sign of kingfishers, and not of rats; and so guided, found plenty of the beautiful eggs of this beautiful bird. Excepting the eggs of swallows and martins, I hardly know any so delicately beautiful as those of the kingfisher, with their slight rose tint and semi-transparent shell. But, alas! when the interior of the egg is removed, the pearly pinkiness vanishes, and the shell becomes of a pure white, very pretty, but not containing a tithe of its former beauty.

The piscatorial propensities of the kingfisher are not the only cause of the slanderous reports concerning the water-vole, and its crime of killing and eating fish. The common house-rat often frequents the water-side; and, it being a great flesh-eater, certainly does catch and eat the fish.

But the water-rat is a vegetable feeder, and I believe almost, if not entirely, a vegetarian in diet. That it is so in individual cases, at all events, I can personally testify, having seen the creature engaged in eating.

In former days, when I thought the water-rats ate fish, I waged war against them, for which warfare there are great facilities at Oxford. However, a circumstance occurred which showed me that I had been wrong.

I saw a water-rat sitting on a kind of raft that had formed from a bundle of reeds which had been cut and were floating down the river. Seeing it busily at work feeding, I took it for granted that it was eating a captured fish, and shot it accordingly, stretching it dead on its reed raft.

On rowing up to the spot, I was rather surprised to find that there was no fish there; and on examining the reeds, I rather wondered at the regular grooves cut by my shot. But a closer inspection revealed a very different state of things; namely, that the poor dead rat was quite innocent of fish eating, and had been gnawing the green bark from the reeds, the grooves being the marks left by its teeth. After this I gave up rat shooting on principle.

Once, though, a rather curious circumstance occurred.

In my possession was a pet pistol, which would throw a ball with great accuracy, and I considered myself sure of an apple at sixteen paces. One day, just as I was standing by a branch of the river Cherwell, I saw a water-rat sitting on the root of a tree at the opposite side of the river, and watching me closely. The river was not above twelve or fourteen yards wide; and the rat presented so good a mark that I fired at him, and of course expected to see him on his back.

But there sat the rat, quite still on the stump, and about two inches below him the round hole where the bullet had struck.

As the creature seemed determined to stay there, I reloaded, and took a good aim, determined to make sure of him. As the smoke cleared away, I had the satisfaction of seeing the rat in exactly the same position, and another bullet-hole close by the former. Four shots I made at that provoking animal, and four bullets did I deposit just under him. As I was reloading for a fifth shot, the rat walked calmly down the stump, slid into the water, and departed.

Now, whether he acted from sheer impertinence, or whether he was stunned by the violent blow beneath him, I cannot say. The latter may perhaps be the case, for squirrels are killed in North America by the shock of the bullet against the bough on which they sit, so that no hole is made in their skins, and the fur receives no damage. Perhaps the rat was actuated by a supreme contempt for me and my shooting powers; and, as the result showed, was quite justified in his opinion.

CHAPTER II.

SHREW-MOUSE—DERIVATION OF ITS NAME—SHREW-ASH—THE SPIRIT AND THE LIFE—WATER-SHREW—ITS HABITS—THE MOLE—MOLE-HILL—A PET MOLE—THE WEASEL.

I HAVE already mentioned that the water-rat has little claim to the title of rat; and there is another creature which has even less claim to the title of mouse. This is the Shrew, or Shrew-mouse, as it is generally called.

SHREW-MOUSE.

This creature bears a very close relationship to the hedgehog, and is a distant connexion of the mole; but with the mouse it has nothing to do.

Numbers of the shrews may be found towards the end of the autumn lying dead on the ground, from some cause at present not perfectly ascertained. If one of these dead shrews be taken, and its little mouth

DERIVATION OF ITS NAME.

opened, an array of sharply-pointed teeth will be seen, something like those of the mole, very like those of the hedgehog; but not at all resembling those of the mouse.

The shrew is an insect and worm-devouring creature, for which purpose its jaws, teeth, and whole structure are framed. A rather powerful scent is diffused from the shrew; and probably on that account cats will not eat a shrew, though they will kill it eagerly.

On examining Webster's Dictionary for the meaning of the word "shrew," we find three things.

Firstly, that it signifies "a peevish, brawling, turbulent, vexatious woman."

Secondly, that it signifies "a shrew-mouse."

Thirdly, that it is derived from a Saxon word, "*screawa*," a combination of letters which defies any attempt at pronunciation, except perhaps by a Russian or a Welchman.

Now, it may be a matter of wonder that the same word should be used to represent the very unpleasant female above-mentioned and also such a pretty, harmless little creature as the shrew. The reason is shortly as follows.

In days not long gone by, the shrew was considered a most poisonous creature, as may be seen in the works of many authors. In the time of Katherine—the shrew most celebrated of all shrews—any cow or horse that was attacked with cramp, or indeed with any sudden disease, was supposed to have suffered in consequence of a shrew running over the injured part. In those days homœopathic remedies were generally resorted to; and nothing but a shrew-infected plant could cure a shrew-infected animal. And the shrew-ash, as the remedial plant was called, was prepared in the following manner.

In the stem of an ash-tree a hole was bored; into the hole a poor shrew was thrust alive, and the orifice immediately closed with a wooden plug. The animal strength of the shrew passed by absorption into the substance of the tree, which ever after cured shrew-struck animals by the touch of a leafy branch.

The poor creature that was imprisoned, Ariel-like, in the tree, was, fortunately for itself, not gifted with Ariel's powers of life; and the orifice of the hole being closed by the plug, we may hope that its sufferings were not long, and that it perished immediately for want of air. Still, our fathers were terribly and deliberately cruel; and if the shrew's death was a merciful one, no credit is due to the authors of it.

For on looking through a curious work on natural history, of the date of 1658, where each animal is treated of medicinally, I find recipes of such terrible cruelty that I refrain from giving them, simply out of tenderness for the feelings of my reader. Torture seems to be a necessary medium of healing; and if a man suffers from "the black and melancholy cholic," or "any pain and grief in the winde-pipe or throat," he can only be eased therefrom by medicines prepared from some wretched animal in modes too horrid to narrate, or even to think of.

We are not quite so bad at the present day; but still no one with moderate feelings of compassion can pass through our streets without being greatly shocked at the wanton cruelties practised by human beings on those creatures that were intended for their use, but not as mere machines. Charitably, we may hope that such persons act from thoughtlessness, and not from deliberate cruelty; for it does really seem a new idea to many people that the inferior animals have any feelings at all.

When a horse does not go fast enough to please the driver, he flogs it on the same principle that he would turn on steam to a locomotive engine, thinking about as much of the feelings of one as of the other.

Much of the present heedlessness respecting animals is caused by the popular idea that they have no souls, and that when they die they entirely perish. Whence came that most preposterous idea? Surely not from the only source where we might expect to learn about souls—not from the Bible; for there we distinctly read of "the spirit of the sons of man;" and immediately afterwards of "the spirit of the beast," one aspiring, and the other not so. And the necessary consequence of the spirit is a life after the death of the body. Let any one wait in a frequented thoroughfare for only one short hour, and watch the sufferings of the poor brutes that pass by. Then, unless he denies the Divine Providence, he will see clearly that unless these poor creatures were compensated in another life, there is no such quality as justice.

It is owing to sayings such as these, that men come to deny an all-ruling Providence, and so become infidels. They don't examine the Scriptures for themselves; but take for granted the assertions of those who assume to have done so; and seeing the falsity of the assertion, naturally deduce therefrom the falsity of its source. If a man brings me a cup of putrid water, I naturally conclude that the source is putrid too. And when a man hears horrible and cruel doctrines, which are asserted by theologians to be the religion of the Scriptures, it is no wonder that he turns with disgust from such a religion, and tries to find rest in infidelity. In such a case, where is the fault?

All created things in which there is life, *must* live for ever. There is only one life, and all living things only live as being recipients; so that as that life is immortality, all its recipients are immortal.

If people only knew how much better an animal

will work when kindly treated, they would act kindly towards it, even from so low a motive. And it is so easy to lead these animals by kindness, which will often induce an obstinate creature to obey where the whip would only confirm it in its obstinacy. All cruelty is simply diabolical, and can in no way be justified.

Supposing that the two cases could be reversed for just one hour, what a wonderful change there would be in the opinion of men; for it may be assumed that the person most given to inflicting pain and suffering is the least tolerant of it himself.

There is, perhaps, hardly one of my readers who does not know some one person who finds an exquisite delight in hurting the feelings of others by various means, such as ridicule, practical jokes, ill-natured sayings, and so on. If so, he will be tolerably certain to find that the same person is especially thin-skinned himself, and resents the least approach to a joke of which he is the subject.

So, if the shrew were to be the afflicted individual, and the human the victim, there would be found no one so averse to the medicinal process as he who had formerly resorted to it under different circumstances.

This principle is finely carried out, in the terrible scene of Dennis, the executioner's, last hours, in "Barnaby Rudge."

These are not pleasant subjects; and we will pass on

to another shrew that is generally found in the water, and called from thence the Water-shrew. It is a crea-

WATER-SHREW.

ture that may be found in many running streams, if the eyes are sharp enough to observe it, and is well worth examination. As it dives and runs along the bottom of the stream, it appears to be studded with tiny silver beads, or glittering pearls, on account of the air-bubbles that adhere to its fur. I have seen a whole colony of them disporting themselves in a little brooklet not two feet wide, and so had a good opportunity of inspecting them.

I may mention here, as has been done in one or two other works, that nothing is easier than to watch animals or birds in their state of liberty. All that is required is perfect quiet. If an observer just sits down at the foot of a tree, and does not move, the most timid creatures will come within a few yards as freely as if no human being were within a mile. If he can shroud himself in branches or grass, or fern, so much the better; but quiet is the chief essential.

It is impossible to form an idea of the real beauty of animal life, without seeing it displayed in a free and unconstrained state; and more real knowledge of natural history will be gained in a single summer spent in personal examination, than by years of book study.

The characters of creatures come out so strongly; they have such quaint comical little ways with them; such assumptions of dignity and sudden lowering of the same; such clever little cheateries; such funny flirtations and coquetries, that I have many a time forgotten myself, and burst into a laugh that scattered my little friends for the next half-hour. It is far better than a play, and one gets the fresh air besides.

These little water-shrews are most active in their sports and their work, for which latter purpose they make regular paths along the banks. And as to their sport, they chase one another in and out of the water, making as great a splash as possible, whisk round roots, dodge behind stones, and act altogether just like a set of boys let loose from the school-room. And then—what a revulsion of feeling to see a stuffed water-shrew in a glass-case!

Now for a few words respecting the distant relation of the shrews, namely, the mole. Of its near relation, the hedgehog, there will not be time to speak.

Every one is familiar with the little heaps of earth

thrown up by the mole, and called mole-hills. But as the animal itself lives almost entirely underground, comparatively little is known of it; at all events, to the generality of those who see the hills. The mole is not often seen alive; and few who see it suspended among the branches by the professional killer would form any conception of the real character of this subterranean animal.

Meek and quiet as the mole looks, it is one of the fiercest, if not the very fiercest, of animals; it labours, eats, fights, and loves as if animated by one of the furies, or rather by all of them together.

MOLE.

Intervals of profound rest alternate with savage action; and, according to the accounts of country folks near Oxford, it works and rests at regular intervals of three hours each.

Useful as these creatures are as subsoil drain-makers, they sometimes increase to an inconvenient extent, and then the professed mole-catcher comes into practice, and destroys the moles with an apparatus apparently inadequate to such a purpose. But the mole is easily killed, and pressure he cannot survive; so the traps are formed for the purpose of squeezing the mole, not of smashing or strangling him.

The mole-catchers are in the habit of suspending their victims on branches, mostly of the willow or similar trees; but their object I never could make out, nor could they give me any reason, except that it was the custom.

When a mole is taken out of the ground, very little earth clings to it. There is always some on its great digging claws; but very little indeed on its fur, which is beautifully formed to prevent such accumulation. The fur of most animals "sets" in some definite direction, according to its position on the body; but that of the mole has no particular set, and is fixed almost perpendicularly on the creature's skin, much like the pile of velvet. Indeed the mole's fur has much the feel of silk-velvet; and so the title of the "Little gentleman in the velvet coat" is justly applied.

Those small heaps of earth that are so common in the fields, and called mole-hills, are merely the result of the mole's travelling in search of the earth-worms, on which it principally feeds; and in their structure there is nothing remarkable.

But the great mole-hill, or mole-palace, in which the animal makes its residence, is a very different affair, and complicated in its structure. In it is found a central chamber, in which the mole resides; and round this chamber there run galleries or corridors in a regular series, so as to form a kind of labyrinth, by means of which the creature may make its escape, if threatened with danger.

The accompanying cut shows the section of the mole-palace.

MOLE-HILL

This palace is formed, if possible, under the protection of large stones, roots of trees, thick bushes, or some such situation; and is located as far as possible from paths or roads.

The food of the mole mostly consists of earthworms, in search of which it drives these tunnels with such assiduity. The depth of the tunnel is necessarily regulated by the position of the worms; so that in warm pleasant days or evenings the run, as it is called, is within a few inches of the surface; but in winter the worms retire deeply into the unfrozen soil, and thither the mole must follow them. For this purpose it sinks perpendicular shafts, and from thence drives horizontal tunnels. It may be seen how useful this provision is, when one thinks of the work that is done by the mole when providing for its own sustenance.

In the cold months, it drives deeply into the ground, thereby draining it, and preventing the roots of plants from becoming sodden by the retention of water above ; and the earth is brought from below, where it was useless, and, with all its properties inexhausted by crops, is laid on the surface, there to be frozen, the particles to be forced asunder by the icy particles with which it is filled, and, after the thaw, to be vivified by the oxygen of the atmosphere, and made ready for the reception of seeds.

The worms have a mission of a similar nature ; but their tunnels are smaller, and so are their hills. Every floriculturist knows how useful for certain plants are the little heaps of earth left by the worms at the entrance of their holes. And by the united exertions of moles and worms, a new surface is made to the earth, even without the intervention of human labour.

Among other pets, I have had a mole—rather a strange pet, one may say ; but I rather incline to pets, and have numbered among them creatures that are not generally petted—snakes, to wit—but which are very interesting creatures, notwithstanding.

Being very desirous of watching the mole in its living state, I directed a professional catcher to procure one alive, if possible; and after awhile the animal was produced. At first there was some difficulty in finding a proper place in which to keep a creature so fond of digging; but the difficulty was surmounted by procuring a tub, and filling it half full of earth.

In this tub the mole was placed, and instantly sank below the surface of the earth. It was fed by placing large quantities of earth-worms or grubs in the cask; and the number of worms that this single mole devoured was quite surprising.

As far as regards actual inspection, this arrangement was useless; for the mole never would show itself, and when it was wanted for observation, it had to be dug up. But many opportunities for investigating its manners were afforded by taking it from its tub, and letting it run on a hard surface, such as a gravel-walk.

There it used to run with some speed, continually grubbing with its long and powerful snout, trying to discover a spot sufficiently soft for a tunnel. More than once it did succeed in partially burying itself, and had to be dragged out again, at the risk of personal damage. At last it contrived to slip over the side of the gravel-walk, and, finding a patch of soft mould, sank with a rapidity that seemed the effect of magic. Spades were put in requisition; but a mole is more than a match for a spade, and the pet mole was never seen more.

I was by no means pleased at the escape of my

prisoner; but there was one person more displeased than myself—namely, the gardener: for he, seeing in the far perspective of the future a mole running wild in the garden, disfiguring his lawn and destroying his seed-beds, was extremely exasperated, and could by no blandishments be pacified.

However, his fears and anxieties were all in vain, as is often the case with such matters, and a mole-heap was never seen in the garden. We therefore concluded that the creature must have burrowed under the garden wall, and so have got away.

Sometimes the fur of the mole takes other tints besides that greyish black that is worn by most moles. There are varieties where the fur is of an orange colour; and I have in my own possession a skin of a light cream colour.

A perpetual thirst seems to be on the mole, for it never chooses a locality at any great distance from water; and should the season turn out too dry, and the necessary supply of water be thus diminished or cut off, the mole counteracts the drought by digging wells, until it comes to a depth at which water is found.

I should like to say something of the Hedgehog, the Stoat, and other wild animals; but I must only take one more example of the British Mammalia, the common Weasel.

Gifted with a lithe and almost snake-like body, a

long and yet powerful neck, and with a set of sharp teeth, this little quadruped attacks and destroys animals which are as superior to itself in size as an elephant to a dog.

WEASEL.

Small men are generally the most pugnacious, and the same circumstance is noted of small animals. The weasel, although sufficiently discreet when discretion will serve its purpose, is ever ready to lay down that part of valour, and take up the other.

Many instances are known of attacks on man by weasels, and in every case they proved to be dangerous enemies. They can spring to a great distance, they can climb almost anything, and are as active as—weasels; for there is hardly any other animal so active: their audacity is irrepressible, and their bite is fierce and deep. So, when five or six weasels unite in one attack, it may be imagined that their opponent has no trifling combat before him before he can claim the victory. In such attacks, they invariably direct their efforts to the throat, whether their antagonist be man or beast.

They feed upon various animals, chiefly those of the smaller sort, and especially affect mice; so that they do much service to the farmer. There is no benefit without its drawbacks; and in this case, the benefits which the weasel confers on farmers by mouse-eating is counterbalanced, in some degree, by a practice on the part of the weasel of varying its mouse diet by an occasional chicken, duckling, or young pheasant. Perhaps to the destruction of the latter creature the farmer would have no great objection.

The weasel is a notable hunter, using eyes and nose in the pursuit of its game, which it tracks through every winding, and which it seldom fails to secure. Should it lose the scent, it quarters the ground like a well-trained dog, and occasionally aids itself by sitting upright.

Very impertinent looks has the weasel when it thus sits up, and it has a way of crossing its fore-paws over its nose that is almost insulting. At least I thought so on one occasion, when I was out with a gun, ready to shoot anything—more shame to me! There was a stir at the bottom of a hedge, some thirty yards distant, and catching a glimpse of some reddish animal glancing among the leaves, I straightway fired at it.

Out ran a weasel, and, instead of trying to hide, went into the very middle of a footpath on which I was walking, sat upright, crossed its paws over its nose, and contemplated me steadily. It was a most humiliating affair.

The weasel has been tamed, and, strange to say, was found to be a delightful little animal in every way but one. The single exception was the evil odour which exudes from the weasel tribe in general, and which advances from merely being unpleasant, as in our English weasels, to the quintessence of stenches as exhibited by the Skunk and the Teledu. A single individual of the latter species has been known to infect a whole village, and even to cause fainting in some persons; and the scent of the former is so powerful, that it almost instantaneously tainted the provisions that were in the vicinity, and they were all thrown away.

The Polecat, Ferret, Marten, and Stoat belong to the true weasels; the Otters and Gluttons claiming a near relationship.

CHAPTER III.

THE COMMON LIZARD — SUDDEN CURTAILMENT — BLIND-WORM — A CURIOUS DANCE — THE VIPER — CURE FOR ITS BITE — THE COMMON SNAKE — SNAKE-HUNTING — CURIOUS PETS — SNAKE AND FROG — CASTING THE SKIN—EGGS OF THE SNAKE — HYBERNATION — THE FROG — THE TADPOLE — THE EDIBLE FROG — THE TOAD — TOADS IN FRANCE — TOAD'S TEETH — VALUE OF TOADS — MODE OF CATCHING PREY — POISON OF THE TOAD — CHANGE OF ITS SKIN.

I HAVE already said that the birds must be entirely passed over in this little work; and therefore we make a jump down two steps at once, and come upon the Reptiles, of whom are many British examples.

The first reptile of which we shall treat is the common little Lizard, that is found in profusion on heaths, or, indeed, on most uncultivated grounds.

It is an agile and very pretty little creature, darting about among the grass and heather, and twisting about with such quickness that its capture is not always easy. Sunny banks and sunny days are its delight; and any one who wishes to see this elegant little reptile need only visit such a locality, and then he will run little risk of disappointment.

There is one peculiarity about it that is rather

startling. If suddenly seized, it snaps off its tail, breaking it as if it were a stick of sealing-wax, or a glass rod. Several lizards possess this curious faculty, and of one of them we shall presently treat.

THE COMMON LIZARD.

The food of this lizard is composed of insects, which it catches with great agility as they settle on the leaves or the ground. If captured without injury—a feat that cannot always be accomplished, on account of the fragility of its tail—it can be kept in a fern-case, and has a very pretty effect there.

One of the chief beauties of this animal is its brilliant eye; and this feature will be found equally beautiful in many of the reptiles, and especially in that generally-hated one, the toad.

In the winter-time, the lizard is not seen; for it is lying fast asleep in a snug burrow under the roots of any favourable shrub, and does not show itself until the warm beams of the sun call it from its retreat.

The next British lizard that I shall mention is one that is generally considered as a snake, and a poisonous

BLIND-WORM.

one; both ideas being equally false. It is popularly known by the name of the Blind-worm, or Slow-worm; and is not a snake at all, but a lizard of the Skink tribe, without any legs.

The scientific name for it is *Anguis fragilis;* and it is called fragile on account of its custom of snapping itself in two, when struck.

Only very lately, I saw an example of this strange propensity, and was the cause of it. Near Dover, there is a small wood, where vipers are reported to dwell; and as I was walking in the wood, I caught a glimpse of a snake-like body close by my foot. I struck, or rather stabbed, it with a little stick,—for it had a very viperine look about it,—and with success rather remarkable, for the very slight blow that the creature could have received from so insignificant a weapon, used in such a manner. The viper was clearly cut into two parts, but how or where could not be seen, owing to the thick leaves and grass that rose nearly knee-high.

On pushing among the leaves, I found with regret that the creature was only a blindworm.

A curious performance was being exhibited by the severed tail, a portion of the animal about five inches long; this was springing and jumping about with great liveliness and agility, entirely on its own account, for by this time the blindworm itself had made its escape, and all search was unavailing.

Some ten minutes or so were consumed in looking for the reptile itself; and by that time the activity of the tail was at an end, and it was lying flat on the ground, coiled into a curve of nearly three-fourths of a circle. I gave it a push with the stick, when I was startled by the severed member jumping fairly into the air, and recommencing its dance with as much vigour as before. This performance lasted for some minutes, and was again exhibited when the tail was roused by another touch from the stick. Nearly half an hour elapsed before the touch of the stick failed to make the ail jump, and even then it produced sharp convulsive movements.

The object of this strange compound of insensibility and irritability may perhaps be, that when an assailant's attention is occupied by looking at the tail, the creature itself may quietly make its escape.

The food of the blindworm is generally of an insect nature, and it seems to be fond of small slugs. The country people declare that it is guilty of various crimes, such as biting cattle and similar offences, of which bite an old author says that, " unless remedy be had, there followeth mortality or death, for the poyson thereof is very strong."

Fortunately for us, we have but one poisonous reptile, the viper; and the slow-worm is as innocent of poison as an earthworm. It is true, that if provoked, it will sometimes bite; but its mouth is so small, and its teeth so minute, that it cannot even draw blood.

The names that are given to it are hardly in accordance with its formation, for it is not very sluggish in its movements, although it can be easier taken than the lizard; while it is anything but blind, and its eyes, though small, are brilliant. Perhaps the epithets ought to have been applied to the givers, and not to the receiver.

As for the real snakes, there are but two species in England, one being called the Viper, or Adder, and the other the Ringed, or Grass-snake. The viper is rather to be avoided, as it is possessed of poison-fangs, and if irritated, is not slow in using them.

Of this latter I have little to say, and would not have mentioned it excepting for two reasons; the one to enable any person to distinguish it from the common snake, and to avoid, as far as possible, the chance of being bitten; and the other to tell how to heal the bite, should so untoward an event happen.

VIPER.

Poisonous snakes may be readily known by the shape of their head and neck; the head being very wide at the back, and the neck comparatively small. Some persons compare the head of a poisonous snake to the ace of spades, which comparison, although rather exaggerated, gives a good idea of the poison-bearing head.

THE RINGED SNAKE.

It has a cruel and wicked look about it also, and one recoils almost instinctively.

Should a person be bitten by the viper, the effects of the poison may be much diminished by the liberal use of olive oil; and the effect of the oil is said to be much increased by heat. Strong ammonia, or hartshorn, as it is popularly called, is also useful, as is the case with the stings of bees and wasps, and for the same reason. The evil consequences of the viper's bite vary much in different persons, and at different times, according to the temperament of the individual, or his state of health.

I may as well put in one word of favour for the viper before it is dismissed. It is not a malignant creature, nor does it seek after victims; but it is as timid as any creature in existence, slipping away at the sound of a footstep, and only using its fangs if trodden on accidentally, or intentionally assaulted.

The second English snake is the common harmless Ringed snake; which does not bite, because it has no teeth to speak of; and does not poison people, because it has no venom at all.

Its only mode of defence is by pouring forth a most unpleasant, pungent odour, which adheres to the hands or clothes so pertinaciously, that many washings are required before it is expelled. Yet it is sparing enough even of this solitary weapon, and may, after a while, be handled without any inconvenience.

To this assertion I can bear personal and somewhat extensive witness; for I have caught and kept numbers of snakes. The worthy villagers must have formed

COMMON SNAKE.

curious ideas of me, and I rather fancy must have accredited me with something of the wizard character; for I contrived to oppose their prejudices—all, by the way, of a cruel character—in so many instances, that they were rather afraid, as well as annoyed. To see them run away, as if from a lighted shell, when I came among them with a snake in each hand, was decidedly amusing, and not less curious was the pertinacity with which they clung to their prejudices.

In vain were arguments used to prove that the snake was not a venomous animal, and ought not to be killed and tortured; in vain did I put my finger into the snake's mouth, and let its forked tongue glide over my very hand or face; they were not to be so taken in, and they remained wise in their own conceit.

They certainly could not deny that the snake did not bite me, and that its tongue did not pierce me, but the conclusion deduced therefrom was simply that my constitution, or perchance my magical art, was such that I was unbiteable and unpoisonable.

No! to them the snake was still poisonous, and its tongue still envenomed.

At one time we had so many snakes that they were kept in the crevices of an old wall, and left to stay or go as they pleased. My boys—I had a school at that time—took wonderfully to snake hunting, and every half holiday produced a fresh supply of snakes. The boys used to devise the strangest amusements in connexion with their snakes, of which they were very proud, each boy exhibiting his particular favourite, and expatiating on its excellences.

One of their fashions, and one which lasted for some time, was to make tunnels in the side of the Wiltshire Downs, and to turn in their snakes at one end, merely for the purpose of seeing them come out at the other.

Then there was a stone-quarry some three miles

distant, which was in some parts of the year nearly filled with water. Thither the boys were accustomed to repair for the purpose of indulging their snakes with a bath. They certainly seemed to enjoy the swim, and were the better for it.

Sometimes there was great excitement; for a snake would now and then act in too independent a manner, and instead of swimming straight across, so as to be caught by a boy on the opposite side, would sink to the bottom, and there lie flat and immoveable. Long sticks could not be found there; and their only mode of making the snake stir was to startle it by throwing stones. Even then there was a difficulty; for if the stones fell too far from the snake they had no effect, and if they fell on him they might hurt him.

To wait until the truant chose to move would have been hopeless, for snakes are able to take so much pure air into their lungs, and they require so little of it for respiration, that the patience of the boys would be exhausted long before the snake felt a necessity for moving.

Sometimes a snake would try to get away, and insinuate his head and part of his body into a crevice: in that case there was sad anxiety, and judicious management was required to eliminate the reptile without damage. It is a very difficult matter to drag a snake backwards, because the creature sets up the edges of

SNAKE AND FROG.

the scales, and each one serves as a point of resistance. So, when the snake is within a crevice, where the scales of the back can act as well as those of the belly, the difficulty is increased.

When such an event took place, the best mode of extracting the snake was to let it glide on, and so lower its scales, and then to pluck it out with a sudden jerk, before it had time to erect them afresh. But as often as otherwise, the snake got the better in the struggle, and by slow degrees was lost to view.

Perhaps the pleasantest portion of snake-keeping was the feeding. It was found that the snakes lost their appetite, and would not eat, though frogs and newts were liberally supplied. So the boys settled the matter by opening the mouth of the snake, and pushing a newt fairly down its throat.

One of the largest snakes that I have seen, was engaged in feeding himself, not trusting to boys for any help. I was walking in a field, and heard a strange cry from a neighbouring ditch. On going towards the spot, I saw there a large snake struggling with a frog. The frog was comparatively as large as the snake, and as it had a plain objection to being swallowed, there was some turmoil.

The snake was stretched along the bottom of the ditch, which at this time was dry, and he held in his mouth both hind feet of the frog, who was also stretched forward at full length, resisting with its forelegs the attempts of the snake to draw it back, and croaking dismally. The strife continued for some time, when I made a sudden movement, and the snake, loosing its hold of the frog, glided up the opposite bank. The frog slowly gathered itself together, sat still for some little time, and then hopped away.

The entire empty skin of the snake may often be found among bushes, where the creature has gone in order to assist itself in casting off its old skin. Snakes, as well as other animals, wear out their coats, and are obliged to change them for others. When the change is about to take place, and a new coat has formed under the old, like new skin under a blister, the creature betakes itself to some spot where is thick grass, reeds, or similar substances. A rent then opens in the neck, and the snake, by wriggling about among the stems, literally crawls out of its skin, which it leaves behind, turned inside out. Even the covering of the eyes is cast away, and in consequence the snake is partially blind for a day or two previously to the moult, if we may call it so.

Eggs laid by the snake are also of frequent occurrence. I have found them in manure heaps, the warmth of which places is attractive to them. The eggs are white, and covered with a strong membrane, but have no shell. They are laid in long strings, from sixteen to twenty eggs being in each chain.

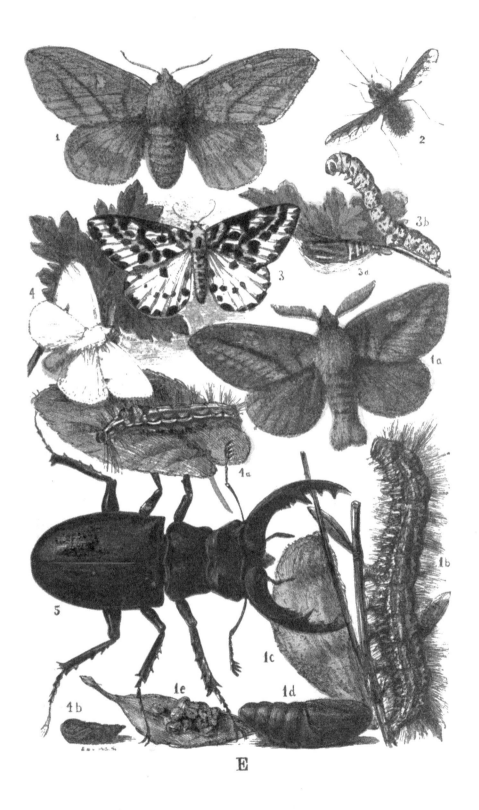

THE FROG.

In the winter the Ringed snake retires to a convenient cell, such as a hollow tree, or a heap of wood, and there it remains in a torpid state until the warm weather. Many individuals have been found collected together in these winter quarters, probably for the sake of affording each other mutual warmth.

The reptiles of which we have just treated live exclusively on land, though they may occasionally be found in water; but those which we shall now inspect belong rather to the water than the land. The most common of these amphibious reptiles, as they are called, is the Frog.

FROG.

A very curious animal is a frog, and well worth examining, as well in its perfect state as in its intermediate state. To begin at the beginning of a frog's existence, we find it exhibited in masses of eggs, fixed to each other by a kind of gelatinous substance, and floating in large quantities in ditches or ponds. Each egg is about the size and shape of a pea, and in the centre is the little black speck from which the young frog proceeds.

In process of time the egg is hatched, and out comes a queer little creature, with a big head and a flat slender tail, called generally a tadpole, and in some places a pollywog. In this state of life the young creature is simply a fish, with fish-like bones, and breathing through gills, after the manner of fish.

Being very voracious, it grows rapidly: little legs begin to show themselves; and, at the proper season, the gills are laid aside, the tail vanishes, and the little frog is then in its usual form. The circulation of the blood can be well exhibited by means of a microscope, if a tadpole be laid on the stage so as to bring its tail within the focus, care being taken to keep the member well wetted.

At the time when the tail is laid aside, the young frog is very small, and in this state is generally found to swarm immediately after rain. The frog-showers, of which we so often hear, are probably occasioned, not by the actual descent of frogs from the clouds, but from the genial influence of the moisture on the young frogs who have already been hatched and developed,

THIRSTINESS OF FROGS.

and who have been biding their time before they dared to venture abroad.

Still I would not venture to say that frogs have not descended *in* the rain, for there are several accredited accounts of fish-showers, both being probably caused in the same way.

For a drawing of the Tadpole, see page 88.

It is not often that frogs are found far from water, for they are the thirstiest of beings, and drink with every pore of their body. If, for example, a wrinkled and emaciated frog is placed in confinement, and plentifully supplied with water, it absorbs the grateful moisture like a sponge, and plumps up in a wonderfully short time.

From the same cause, it parts with its moisture with equal rapidity; and if a dead frog be laid in the open air on a dry day, it speedily shrinks up, and becomes hard as horn. The skin and lungs co-operate in respiration, but only when the former is moist. So, in order to secure that object, the frog is furnished with an internal tank, so to speak, which receives the superabundance of the absorbed water, and keeps it pure until it is required for use. So great is the power of absorption, that a frog has been known to absorb a quantity of water equal to itself in weight merely through the pores of the abdominal surface, and this in a very short time.

In England we don't eat frogs, for what reason I know not. One species of frog is very excellent food, and it is but natural to suppose that another may be so, *i.e.* if properly cooked. However, the old belief still keeps its ground, that the French are the natural foes of the English, and we ought to hate them, because they "eat frogs and are saddled with wooden shoes." Still I cannot but think that to eat frogs is better than to starve or to steal, and that to wear wooden shoes is not more humiliating than to wear no shoes at all.

After its fashion, the frog sings, though it is but after a fashion. We call the frog's song a croak: I wonder what name the frog would give to our singing. When the frog sings, it generally sinks itself under water, with the exception of its head, opens its mouth, lays its lower jaw flat on the water, and sets to work as if it meant to make the best of its time. Even in England we have fine specimens of frog concerts, though not to such an extent as in many other countries. In France the frogs make such a croaking, that we hardly wonder at the rather tyrannous conduct of the noblesse just before the great Revolution. When the nobility or courtiers spent any time in the country, the miserable peasants were forced to flog the water all night, on purpose to keep the frogs quiet, for their croaking was so noisy that the fastidious senses of the fashionables could not be lulled to sleep.

Now-a-days, the people don't seem to be satisfied with the country croakings, but they import the horrid sounds into the city by means of a toy called a "grenouille," which, when set in motion, makes a croaking sound, just like that of a frog.

As a general fact, frogs are just endurable, and people will inspect them—from a distance—without much ado. But the case is widely altered when they see the frog's first-cousin, the Toad.

A large volume might easily be filled with tales respecting this much-calumniated creature; in which tales the toad appears to be a very incarnation of malignity, and to be wholly formed of poison. If it burrowed near the root of a tree, every one who ate a leaf of that tree would die; and, if he only handled it, would be struck with sudden cramp. And the cause of this poisonous nature was its liver, which was "very vitious, and causeth the whole body to be of an ill temperament."

Fortunately, toads had two livers; and although both of them were corrupted, yet one was full of poison, and the other resisted poison. As for remedies, the only effectual one was of rather a complicated nature, and consisted of plantain, black hellebore, powdered crabs, the blood of the sea-tortoise mixed with wine, the stalks of dogs' tongues, the powder of the right horn of a hart, cummin, the vermet of a hare, the quintessence of treacle, and the oil of a scorpion, mixed and taken *ad libitum*.

Even in the days when this prodigious prescription was invented, some good was acknowledged to exist in a toad, the one being the precious jewel in its head, and

THE COMMON TOAD.

the other its power as a styptic. Supposing any one to fall down and knock his nose against a stone, he could instantly stop the bleeding if he only had in his pocket a toad that had been pierced through with a piece of wood and dried in the shade or smoke. All that was requisite was to hold the dried toad in the hand, and the bleeding would immediately cease. The reason for this effect is, that "horror and fear constrained the blood to run into his proper place, for fear of a beast so contrary to humane nature."

And, as a concluding instance of the wonderful

things that happened whenever toads were the subject, we are told that at Darien, where the household slaves water the door-steps in the evening, all the drops that fall on the right hand turn into toads.

These poor creatures fare little better even now, as far as public opinion goes; and in France worse than in England.

I was once walking in the forest at Meudon with a party of friends, and was brought to a check by a sudden attack made on a large toad that was walking along the pathway. I succeeded in stopping a blow that was aimed at it; and was stooping down, intending to remove it to a place of safety, when I was hastily pulled away, and horror was depicted on the countenances of all the spectators.

"It will bite you," cried one.

"Pouah!" exclaimed another, "it will spit poison at you."

"In France, every one kills toads," said a third.

I objected that it could not bite, because it had no teeth.

"No teeth!" they all exclaimed. "In France, toads *always* have teeth."

"Well, then," I said, "I will open its mouth, and show you that it has none."

But before I could touch it, I was again dragged away.

"Teeth come when the toads are fifty years old," was the explanation that was given; but still the death-sentence had passed in every mind, and I knew that when I moved the poor toad would be killed.

Just then, some one remarked that tobacco killed toads, if put on their backs. So I took advantage of the assertion, and made a compromise, that, on my part, I would not handle the toad; and that, on theirs, the only mode by which they might kill it was by putting tobacco on it.

The terms being thus arranged, plenty of tobacco was produced—and very bad tobacco, too, as is generally the case in France; and, as no one but myself dared come so near, I put about half an ounce of the weed on the back of the toad, as it sat in a rut. For a minute or more, the creature sat quite still, and all the party began to exclaim—

"See! the toad is quite dead!"

"Ah! the nasty animal!"

"Monsieur Ool!"—(no one ever made a better shot at my name than Ool)—"Monsieur Ool! the toad is dead!"

However, the toad rose, shook off all the tobacco, and recommenced his march along the road. The only good that was done was the saving of that individual toad's life, for all the party retained their faith in toads' teeth, and probably thought that the creature

VALUE OF THE TOAD.

would not touch me because I was a trifle madder than the rest of my nation, who are always very mad on the French stage.

Afterwards, I found that the belief in toads' teeth was quite general; and one person offered to show me some, half an inch in length, which he kept in a box at home. But I was never fortunate enough to see them.

In England, toads are sometimes valued for the good which they do; and the market-gardeners, whose trim grounds surround London, actually import toads from the country, paying for them a certain sum per dozen. For toads are voracious creatures, feeding upon slugs, worms, grubs, and insects of various kinds, and so devour great numbers of these little pests to the gardener.

The mode in which a toad catches its prey is curious enough. Its tongue is fastened into its mouth in a very peculiar way, the base of the tongue being fixed at the entrance of the mouth, the tip pointing down its throat when it is at rest. When, however, the toad sees an insect or slug within reach, the tongue is suddenly shot out of the mouth, and again drawn back, carrying the creature with it.

So rapidly is this operation performed, that the insect seems to disappear by magic. The frog feeds in the same manner

For the poisonous properties attributed to the toad, there is some foundation, though a small one. But a very small foundation is generally found strong enough to bear a very large superstructure of calumny; though the reverse is the case when the report is a favourable one. The skin of the toad is covered with small tubercles, which secrete an acid humour sufficiently sharp and unpleasant to prevent dogs from carrying a toad in their mouths, though not so powerful as to deter them from attacking toads and killing them.

A rather curious advantage has been taken of the insect-eating propensities of the toad. A gentleman had killed a toad at a very early hour one morning, and after skinning it, for the purpose of stuffing the skin, he dissected its digestive system. The contents of the stomach he turned out into a basin of water, and found there a mass of insects, some of them very rare and in good preservation.

Afterwards, he was accustomed to kill toads for the express purpose of collecting the insects that were found within them, and which, being caught during the night, were often of such species as are not often found.

The same experiment elicited another curious fact, namely, the great tenacity of life possessed by some insects. Before pinning out the insects that were found, and which were mostly beetles, they had been allowed to remain in the water for several days, and were apparently dead. Yet, when they were pinned on cork, they revived; and, when they were visited, were found sprawling about in quite a lively style.

Like all the reptiles, the toad changes its skin, but the cast envelope is never found, although those of the serpents are common enough. The reason why it is not found is this: the toad is an economical animal, and does not choose that so much substance should be wasted. So, after the skin has been entirely thrown off, the toad takes its old coat in its two fore-paws, and dexterously rolls it, and pats it, and twists it, until the coat has been formed into a ball. It is then taken between the paws, pushed into the mouth, and swallowed at a gulp like a big pill.

CHAPTER IV.

NEWTS—A FISH WITH LEGS—NEWTS FEEDING—NEWT-CANNIBALS—CASTING THE SKIN—STRANGE STORIES—ANOTHER NEWT STORY—HATCHING OF YOUNG—TENACITY OF LIFE—THE STICKLEBACK—ITS PUGNACITY—ITS COLOURS—ACCLIMATIZATION—THE LAMPERN—A RUSTIC PHILOSOPHER—THE CRAY-FISH—HOW WE CAUGHT IT—REPRODUCTION OF LIMBS—FRESH-WATER SHRIMP—WOODLOUSE AND ARMADILLO.

THE Newts, or Efts, or Evats, as they are called in different parts of England, can be easily distinguished from the lizard by the flattened tail, which being intended for swimming, is formed accordingly.

Two species of these creatures are found in this country, the common Water-Newt and the Smooth Newt. These beautiful creatures may be found in almost every piece of still water, from ponds and ditches up to lakes. The full beauty of the newt is not seen until the breeding season begins to come on, and even then only in the male.

At this time the green back and orange belly attain a brighter tint, and the back is decorated with a wavy crest, tipped with crimson. This crest is continually waving from side to side as the creature moves, and forms graceful curves. The newts are equally at home in water and on land, and in the latter case have often been mistaken for lizards.

THE COMMON NEWT.

One of these animals, when taking a walk, alarmed an acquaintance of mine sadly. He was rather a tall man than otherwise, and did not appear particularly timid; but one day he came to me looking rather pale, and announced that he had just been terribly frightened.

"A fish, with legs!" said he, "*four* legs! got out of the water and ran right across the path in front of me! I saw it run!"

"A fish with legs!" I replied; "there are no such creatures."

"Indeed there are, though, for I saw them. It had

FOUR LEGS, and it waggled its tail! It was horrible, horrible!"

"It was only a newt," I replied, "an eft. There is nothing to be afraid of."

"It was the *legs*," said he, shuddering, "those dreadful legs. I don't mind getting bitten, or stung, but I can't stand legs."

Newts are very interesting animals, though they have legs, and can easily be kept in a tank if fed properly. Little red worms seem to be their favourite food, and the newt eats them in a rather peculiar style. I have had numbers of newts of all sizes and in all stages of their growth, and always found them eat the worm in the same way. As the worm sank through the water, the newt would swim to it, and by a sudden snap seize it in the middle. For nearly a minute it would remain with the worm in its mouth, one end protruding from each side of its jaws. Another snap would then be given, and after an interval a third, which generally disposed of the worm.

When they have been swimming freely in a large pond, I have often seen large newts attack the smaller, and try to eat them; but I never saw the attempt successful, though I hear that they have been seen to devour the younger individuals. They always came from behind, as if trying to avoid observation, and then made a sudden dart forward, snapping at the tail of their intended victim. In confinement I never saw even an attempt at cannibalism.

Whether it is invariably the case, I cannot say, but every newt that I took cast its skin within a few hours from the time that it was placed in the glass jar. The general surface of the skin came off in flakes, but that from the paws was drawn off like gloves, retaining on their surface all the markings and creases which they exhibited when in their proper place.

How the drawing off of their tiny gloves was effected, I could not see, though I watched carefully. They looked beautiful as they floated in the water, being delicate as gossamer, white, and almost transparent. They might have been made for Queen Mab herself, and were so delicate that I never could preserve any of them so as to give a proper idea of their form.

It may be that the change of water might cause the change of skin, for the water in which they were kept was drawn from a pump, and that in which they formerly lived was the ordinary soft water found in ponds.

Pretty as is the newt, it is as harmless as pretty, and notwithstanding has suffered under the reputation of being a venomous creature. The absurd tales that I have heard of this creature could scarcely be believed; and how people with any share of sense could receive such absurdities, is matter of wonder. And as usual, the moral of the stories is, that newts are to be killed wherever found. The belief of the poisonous character of the newt is of long standing, as may be seen in the ancient works on natural history. In one of these it is said that its poison is like that of vipers; and there is a description of the formation of its tail which is rather beyond my comprehension:—

"The tail standeth out betwixt the hinder-legs in the middle, like the figure of a wheel-whisk, or rather so contracted as if many of them were conjoined together, and the void or empty places in the conjunctions were filled."

The capture and domesticating of newts gave dire offence in a village where I lived for some time; and the expressions used when I took a newt in my hands were not unlike those of the Parisians respecting the toad. Sundry ill-omened tales of effets were told me. For example: A girl of the village was filling her pitcher at a stream which runs near the village, when an effet jumped out of the water, sprang on her arm, bit out a piece of flesh, spat fire into the wound, and, leaping into the water, escaped. The girl's arm instantly swelled to the shoulder, and the doctor was obliged to cut it off.

This was told me with an immensity of circumstantial details common to such narrators, and was corroborated by the bystanders. The wounded lady

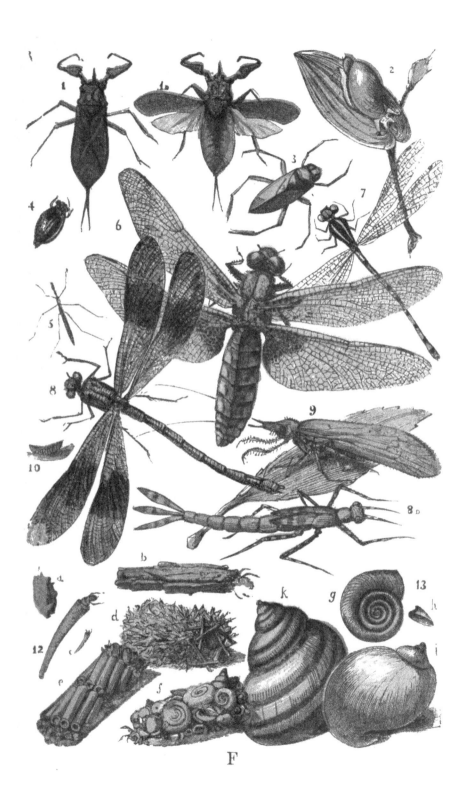

herself was not to be found, and cross-questions elicited that it "weir afoor their time." I asked them how the effet which lived in the water, and had just leaped out of it, was able to keep a fire alight in its interior; but they were not in the least shaken, except perhaps in their heads, which were wagged with a Lord Burleigh kind of emphasis.

Then there was the sexton-clerk-gardener-musician and general factotum, who had a newt tale of his own to tell. He had been cutting grass in the churchyard, and an effet ran at him, and bit him on the thumb. He chopped off the effet's head with his knife, but his thumb was very bad for a week.

Once they got the better of the argument, at all events in the eyes of the owner of the farming stock, and my poor newts were ejected. It happened thus:—

Two or three specimens I kept in my own room in a glass vase, in order to watch them more closely; and some six or seven others lived as stock in the large horse-trough, from whence they could be taken when required.

One day the proprietor came to me and ordered the destruction of my newts, for they had killed one of his calves.

"But," I remonstrated, "they cannot kill a calf or even a mouse, for they have no fangs and very little mouths. Besides, the calf has not come near this trough."

So saying, I took up several of the newts, opened their mouths—no easy matter, by the way—and showed that they had no fangs. And I urged, that even if they had been as poisonous as rattlesnakes, it would not have made any difference to the calf, which had never left the cowhouse, and was at the opposite end of the farm-yard, separated by a barn and several gates. But all was useless.

"There are the newts, and there is the dead calf!" was the answer; and so the newts had to go. However, I would not suffer them to be killed, but put them into a bag and took them back to the pond whence they had come.

Afterwards the proprietor said that the calf died because its mother had drunk at the trough in which the poisonous newts were.

Now, the funniest part of the story is, that there was not a horse-pond that did not swarm with efts, and consequently all the foals and calves ought to have died. Only they didn't.

The care which the female newt takes in depositing her eggs is very remarkable.

Each egg is taken separately, and by the aid of the fore-paws is regularly tied or twisted up in the leaves of water-plants, for which process different people have different reasons. Some think that it is for the purpose of preventing too ready an access of water, and so to

THE FEMALE NEWT.

retard their hatching; while some say that it is to guard the egg against voracious water-animals. To the latter opinion I rather incline; perhaps both may be right.

When hatched, the young newt is very like a tadpole, breathing by gills outside its neck. After a while the gills vanish and the legs appear; but it keeps its tail. It is rather curious that the frog tadpole puts forth its hinder legs first; while in the tadpole of the newt, the fore-legs are the first to show themselves.

After the gills are lost, the newt breathes by means of lungs; and if it is in the water, is forced to rise at intervals for the purpose of breathing.

The tenacity with which these creatures cling to life is quite surprising. Experiments have been tried purposely to see to what degree a body could be mutilated, and yet retain life. They have even been frozen up into a solid block of ice, and, after the thawing of their cold prison, revived, and seemed none the worse for it. I may as well mention that none of these experiments were tried by myself, for I am not scientific enough to care nothing for the infliction of pain; but on one occasion I did try an experiment, and, as it turned out, a very cruel one, although it was not intended for an experiment.

I was studying the anatomy of the frogs and newts; and having eight or ten fine specimens of the latter creature, determined to take advantage of the opportunity. The first thing was, of course, to kill the creature without injuring its structure, and I thought that the best mode of so doing would be to put it into my poison-bottle. This was a large glass jar filled with spirits of wine, in which was held corrosive sublimate in solution. This mixture generally killed the larger insects almost immediately, and seemed just the thing for the newts.

So they were put into the jar—but then there was a scene which I will not describe, which I trust never to see again, and of which I do not even like to think. Suffice it to say, that nearly a quarter of an hour elapsed before these miserable creatures died, though in sheer mercy I kept them pressed below the surface.

Changing our post of observation from the banks of the ponds to those of the running streams, we shall find there many creatures worthy of observation; so many, indeed, that it would be a hopeless task to attempt to give even a slight account of one-fiftieth of them. I shall, therefore, only mention two creatures, as examples of the fish; and these two are chosen because they are exceedingly common, and very different from each other in colour and habits.

The first of these creatures is the common Stickleback, or Tittlebat, as it is sometimes called. There are

THE STICKLEBACK.

several species of British sticklebacks; but the commonest, and I think the most beautiful, is the three-spined stickleback.

These little fish derive their name from the sharp

spines with which they are armed, and which they can raise or depress at pleasure—as I know to my cost. For being, as boys often are, rather silly, I made a wager that I would swallow a minnow alive; and having made the bet, proceeded to win it. Unfortunately, instead of a minnow, a stickleback was handed to me, which having its spines pressed close to the body, was very like a minnow. Just as I swallowed it, the creature stuck up all its spines, and fixed itself firmly.

Neither way would it go, and the torture was horrid. At last, a great piece of apple that I swallowed gave it an impetus that started it from its position; but it was not for some time, that to me appeared hours, that the fish was disposed of. And even then it left its traces; and if it would be any satisfaction to the fish to know that ample vengeance was taken for its death, it must have been thoroughly gratified.

There are few fish more favoured in point of decoration than the stickleback; although the decoration, like that of soldiers, is only given to the gentlemen, and of them only to the victors in fight.

They are most irritable and pugnacious creatures, that is, in the early spring months, when the great business of the nursery is in progress. And the word nursery is used advisedly; for the stickleback does not leave her eggs to the mercy of the waters, but establishes a domicile, over which her husband keeps guard.

The vigilance of this little sentry is wonderful; and I have often seen fierce fights taking place. Not a fish passes within a certain distance of the forbidden spot, but out darts the stickleback like an arrow, all his spines at their full stretch, and his body glowing with green and scarlet. So furious is the fish at this time, that I have sometimes amused myself by making him fight a walking-stick.

If the stick were placed in the water at the distance of a yard or so, no notice was taken. But as the stick was drawn through the water, the watchful sentinel issued from his place of concealment, and when the intruding stick came within the charmed circle, the stickleback shot at it with such violence that he quite jarred the stick.

His nose must have suffered terribly. If the stick were moved, another attack would take place, and this would be continued as long as I liked.

Sometimes a rival male comes by, with all *his* swords drawn ready for battle, and his colours of red and green flying. Then there is a fight that would require the pen of Homer to describe. These valiant warriors dart at each other; they bite, they manœuvre, they strike with their spines, and sometimes a well-aimed cut will rip up the body of the adversary, and send him to the bottom, dead.

When one of the combatants prefers ignominious flight to a glorious death, he is pursued by the victor with relentless fury, and may think himself fortunate if he escapes.

Then comes a curious result. The conqueror assumes brighter colours and a more insolent demeanour; his green is tinged with gold, his scarlet is of a triple dye, and he charges more furiously than ever at intruders, or those whom he is pleased to consider as such. But the vanquished warrior is disgraced; he retires humbly to some obscure retreat; he loses his red, and green, and gold uniform, and becomes a plain civilian in drab.

Sometimes I have brought on a battle royal between the guardians of several palaces, by dropping in the midst of them a temptation which they could not resist. This was generally a fine fat grub taken from a caddis case. The caddis is large and white, and so can be seen to a considerable distance.

As this sank in the water, there would be a general rush at it, and the ensuing contention was amusing in the extreme. First one would catch it in his mouth and shoot off; half a dozen others would unite in chase, overtake the too fortunate one, seize the grub from all sides, and tug desperately, their tails flying, their fins at work, and the whole mass revolving like a wheel, the centre of which was the caddis worm.

It would be swallowed almost immediately; but the mouth of the stickleback is much too small to admit an entire caddis, and the skin of the grub is too tough to be easily pierced or torn. Half an hour often elapses before the great question is settled, and the caddis eaten.

The rapidity of the evolutions and the fierceness of the struggle must be seen to be appreciated—and it is a spectacle easily to be witnessed; wherever there are sticklebacks, caddis worms are nearly certainly found, and it only needs to extract one of these from its case and deposit it judiciously in the water.

The stickleback is a hardy little fish, and can easily be kept in the aquarium, if plenty of room be given to it. It has even been trained to live in sea-water, by adding bay-salt to the water in which it dwelt; so that the plan of pickling salmon alive, by a judicious admixture of vinegar and allspice with the water, has something to which to appeal as collateral evidence.

The other representative of the fishes is a very curious one, and can be easily observed. It is called the "Lampern," and is shown in the accompanying figure.

In some parts of England the lampern goes by the name of "Seven-eyes," in allusion to the row of eye-like holes that may be seen extending along the side of the throat. These apertures are the openings by which the water passes from the gills.

The chief external peculiarity in this creature is the

THE LAMPERN.

mouth, which, instead of being formed with jaws like those of other fishes, resembles none of them, not even those of the eel, which it most resembles externally. Indeed, on looking at the mouth of a lampern, one is forcibly reminded of the leech, for it is possessed of no jaws, and adheres firmly to the skin by exhaustion of the air.

Very delicate food are these lamperns, quite as good as the lampreys themselves, whose excellence is reported to have cost England one of her kings; yet I never knew but one person who would eat them, and very few who would even touch them, they also being called poisonous.

In Germany they know better, and not only eat the lamperns themselves, but packing them up in company with vinegar, bay leaves, and spices, export them as an article of sale.

A RUSTIC PHILOSOPHER.

The solitary sensible individual of whom I have made mention, was truly a wise man. He used to offer the young urchins of the neighbourhood a reward for bringing lamperns, at the rate of a halfpenny per wisket full.

A wisket, I may observe, is a kind of shallow basket, made of very broad strips of willow; and a wisket filled with lamperns would be a tolerable load for a boy.

So for the sum of one halfpenny, that philosopher was furnished with provisions for a day or more.

Really, the prejudice against the lampern is most singular. Even near London, when lamperns lived in vast numbers in the Thames, they were only used as bait, being sold for that purpose to the Dutch fishermen. In one season, four hundred thousand of these creatures have been sold merely for bait for cod-fish and turbot.

The scientific name for the lampreys is "*Petromyzon*," a word signifying "stone-sucker." The name is rightly applied; for when the lampern wishes to remain still in one place, it applies its mouth to a stone, sticks tightly to it by suction, and there remains firmly at anchor, and defying the power of the stream. In favourable spots, thousands of these fish **may** be seen together, quite blackening the bottom of the stream with their numbers. They seem specially to affect shallow mountain streams; and, in spite of the rapid current, wriggle their devious way up the stream with great rapidity. When they are not quite pleased with the spot on which they settle down for the time, they scoop it out to their minds in a very short time. This task is accomplished by means of the sucker-like mouth. If a stone is placed in a position that incommodes them, they affix their mouths to it, and drag it away down the stream. In this way they will remove stones which are apparently beyond the power of so small a creature. By perseverance they thus scoop out small hollows, about eighteen inches long, and a foot wide, in which they lie in groups so thick that I have more than once mistaken them for dark logs lying in the stream, and was only undeceived by the waving of the multitudinous tails. Year after year the lamperns followed the same course, and chose the same positions, so that we could at any time tell where these creatures would be found by the thousand, where they would be found singly, and where none would be seen at all.

The general thickness of this creature is that of a large pencil, but it varies according to the individual. The length is from one foot to fifteen inches or so.

There is a much smaller species of lampern called the Pride, Sand-pride, or Mud Lamprey, which is not more than half the length of the lampern, and only about the thickness of an ordinary quill. This creature has not

THE CRAY-FISH.

the power of affixing itself like the lampern, on account of the construction of its mouth.

Having now taken a hasty glance at the vertebrated animals, we pass to those who have no bones at all, and whose skeleton, so to speak, is carried outside. Our representation of aquatic crustacea, as such creatures are called, will be the Cray-fish and the Water-Shrimp.

Every one knows the Cray-fish, because it is so like a lobster, turning red when boiled in the same way.

THE CRAY-FISH.

This red colour is brought out by heat even if applied by placing the shell before a fire, and spirits of wine has the same effect. The last fact I learned from experience, and was very sorry that it *was* a fact, for the red shell quite spoiled the appearance of a dissected cray-fish that was wanted to look nice in a museum.

Being very delicate food, and, in my opinion, much better than the native lobster, they are much sought after at the proper season, and are sold generally at the rate of half-a-crown for one hundred and twenty.

There are many modes of catching them, which may be practised indifferently. There are the "wheels," for example, being wicker baskets made on the wire mouse-trap principle, which the cray-fish enters and cannot get out again. Also, there is a mode of fishing for them with circular nets baited with a piece of meat. A number of these nets are laid at intervals along the river bank, and after a while are suddenly pulled out of the water, bringing with them the cray-fish that were devouring the meat.

But the most interesting and exciting mode of cray-fish catching is by getting into the water, and pulling them out of their holes.

Cray-fish take to themselves certain nooks and crannies, formed by the roots of willows or other trees that grow on the bank; and they not unfrequently take possession of holes which have been scooped by the water-rat. The hand is thrust into every crevice that can be detected, and if there is a cray-fish, its presence is made known by the sharp thorny points of the head,—for the cray-fish always lies in the hole with its head towards the entrance.

The business is, then, to draw the creature out of its stronghold without being bitten—a matter of no small difficulty. If the hole is small, and the cray-fish large,

I always used to draw it forward by the antennæ or horns, and then seize it across the back, so that its claws were useless.

The power of the claws is extraordinary, considering the size of the creature that bears them. They will often pinch so hard as to bring blood; and when they have once secured a firm hold, they do not easily become loosened. Still, the risk of a bite constitutes one of the chief charms of the chase.

The legitimate mode of disposing of the cray-fish, when taken, is to put them into the hat, and the hat on the head; but they stick their claws into the head so continually, and pull the hair so hard, that only people of tough skin can endure them.

Sometimes, when the bed of the river is stony, the cray-fish live among and under the stones, and then they are difficult of capture; for with one flap of their tail they can shoot through the water to a great distance, and quite out of reach.

It is not unfrequent to find a cray-fish with one large claw and the other very small. The same circumstance may be noted in lobsters. The reason of this peculiarity is, that the claw has been injured, generally in single combat; for the cray-fish are terrible fighters, and the mutilated limb has been cast off. Most wonderfully is this managed.

The blood-vessels of the crustaceans are necessarily

so formed, that if wounded, they cannot easily heal; and if there were no provision against accidents, the creature might soon bleed to death.

But when a limb, say one of the claws, is wounded, the limb is thrown off—not at the injured spot, but at the joint immediately above. The space exposed at the joints is very small in comparison with that of an entire claw; and as the amputation takes place at a spot where there is a soft membrane, it speedily closes. In process of time, a new limb begins to sprout, and takes the place of the member that had been thrown off.

The eyes of the cray-fish are set on footstalks, so as to be turned in any direction, and they can also be partially drawn back, if threatened by danger. If the eye is examined through a magnifying glass of tolerable power, it will be seen that it is not a single eye, but a compound organ, containing a great number of separate eyes, arranged in a wonderful order. As, however, a description of an insect's eye will be given at a succeeding page, we at present pass over this organ.

At the proper season of the year, the female cray-fish may be seen laden with a large mass of eggs, which she carries about with her, and by the movement of the false legs that are arranged in double rows on the under surface of the tail, keeps them supplied with fresh streams of water. In process of time, the eggs are hatched; but very few, in comparison, reach

maturity. Even the mother herself is apt to eat her own young, when they have set themselves free from her control. I have known this to take place when we were trying to breed cray-fish in a tank. Only one attained to any size, and even that was not so large as a house-fly when we took it from the water.

The fresh-water Shrimp may generally be found in plenty in any running stream. Its appearance and

FRESH-WATER SHRIMP.

habits very much resemble the Sandhopper, a little creature that every one must have seen who has walked on a sandy sea-shore. Like the cray-fish, this little creature carries its eggs about until they are hatched. It is a carnivorous animal, and is one of the numerous scavengers of the water, without whose help every stream would soon become putrid and loathsome.

Certain species of crustacea inhabit the land; two of which are well known under the titles of Woodlouse and Armadillo. They belong to the class of crustaceans called "*Isopod*," or equal-footed, because the legs are all of the same nature; whereas in the other crustacean, some legs are used for walking, and others are turned into claws, &c. The woodlouse is to be found in myriads under the scaly bark of trees, under stones,

TADPOLES AND YOUNG FROG.

and, in fact, in almost every crevice. It feeds mostly on decayed vegetable matters, but also eats animal substances, and vegetables that are not decayed. Some gardeners hold the woodlouse in great horror, and say that nothing is so hard or so bitter that a woodlouse will not eat it. If the bark is removed from an ancient willow tree, any number of these creatures may be discovered, in every stage of existence, scuttling about in great fear at the unwelcome light, and sticking close to the wood in hopes that they may not be seen. Dried coats of the woodlouse may be also seen, empty and bleached to an ivory whiteness. They are night-feeders; and, although they can run fast enough if disturbed, walk very deliberately when only employed in feeding.

WOODLOUSE, ARMADILLO, AND PILL MILLEPEDE.

The Armadillo-woodlouse is very curious, and easily recognised from its habit of rolling itself into a round ball when alarmed, just like the quadruped armadillo. Its habits are much the same as those of the common woodlouse. Formerly the armadillo was used in medicine, being swallowed as a pill in its rolled-up state. I have seen a drawer half full of these creatures, all dry and rolled up, ready to be swallowed.

On the preceding cut are two armadillo-like animals, much resembling each other, but belonging to different orders. Fig. *a* is the Woodlouse; *b*, the Pill Millepede, walking; *c*, the same rolled up; *d* is the true Armadillo, walking; and *e*, the same creature rolled up.

CHAPTER V.

A SHORT ESSAY ON LEGS — TAKING A WALK — BRITISH FAKIRS — INSECT LIFE — DEVELOPMENT — THE TIGER MOTH — GROWTH OF THE CATERPILLAR — HOW TO DISSECT INSECTS — PLAN OF CATERPILLAR ANATOMY — SILK ORGANS — ORGANS OF RESPIRATION — SPIRACLES AND THEIR USE — WONDERS OF NATURE — THE CHRYSALIS — SCIENTIFIC LANGUAGE.

As, in common with many other animals, mankind are furnished with legs, and the power to move them, it is universally acknowledged that those limbs ought to be put to their proper use. But while men agree respecting the importance of the members alluded to, they differ greatly in the mode of employing them.

To the tailor, for example, legs are chiefly valuable as cushions, whereon to lay his cloth. For the jockey, the same members form a bifurcated or pronged apparatus, by the help of which he sticks on a horse. The legs of the acrobat are mostly employed to show the extent of ill-treatment to which the hip-joint can be subjected without suffering permanent dislocation. The dancer values his leg solely on account of the "light fantastic toe" which it carries at its extremity. The turner sees that two legs are absolutely necessary to mankind,—*i.e.* one to stand upon, and the other to make a wheel run round. The surgeon views legs—on other people—as objects affording facilities for amputation. The boxer professionally regards his legs as "pins," upon which the striking apparatus is kept off the ground. The soldier's opinion of his legs is modified according to the temperament of the individual, and the position of the enemy. Some people employ their legs in continually mounting the same stairs, and never getting any higher; while others use those limbs in continually pacing the same path and never going any farther.

And of all these modes of employing the legs, the last, which is called "taking a walk," is the dreariest and least excusable.

For, in the preceding cases, the owners of the legs gain their living, or at all events their life, by such employment of those members; and in the case of the interminable stairs, the individual is not acting by his own free will. But it does seem wonderful, that a being possessed of intellectual powers should fancy himself to be the possessor of a right leg and a left one, merely that the right should mechanically pass the left so many thousand times daily and in its turn be passed by the left; while the sentient being above was occupied in exactly the same manner as if both legs were at rest, snugly tucked under a table. .

Sad to relate, such is the general method of taking recreation.

A man who has been over-tasking his brain all the early part of the day, rises corporeally from his work at a certain time, places his hat above his brain, buttons his coat underneath it, and sallies forth to take a walk.

Whatever subject he may be working upon, he takes with him, and on that subject he concentrates his attention. Supposing him to be a mathematician, and that the prevalent idea in his mind is to prove that $\triangle A B C = (\angle D E F + \angle G H I)$. He takes one final look at his Euclid while drawing on his gloves, and sets off with A B C before his eyes.

As he walks along, he sees nothing but A B C, hears nothing but D E F, feels nothing but G H I, and thinks of nothing but the connexion of all three.

An hour has passed away, and he re-enters his room without any very definite recollection of the manner in which he got there. He has mechanically paced to a certain point, mechanically stopped and turned round, mechanically retraced his steps, and mechanically come back again.

He has not the least recollection of anything that happened during his walk; he don't know whether the sky was blue or cloudy, whether there was any wind, nor would he venture to say decidedly whether it was night or day. He *does* recollect seeing a tree on a hill and a spire in a valley, because, together with himself, they formed an angle that illustrated the proportions of the triangle, A B C; but whether the tree had leaves, or not, he could not tell. But he is happy in the consciousness of having performed his duty;—he has taken a walk, he has been for a "constitutional."

O deluded and misguided individual! The walking powers are meant to carry yourself—not only your corporeal body—into other scenes, to give a fresh current to your thoughts, and to give your brain an airing as well as your nose. The mind requires variety in its food, as does the body; and to obtain that change of nutriment is the proper object of taking a walk.

That a rational being can condemn himself to walk three miles along a turnpike road, and three miles back again, at one uniform pace, his eyes directed straight ahead, and his thoughts at home with his books, seems incredible to ordinary personages.

Yet such British fakirs may be seen daily in all weathers, on the roads leading from university towns, going at a rate of four miles per hour, their hats tilted towards the back of their heads, their bodies inclining forward at an angle of 80°, their lips muttering polysyllabic language, and their eyes as beaming as those of a boiled cod-fish.

Now the real use of taking a walk is, to get away from one's self, and to change the current of the thoughts for a while, by changing the locality of the individual.

In order so to do, he should cast his senses abroad, instead of concentrating them all within himself; and from sky, air, water, and earth draw a new succession of images wherewith to relieve the monotony within. There are various modes of attaining this object; and each man will follow that mode which most accords with his own character.

For example, if he is an astronomer, he will look to the heavenly bodies; if a geologist, his eyes will be directed to the earth; if a botanist, his mind seeks employment among the vegetable productions; if a meteorologist, the wind's temperature and atmospheric phenomena will claim his attention; if an entomologist, he will find recreation in watching the phases of insect life, and so on.

It is evident enough that to treat of all these subjects would render necessary a volume that numbered its pages by thousands, and its volumes by at least tens; and therefore, in a work of this nature, it must be sufficient to lay particular stress on one portion, to treat slightly of others, and to leave many entirely untouched. And that portion on which I shall lay the chief stress, is that which is brought more constantly before the eye and ear than any other, namely, the entomological department.

As when approaching cities, the "busy hum of men" is the first indication that meets the ear, so in the country the busy hum of insects is, next to the song of the birds, the sound that gives strongest evidence of a life untrammelled by the artificial rules of society.

Not only do insects make their presence known to the ear, but they also address themselves to the eye. Their forms may be seen flitting through the air, running upon the ground, or making their abode on the various examples of vegetable life. Comparatively small as insects are, they are of vast importance collectively; and there is hardly a leaf of a tree, a blade of grass, or a square inch of ground, where we may not trace the work of some insect. Nearly all strange and curious objects that are noticed by observant eyes in the woods or fields, are caused by the action of insects, and are often the insects themselves, in one or other of the phases of their varied life. Certain examples of insect life, and its effects, will now be given. No particular order will be observed, no long scientific terms will be used, and every creature that is mentioned will be so common that it may be found almost in every field.

The first creature that we will notice is that caterpillar which is so abundantly found at several seasons of the spring and summer, and, from the long hairy

DEVELOPMENT OF INSECTS.

skin in which it is enveloped, goes by the popular name of the "Woolly Bear!" A figure of this creature may be seen in plate B, fig. 5 a. This creature is the larva of the common Tiger-moth, which is represented on the same plate, fig. 5.

It will be necessary to pause here a little, before proceeding to the description and histories of the various insects, because in the course of description certain terms must be used, which must be explained in order to make the description intelligible.

In the first place, let it be laid down as a definite rule, that

INSECTS NEVER GROW.

Many people fancy that a little fly is only little because it is young, and that it will grow up in process of time to be as big as a blue-bottle. Now this idea is entirely wrong; for when an insect has once attained to its winged state, it grows no more. All the growing, and most part of the eating, is done in its previous states of life; and, indeed, there are many insects, such as the silkworm-moth, which do not eat at all from the time that they assume the chrysalis state to the time when they die.

It is a universal rule in nature, that nothing comes to its perfection at once, but has to pass through a series of changes, which if carefully examined can mostly be reduced to three in number. Sometimes these changes glide imperceptibly into each other, but mostly each stage of progress is marked clearly and distinctly. Such is the case with the insect of which we are now considering; and when we have examined the development of the Tiger-moth through its phases of existence, we have the key to the remainder of the insects.

After an insect has left the egg, and entered upon the world as an individual being, it has to pass through three stages, which are called larva, pupa, and imago.

The word "larva," in Latin, signifies "a mask," and this word is used because the insect is at that time "masked," so to speak, under a covering quite different from that which it will finally assume. In the present instance, the Tiger-moth is so effectually masked under the Woolly Bear, that no one who was ignorant of the fact would imagine two creatures so dissimilar to have any connexion with each other.

Throughout this work the word "larva" will be always understood to signify the first of the three states of insect life, whether it be a "caterpillar," a "grub," or a "worm."

In its next stage the insect becomes a "pupa," which word means a "mummy," or a body wrapped in swaddling clothes. This name is employed, because in very many insects the pupa is quite still, is shut up

H

without the power of escape, and looks altogether much like a mummy, wrapped round in folds of cloth. In the moths and butterflies the insect in this stage is called a "chrysalis," or "aurelia," both words having the same import, the first Greek and the other Latin, both derived from a word meaning "gold." Several butterflies—that of the common cabbage butterfly, for example—take a beautiful golden tinge on their pupal garments, and from these individual instances the golden title has been universally bestowed.

The last, and perfected state, is called the "imago," or image, because now each individual is an image and representative of the entire species.

The Woolly Bear, then, is the larva of the Tiger-moth; and if any inquiring reader would like to keep the creature, and watch it through its stages, he will find it an interesting occupation. There is less difficulty than with most insects; for the creature is very hardy, and the plant on which it mostly feeds is exceedingly common.

Generally, the Woolly Bear is found feeding on the common blind nettle; but it may often be detected at some distance from its food, getting over the ground at a great rate, and reminding the spectator of the porcupine. In this case it is usually seeking for a retired spot, whither it resorts for the purpose of passing the helpless period of pupa-hood.

If it is captured on such an occasion, there will be

little trouble in feeding, as it will generally refuse food altogether, and, betaking itself to a quiet corner, prepare for its next stage of existence.

If taken at an earlier period of its life, it feeds greedily on the nettle above-mentioned, and the amount of nutriment which one caterpillar will consume is perfectly astounding. I once had nearly four hundred of them all alive at the same time, and they used to be furnished with nettles by the armful. Of course so large a number is not necessary for ordinary purposes; but this regiment was required for the purpose of watching the development and anatomy of the creature through its entire life.

As the skins of caterpillars are not capable of growth, and the creature itself grows with singular rapidity, it is evident that the skins themselves must be changed, as is the case with many other animals of a higher class, such as the snakes, newts, &c.

For this purpose the skin of the caterpillar splits along the back of the neck, and by degrees the creature emerges, soft, moist, and helpless. A very short time suffices for the hardening of the new envelope; and as the caterpillar has been obliged to fast for a day or two, previously to changing the skin, it sets to work to make up for lost time, and does make up effectually.

In the case of the Woolly Bear, and several others, the cast skin retains nearly the same shape and appearance as when it formed the living envelope of the caterpillar; and, consequently, if any number of these insects are kept, the interior of their habitation soon becomes peopled with these imitation caterpillars. Each individual changes its skin some ten or eleven times, each time leaving behind it a model of its former self, so that caterpillars seem to multiply almost miraculously.

Although even the exterior appearance of an insect is very wonderful, yet its interior anatomy is, if possible, even more wonderful, and, if possible, should be examined. The mode of doing so is simple and easy. If the Woolly Bear, for example, is to be dissected, the easiest mode of doing so is as follows:—

Get a shallow vessel, glass if possible, about an inch or so in depth; load a flat piece of cork with lead, put it at the bottom of the vessel, and fill it nearly to the top with water. Now take the caterpillar, which may be killed by a momentary immersion in boiling water, or by being placed in spirits of wine, and with a few minikin pins fasten it on its back on the cork. The pins of course must only just run through the skin, and two will be sufficient at first, one at each end.

Now take a pair of fine scissors, and carefully slit up the skin the entire length of the creature, draw the skin aside right and left, and pin it down to the cork. The creature will now exhibit portions of organs of

CATERPILLAR ANATOMY.

different shapes and characters, the remainder being concealed under the mass of fat that is collected in the interior. This fat must be carefully removed in order to show the vital organs; and this object is best attained by using a fine needle stuck into a handle. I generally use a common crochet-needle handle, so that needles of various sizes can be used at pleasure.

Now will appear a number of organs closely packed together, and mostly stretching along the entire length of the creature. In order to assist the inquirer, I here present a plan or chart of the interior of the caterpillar when thus opened. It must be understood that the drawing is not meant to represent the particular anatomy of any one species, but to give a general view, by means of which the anatomical details of any caterpillar may be recognised. And

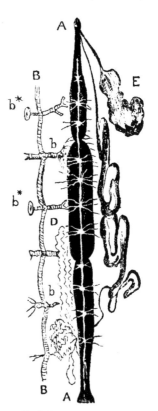

INTERIOR OF CATERPILLAR.

in order to give greater distinctness, only one of each organ is seen, though, with the exception of the intestinal canal, there is a double set of each organ, one on each side.

Running in a straight line from head to tail is seen the digestive apparatus, consisting of throat, stomach, and intestines, with their modifications; and this apparatus is marked A A in the cut.

On the surface of the digestive apparatus, and straight along its centre, lies the nervous system, represented by tiny white threads, dotted at regular distances by rather larger spots of the same substance. If the nerve is examined closely, it will be seen to be composed of *two* very slender threads, lying closely against each other, but easily separable: in which state they are shown. And the little knobs are called "ganglia," each forming a nervous centre, from which smaller nerves radiate to the different portions of the body.

As for brains, the caterpillar dispenses with them almost entirely; and instead of wearing one large brain in the head, is furnished with a row of lesser brains, or ganglia, extending through its whole length. This is the reason why caterpillars are so tenacious of life. If a man loses his head, he dies immediately; but an insect is not nearly so fastidious, and continues to live for a long time without any head at all. Indeed, there are some insects, which. if beheaded, die, not so much on account of the head, but of the stomach: for, having then no mouth, they cannot eat, and so die of hunger. And some insects there are which positively live longer if decapitated than if left in possession of their head.

H

SILK-PRODUCING ORGANS.

On the right hand may be seen a curiously twisted organ, marked c, swelling to a considerable size in the middle, and diminishing to a mere thread at each end. This is one of the vessels that contain the silk, or rather the substance which becomes silk when it is spun.

If this organ be cut open in the middle, it will be seen filled with a gummy substance of curious texture, partly brittle and partly tough. From this substance silk is spun, by passing up the tube, through the thread-like portion, and so at last into a tiny tube, called the spinneret, which opens from the mouth, and wherefrom it issues in a fine thread.

There are two of these silk-making organs, and both unite in the spinneret. Consequently, if silk is examined in the microscope, the double thread can clearly be made out, both threads adhering to each other, but still distinguishable. If the threads lie parallel to each other, the silk is good; if not so, it is of an inferior quality, and liable to snap.

Most caterpillars possess this silk-factory, but some have it much more largely developed than others—the silkworm, for instance. It is of considerable size in the larva which we are examining, because the Woolly Bear has to spin for itself a silken hammock in which to swing while it is in the sleep of its pupal state. Just before it begins to spin, the organ is of very large size, and distended with the liquid silk; but after the hammock is completed, the organ diminishes to a mere thread, and is soon altogether absorbed.

At the left hand of the drawing may be seen a curious structure, marked B B. This is the chief portion of the respiratory system, and may be at once recognised by the ringed structure of the tube. Indeed, it is quite analogous to that of the windpipe in animals.

The mode in which insects breathe differs much from that of the higher animals. In them the breathing apparatus is gathered into one mass, called lungs or gills, as the case may be; but with insects, the respiratory system runs entirely over, round, and through the body, even to the tips of the claws, and the end of the feelers or antennæ.

Every internal organ is also surrounded and enveloped by the breathing tubes; and this often to such an extent, that the dissector is sadly perplexed how to remove the tracheal tubes, as they are called, without injuring the organs to which they so tightly cling. Sometimes they are so strongly bound together, that they may be removed like a net, but mostly each must be taken away separately. The mode in which these tracheal tubes supply the digestive apparatus may be seen at bb; and as there is a double set of them, it may be seen how closely they envelop the organ to which they direct their course.

USE OF THE SPIRACLES.

The ringed structure runs throughout the entire course of the air tubes, and is caused by a thread running spirally between the two membranes of which the tube is composed. The object of this curious thread is to keep the tube always distended, and ready for the passage of air. Otherwise, whenever the insect bends its flexible body, it would cut off the supply of air in every tube which partook of the flexure of the body.

The structure is precisely similar to that of a spiral wire bell-spring; and so strong is the thread, that I have succeeded in unwinding nearly two inches of it from the trachea of a humble-bee.

The air obtains entrance into these tubes, not through the mouth or nostrils, but through a set of oval apertures arranged along the sides of the insect, which apertures are called "spiracles;" and two of them are indicated at $b^* b^*$.

In order to prevent dust, water, or anything but air, from entering, the spiracles are defended by an elaborate *chevaux de frise* of hair, or rather quill, so disposed as to keep out every particle that could injure. So powerful are these defences, that, even under the air-pump, I was unable to force a single particle of mercury through them, though a stick will be entirely permeated by the metal, so that if cut it starts from every pore. I kept the creature in a vacuum for three days, then plunged it under mercury, and let in the air. Even then, no effect was produced, except that the whole of the stomach and intestinal canal were charged with mercury.

But, though the spiracles are such excellent defences against obnoxious substances, they are not capable of throwing off any substance that may choke them. Consequently, nothing is easier than to kill an insect humanely, if one only knows how; and few things more difficult, if one does not know.

For example, if ladies catch a wasp, they proceed to immolate it, by snipping it in two with their scissors; a dreadfully cruel process, for the poor creature has still some four or five brains left intact, and lives for many hours. But if a feather is dipped in oil and swept across the body of the creature, it collapses, turns on its back, and dies straightway. For the oil has stopped up the spiracles, and so the supply of air is cut off from every portion of the body at once. The same rule holds good with all insects.

There is yet one more organ to which I must draw attention, and that is the curious bag-shaped object marked E.

Just as the silk is contained in the vessel C, so the saliva is contained in E, and is developed according to the character and habits of the insect. Some insects require a large supply of that liquid, which is used for various purposes, and others require comparatively little. The caterpillar in which these receptacles may be found best developed is the larva of the Goat-moth, which may be easily found within the substance of decaying trees. Of the Goat-moth we may speak in a future page.

If the reader will again refer to the engraving on p. 101, he will see that between the tracheal tube and the digestive apparatus is a curiously waved line, forming two loops in its upper portion, and running into a confused entanglement below. This entanglement, however, is only apparent, for in nature there is no entangling; all is perfect in order.

This wavy line represents one of the numerous thread-like vessels that surround this portion of the digestive apparatus, and are called the biliary vessels, being, in fact, the insect's liver. There is a large mass of these biliary vessels, and they are found so closely entwined among each other, and so encircled with the air tubes, that to separate them is no easy matter. Their microscopic structure is curious, and will repay a careful examination.

In examining the creature for the first time, the dissector will be tolerably sure to damage the organs and unfit it for preservation, and therefore it is best to take such a course for granted, and to make the best of it.

Removing all these vital organs, he should then

examine the wonderful and most complicated muscular structure, by which the caterpillar is enabled to lengthen, shorten, twist, and bend its body in almost any direction, and that with such power that many caterpillars are enabled to stretch themselves horizontally into the air, and there to keep themselves motionless for hours together.

Few people have any idea of the wonders that they will find inside even so lowly a creature as a caterpillar—wonders, too, that only increase in number and beauty the more closely they are examined. When the outer form has been carefully made out, there yet remains the microscopical view, and after that the chemical, in either of which lie hidden innumerable treasures.

A very forcible and unsophisticated opinion was once expressed to me, after I had dissected and explained the anatomy of a silkworm to an elderly friend. He remained silent for some time, and then uttered disconnected exclamations of astonishment.

I asked him what had so much astonished him.

"Why," said he, "it's that caterpillar. It is a new world to me. I always thought that caterpillars were nothing but skin and squash."

Having now seen something of the exterior and interior of the caterpillar, we will watch it as it prepares for its next state of existence.

THE CHRYSALIS.

Hitherto it has been tolerably active, and if alarmed while feeding, it curls itself round like a hedgehog and falls to the ground, hoping to lie concealed among the foliage, and guarded from the effects of the fall by its hairy armour, which stands out on all sides, and secures it from harm. But a time approaches wherein it will have no defence and no means of escape, so it must find a means of lying quiet and concealed. This object it achieves in the following manner.

It leaves its food, and sets off on its travels to find a retired spot where it may sling its hammock and sleep in peace. Having found a convenient spot, it sets busily to work, and in a very short time spins for itself a kind of silken net, much like a sailor's hammock in shape, and used in the same manner. It is not a very solid piece of work, for the creature can be seen through the meshes; but it is more than sufficiently strong to bear the weight of the inclosed insect, and to guard it from small foes.

On plate B, and fig. 5 *b*, the silken hammock is represented, the form of the pupa inside being visible. It casts off its skin for the last time, and instead of being a hirsute and active caterpillar, becomes a smooth and quiescent chrysalis. In this state it abides for a time that varies according to the time of year and the degree of temperature, and at last bursts its earthy holdings, coming to the light of the sun a perfect insect.

When first the creature becomes a chrysalis, its colour is white, and its surface is bathed in an oily kind of liquid, which soon hardens in the air, and darkens in the light.

On one occasion, I watched a Woolly Bear changing its skin, and, seizing it immediately that the task was accomplished, put it into spirits of wine, intending to keep it for observation.

Next day, the spirit was found to have dissolved away the oily coating, and all the limbs and wings of the future moth were standing boldly out.

Before closing this chapter, I must just remark that the absence of scientific terms throughout the work will be intentional, from a wish to make the subject intelligible, instead of imposing. It would have been easy enough to speak of the Woolly Bear as the larva of Arctia Caja; to describe it as a chilognathiform larva, with a subcylindrical body, and no thoracic shield: passing through an obtected metamorphosis, and becoming a pomeridian lepidopterous imago; and to have proceeded in the same style throughout. But as nearly every one who has taken a country walk has seen Woolly Bears, and hardly any one knows what is meant by "chilognathiform," the subject is treated of for the benefit of the many, even at the risk of incurring the contempt of the few.

CHAPTER VI.

THE PUSS-MOTH—CURIOUS CATERPILLAR—A STRONG FORTRESS—THE BURNET-MOTH—OAK EGGER—HOW TO KILL INSECTS—TWOFOLD LIFE—VICTIMS OF LOVE—ACUTE SENSES—THE STORY OF INSECT LIFE—DRINKER MOTH—CATERPILLAR BOX—EMPEROR MOTH—TYPE OF THE MOUSETRAP.

JUST at the right-hand of the Tiger-moth, on plate B, may be seen a caterpillar of a very strange and eccentric form, and marked by the number 4 *a*. This is the larva or caterpillar of the Puss-moth, and is no less beautiful in colouring than fantastic in form. Its attitude, too, when it is at rest, is quite as curious as its general appearance.

While eating, it sits on the leaves and twigs much as any other caterpillar; but when it ceases to feed, and reposes itself, it grasps the twig firmly with the claspers with which the hinder portion of its body is furnished, and raises the fore-part of its body half upright. In this attitude it much resembles that of the Egyptian Sphinx, and from this circumstance the moth itself is called a Sphinx. An old gardener was once quite put out of temper by seeing several of these caterpillars for the first time, because they had so consequential an air.

The colouring of this creature varies according to

the time of year; but it may be easily recognised by its form alone, which is very peculiar.

One of the most remarkable points in the creature is the forked apparatus at the end of the tail, and which frightens people who do not know the habits of the caterpillar. These forks are black externally, and rather stiff, but are only sheaths for two curious rose-coloured tentacles, which are usually kept hidden, but which may be seen by touching the caterpillar with the point of a needle. When the creature is thus irritated, it will protrude these tentacles from their sheath, and will then strike the part that had been touched.

It is supposed that this apparatus is intended as a kind of whip, wherewith to drive away the ichneumon flies, and other parasites, that inflict such annoyance on many caterpillars.

When this caterpillar proceeds to its pupal state, it makes itself a wonderful fortress—not suspended like that of the Tiger-moth, nor hidden in a dark spot; but it boldly fixes its residence on the exterior of the tree on which it feeds, trusting to its similitude to the bark for concealment, and to the strength of its habitation for safety, even if discovered.

It is furnished with a gummy substance, something after the manner of the silk of the Tiger-moth; but instead of spinning that substance into threads, it uses it in the following manner.

A STRONG FORTRESS.

Biting little chips of wood from the bark of the tree, the caterpillar glues them together with this natural cement; and so builds an arched house for itself, much about the size and shape of half a walnut-shell. So strongly compacted is this residence, that rain and wind have no effect on it, and a penknife does not find an easy entrance.

One or two of these caterpillars which I brought home modified their dwellings in a curious manner. One of them nibbled to pieces a portion of a cardboard box, and so made a kind of papier-maché house; while others, who were placed under a glass-tumbler, and upon a stone surface, simply made their house of the hardened gum. In this state, it appeared as if it had been made of thin horn, and was so transparent that the chrysalis could be seen through the walls.

The caterpillar is common enough, and may be found on the willow or poplar. And a sharp eye will soon learn to detect the winter house, which to an unpractised eye looks as if it were merely a natural excrescence on the bark.

If one of these habitations is found, the best mode of removing it is to avoid touching the dwelling itself, but to cut away the bark round it; and then, by inserting the point of a stout knife, gently raise up the house, together with the bark on which it is placed. This is one of the modes by which an entomologist may find employment even during the winter months, and others will be mentioned in the course of this work.

The moth itself may be seen figured on plate B, fig. 4. It is called the Puss-moth, on account of the soft furry down with which its body is covered, and it is fancifully thought to resemble the fur of the cat.

It is rather a difficult moth to preserve effectually, as it is apt to become "greasy"—that is, to have its whole beauty destroyed by an oiliness that exudes from the body, and gradually creeps even over the wings. The best preservative is to remove the contents of the abdomen, and stuff it with cotton-wool that has been scented with spirits of turpentine. But even that plan is rather precarious, and the delicate, downy plumage is apt to be sadly damaged during the process of stuffing.

Still keeping to the same plate, and referring to the right-hand corner at the top, a moth of strange aspect will be seen; and immediately below it an object that somewhat resembles the hammock of the Tiger-moth, affixed in a perpendicular instead of an horizontal direction. This moth is called the Burnet-moth, and the hammock is the pupa case of the same insect.

The colouring of this moth is very rich and beautiful. The two upper wings are green, and of a tint so deep that, like green velvet, they almost appear to be black. On each of these wings are several red spots, varying in number according to the species; some wearing six spots, and others only five. The two under-wings are of a carmine red, edged with a border of black, in which is a tinge of steely blue. The body is velvety-black, with the same blue tint.

The moth is rather local; but when one is found in a field, hundreds will certainly be near.

At the best of times it is not an active insect, and on a cold or a dull day hundreds of them may be seen clinging to the upright grass stems, from which they can be removed at pleasure.

The caterpillar of this beautiful moth keeps close to the ground, and feeds on grasses, the speedwell, dandelion, and other plants. When it is about to become a pupa, it ascends some slender upright plant, generally a grass stem, and then spins for itself the residence which is represented on the plate.

In this state it may be gathered, and placed under a glass shade; and in the summer months the perfect insect will make its appearance. There are some places which it specially favours, and where it may be found in great profusion. At Hastings, for example, the fields about the cliffs were so populated by these moths, that hardly a grass stem was without its Burnet-moth's habitation.

Feeding on the same plant as the Tiger-moth caterpillar, may often be found another caterpillar of a very different aspect. It is very much larger, and instead of presenting an array of stiff bristles, is covered with thick soft hair of a yellowish-brown colour, diversified with stripes of a deep velvety black, arranged so as to resemble the slashed vestments that were so fashionable some centuries ago.

This caterpillar is the larva of the Oak-Egger moth, and is not so remarkable as a caterpillar as for the house which it builds for its pupal residence.

After changing its skin the requisite number of times, the caterpillar ceases to feed, and, proceeding to some convenient spot (generally a faded thorn-branch), spins its temporary habitation. This cocoon, as it is called, is about an inch in length; and into that narrow space the creature contrives to push, not only itself, but also its last and largest skin.

The substance of the cocoon is hard and rather brittle when dry; and in texture somewhat resembles thin brown cardboard. In its substance, and on its surface, are woven many of the hairs with which the caterpillar is furnished. If the cocoon is carefully opened, the chrysalis will be found within, its head towards the spot where the moth is to emerge, and the cast caterpillar-skin crumpled down by its tail.

In course of time, the chrysalis passes through its development, and the egger-moth itself pushes its way out of the cocoon, with wings and body wet and

HOW TO KILL A BEETLE.

wrinkled, but soon to assume their proper form and strength. The cocoon is shown at plate I, fig. 5 a.

Sometimes the cocoon remains unbroken beyond the proper season; and if it is examined, one or two little holes will generally be found in it. These are signs that the egger has met with an untimely fate, and that it has fallen a victim to those scourges of the insect world, the ichneumon flies. Of these creatures we shall speak in a future page, and therefore omit to describe them here. The moth is shown at fig. 5.

If the moth is intended to be killed, and then placed in a cabinet, the use of sulphur must be avoided. It kills the moth, certainly; but it kills the colours also, and quite ruins its appearance. Sulphur is always a dangerous instrument in insect killing, and should on no account be used. There are many ways of destroying insects humanely, and extinguishing their life as if by a lightning flash; but these modes vary according to the size, sex, and nature of the insect. Some of them I will here mention.

If the insect is a beetle, it may be plunged into boiling water, or into spirits of wine, in which a very little corrosive sublimate has been dissolved. Both modes will destroy the life rapidly, but the former is the better of the two. When walking in the fields or woods, a wide-mouthed, strong bottle, about half full of spirits of wine, is a useful auxiliary, as all kinds of beetles, and even flies and bees, can be put into it; and if dried in a thorough draught, will look as well as before. If this precaution be not taken, all the insects that have long hair, as the humble-bee and others, will lose their good looks, and their hair will be matted together in unseemly elf-locks.

Butterflies, and most of the Diptera, or two-winged flies, can be instantaneously killed by a sharp pinch on the under-surface of the thorax among the legs, as the great mass of nerves is there collected. Many people seem to fancy that the head is the vital part in an insect; and having pinched or run a pin through its head, they think that they have effectually slain the creature, and marvel much to see it lively some twenty-four hours afterwards.

Especially is this the case with the large-bodied moths, whose vitality is quite astonishing. You may even stamp upon them, and yet not crush the life out of that frail casket. If you drive the life out of one-half of the creature, it only seems to take refuge in the other; and then retain a more powerful hold, like a garrison driven into a small redoubt.

It is not at all uncommon to find one of these moths dead and dry as to its wings and limbs, which snap like withered sticks if touched, and yet with so much life in it as to writhe its abdomen if irritated, and to deposit its eggs just as if it were in full activity.

I

HOW TO KILL MOTHS.

Indeed, so strong is this power that the creature seems to be gifted with a double life, one for itself and the other for its progeny. The former is comparatively weak, and but loosely clings to its home; but the latter intrenches itself in every organ, penetrates every fibre, and, until its great work is completed, refuses to be expelled. So, unless the entire mechanism of the insect be killed, the poor creature may live for days in pain.

Fortunately, there is a mode of so doing; and this is the way of doing it:—

Make a strong solution of oxalic acïd, or get a little bottle of prussic acid—it is the better of the two, if you have discretion as beseems a naturalist. Also make a bone or iron instrument, something like a pen, but without a nib. Dip this instrument into the poison as you would a pen, and then you have a weapon as deadly as the cobra's tooth, and infinitely more rapid in its work. Now hold your moth delicately as entomologists hold moths, near the root of the wings. Keep the creature from fluttering; plunge the instrument smartly into the thorax, between the insertion of the first and second pair of legs; withdraw it as smartly, and the effect will be instantaneous. The moth will stretch out all its legs to their full extent; there will be a slight quiver of the extremities; they will be gently folded over each other; and you lay your dead moth on the table.

The reason of this rapid decease is of a twofold nature.

In the first place, the chief nerve mass is cut asunder, and even thus a large portion of the life is destroyed. But the chief breathing tubes are also severed, and a drop of poison deposited at their severed portions. Consequently, at the next inspiration, either the poison itself or its subtle atmosphere rushes to every part, and to every joint of the insect, thus carrying death through its whole substance.

The male insect is very different in appearance to the female, and in general is hardly more than two-thirds of her size. The colours, too, are very different; for in the male insect the wings are partially of a dark chesnut brown, with a light band running round them, as may be seen in the engraving; while in the female the wings are almost entirely of a uniform yellowish brown.

The antennæ, too, of the male are deeply cleft, like the teeth of a comb; while those of the female are narrow, and comparatively slightly toothed.

As is the case with several other moths, the male oak eggers are sad victims to the tender passion, and fall in love not only at first sight, but long before they see the object of their affection at all.

If a female egger is caught immediately after her entrance into the regions of air, and placed in a perforated box near an open window, her unseen charms will be so powerfully felt by gentlemen of her own race that they will flock to the casket that contains their desired treasure, and fearlessly run about it, fluttering their wings, and striving to gain admission. So entirely do they abandon themselves to the captivity of love, that they do not fear the risk of a bodily captivity, and will suffer themselves to be taken by hand, without even an endeavour to escape.

Carry the imprisoned moth into the fields, and even there the eager suitors will arrive from all quarters, and boldly alight on the box while in the hand of the entomologist.

Most wonderful must be the influence that can emanate from so small a creature, and extend to so great a distance—an influence which, although entirely inappreciable by any human sense, exercises so potent a sway on all sides, and to so great a distance.

The conditions, too, of this mysterious influence are singularly delicate; for after the moth has once found her mate, she may be placed amid a crowd of gentlemen, and not one will take the least notice of her.

Like the young beauty of the ball-room, who whilom attracted to herself crowds of beaux, that fluttered around her, and contended with each other for a look or a smile of their temporary divinity, but who finds herself deserted by the fickle crowd when her election is made; so our Lady Lasiocampa Quercus, after setting all hearts ablaze for a time, makes happy one favoured individual, is deserted by the many rejected, and left in quiet to the duties of a wife and a mother.

Her married life is but short, for her husband rarely survives his happiness more than a few hours, and she, after making due preparation for the welfare of her numerous family, whom she is never to see, feels that she has fulfilled her destiny, and gives up a life which has now no further object.

There is really something very human in the life even of an insect. Many a life story have I watched in the insect world, which, if transferred to the human world, would be full of interest. There is also one great advantage in the insect life, namely, that as it only consists of a year or two, the events of several successive generations come under the observation of a single historian.

First, a number of tiny, purposeless beings come into the world, spreading about much at random, and seeming to have no other object except to eat. It is but just to them to say that they don't cry, and are always contented with the food that is given them.

They rapidly increase in size, pass through a regular series of childish complaints, which we mass together under this single term, "moulting," but which are probably to their senses as distinct as measles, and chicken-pox, and hooping-cough.

They outgrow a great many suits of clothes in a wonderfully short period; they retire for a time to finish their education; and then come before the world in all the glory of their new attire.

Up to this time they are nearly exactly alike in habits and manners; but, when freed from the trammels that held them, they diverge, each in his appointed way, each exulting for a short space in the buoyancy of youth, and fluttering indeterminately in the new world, but soon settling down to the business for which they were made.

So even in insects a human soul can find a companionship, and a solitary man need never feel entirely alone as long as he can watch the life of a humble moth, and see in that despised creature some manifestations of the same feelings which actuate himself.

And it even seems that, through this companionship, the higher nature communicates itself in some degree to the lower, as is shown by the many instances of men who have tamed spiders and other creatures quite as far removed from the human nature. In such a case it seems very clear that either the higher nature gives to the lower an intelligence not its own, or that it develops powers which would have lain dormant had they not been called forth by the contact of a superior being.

DRINKER MOTH.

This subject is a very wide one, and well worth following up. But as it runs through the whole creation, and this book is only to consist of a few pages, it must suffice merely to put forth the idea.

To pass to another insect.

On plate E, and fig. 1 and 1 *a*, may be seen an insect which somewhat resembles the oak-egger moth, and is often mistaken for it by inexperienced eyes. This is the "Drinker" moth, remarkable for the thick furry coat which it wears, as a caterpillar and as a moth, and which it employs in the construction of its cocoon. This moth is one of my particular friends; and I have had hundreds of them from the egg to their perfect state. I had quite a large establishment for the education and development of lepidoptera, and especially favoured the tiger-moth, the oak egger, and the drinker.

The caterpillar of this moth is entirely covered with dense hair, even down to the very feet; and by means of this protection it is enabled to brave the winter frost, needing not to pass the cold months in a torpid state. It is a pretty caterpillar, and very easily recognised by the figure. Its chief peculiarities are the two tufts of hair that it bears at its opposite extremities, and the double line of black spots along its sides.

Generally, it feeds on various grasses, but it is not dainty, as are many caterpillars; and I have always found it to eat freely of the same food as the oak egger larva. This caterpillar is seen at fig. 1 *b*.

When alarmed, it loosens its hold of the plant on which it is feeding, rolls itself into a ring, and drops to the ground, hoping to evade notice among the foliage. This habit used to be rather perplexing to me, not because the creature could escape by so well-known a trick, but because it would not go into the box prepared for its reception.

It is necessary to have a box of a peculiar form for the collection of caterpillars. If the lid is raised every time that a fresh capture is made, difficulties increase in proportion to the number of caterpillars. For, when some thirty larvæ are in the box, they all begin to crawl out when the lid is opened; and Hercules had hardly a more bewildering task among the hydra's heads than the entomologist among his captives.

No sooner is the light admitted, than a dozen heads are over the side; and as fast as one is replaced, six or seven more make their appearance. The only remedy is to sweep them all back with a rapid movement of the hand, to shake them all to the bottom, and then to replace the lid as fast as possible. Even with all precaution, caterpillars are crushed; and, besides, they are delicate in their constitutions, and require gentle handling.

So the best plan is to have a tin box made with a short tube, through which the caterpillars can be introduced, and which can be stopped by a cork when the creatures are fairly inside.

Now, although this is a capital contrivance for caterpillars that hold themselves straight, it fails entirely when they curl themselves into a ring and refuse to be straightened. It is as impossible to straighten a rolled-up hedgehog as a caterpillar in a similar attitude; and if force is used in either case, the creature will be mortally injured. However, gentle means succeed when violence fails, with insects as with men. A Bheel robber will steal the bedding from under a sleeping man without waking him; and, by an analogous process, the refractory caterpillar is lodged in his prison before he is fairly awake to his condition.

The entomologist feels a justifiable pride in executing similar achievements; for there is quite as much force of intellect needed to outwit a caterpillar as a quadruped.

COCOON OF THE EMPEROR MOTH.

When the Drinker Caterpillar passes into its pupal state, it makes for itself a very curious cocoon, not unlike a weaver's shuttle in shape, being large in the middle, and tapering to a point at each end. The texture is soft and flexible, as if the cocoon were made of very thin felt, and the larval hairs are quite distinguishable on its surface. The moth leaves the cocoon about August. For the cocoon, see fig. 1 c.

I found that few caterpillars are so liable to the attack of ichneumon flies as those of the Drinker moth. A cocoon now before me is pierced with thirteen holes from which ichneumon flies have issued, having eaten up the caterpillar. The eggs are shown at fig. 1 e.

If the reader will now refer to plate C, the central figure will be found to represent a strikingly handsome moth, called, from its gorgeous plumage, the "Emperor Moth."

Its body is covered with a thick downy raiment, and the wings are clothed with plumage of a peculiarly soft character, which is well represented in the figure. The antennæ, too, are elaborately feathered.

Although the beauty of this insect would entitle it to notice in its perfect state, and the peculiar shape of its larva—(see plate C, fig. 4 a)—would draw attention, yet its chief title to admiration lies in the cocoon which it constructs for its pupal existence.

Externally, there is nothing remarkable in the cocoon; and, as may be seen in the same plate, fig. 4 b, it is a very ordinary, rough, flask-shaped piece of workmanship. But if the outer covering be carefully removed, or if the cocoon be divided lengthways, a very wonderful structure is exhibited.

The inventor of lobster-pots is not known, and

history has failed to record the name of the man who first made wire mouse-traps with conical entrances, into which the mice can squeeze themselves, but exit from which is impossible.

But, though the principle had not been applied to lobsters or mice, it was in existence ages upon ages ago. Before human emperors had been invented, and very probably long before mankind had been placed on our earth, the caterpillar of the Emperor Moth wove its wondrous cell, and thereby became a silent teacher to the cunning race of mankind how to make mouse-traps and lobster-pots.

For inside the rough outer case, which is composed of silken threads, woven almost at random, and very delicate, is a lesser case, corresponding in shape with its covering, but made of stiff threads laid nearly parallel to each other, their points converging at the small end of the case. See the cut on page 126.

It will now be seen that the moth when it leaves its chrysalid case can easily walk out of the cocoon, but that no other creature could enter. So within its trapped case the chrysalis lies secure, until time and warmth bring it to its perfection. It breaks from its pupal shell, walks forward, the threads separate to permit its egress, and then converge again so closely that to all appearance the cocoon is precisely the same as when the moth was within.

Now, any observant member of the human race, who had been meditating upon traps, and happened in a contemplative mood to open one of these cocoons, would feel a new light break in upon him, and, Archimedes-like, he would exclaim "Eureka," or its equivalent, " I have found my trap!" Reverse the process, make the converging threads to lead into instead of out of the trap, and the thing is done. " I will make it of wire, put it on my shelf, and I catch mice and rats. I will make it of osier, sink it to the bottom of the sea, and I catch lobsters and crabs. I will lay it in a rapid, and I catch roach and dace ; I will place it under the river banks, and then I have cray-fish."

So might he soliloquize on the future achievements of his newly-discovered principle. But unless he had the prophetic afflatus strong within him, never would he imagine that in future times his discovery would catch a monarch, and an Elector to boot.

CHAPTER VII.

ELEPHANT HAWK-MOTH — PRIVET HAWK-MOTH — DIGGING FOR LARVÆ—BUFF-TIPP MOTH—GOLD-TAILED MOTH—CASE FOR ITS EGGS—CURIOUS PROPERTY OF ITS CATERPILLAR—VAPOURER MOTH — LEAF-ROLLERS — GREEN OAK MOTH — ITS CONSTANT ENEMY—LEAF MINERS—LACKEY MOTH—EGG BRACELETS.

It will be noticed that the insects mentioned in the preceding chapter are mostly remarkable for the cocoons which they construct, and that the peculiarities of the larva and the perfect insect are but casually mentioned. Those, however, which will be noticed in this chapter are chosen because there is "something rare and strange" in the habits and manners of the creatures themselves.

As it will be more convenient to keep to the same plate as much as possible, we still refer to plate C. On casting the eye over the objects there depicted, the strangest and most fantastic shape is evidently that creature which is marked 5 *a*.

The aspect of the creature is almost appalling, and it seems to glare at us with two malignant eyes, threatening the poisoned blow which the horrid tail seems well able to deliver.

PRIVET HAWK-MOTH.

Yet this is as harmless a creature as lives, and it can injure nothing except the leaves of the plant on which it feeds. The eye-like spots are not eyes at all, but simply markings on the surface of the skin, and the formidable horn at the tail cannot scratch the most delicate skin.

The creature is in fact simply the caterpillar of a very beautiful moth, represented in fig. 5, and called the Elephant Hawk-moth—elephant, on account of its long proboscis, and hawk on account of its sharp hawk-like wings and flight. The caterpillar may be found in many places, and especially on the banks of streams, feeding on various plants, such as the willow-herbs.

Another kind of Hawk-moth is much more common than the elephant, and is represented on plate A; the moth itself at fig. 5, and its caterpillar at fig. 5, a.

This is called the Privet Hawk-moth, because the caterpillar feeds on the leaves of that shrub. The colours of both moth and caterpillar are very beautiful, and not unlike in character.

The bright leafy green tint of the caterpillar, and the seven rose-coloured stripes on each side, make it a very conspicuous insect, and raise wishes that tints so beautiful could be preserved. But as yet it cannot be done, for even in the most successful specimens the colours fade sadly in a day or two, and after a while there is a determination towards a blackish brown tint that cannot be checked.

Any one, however, who wishes to try the experiment may easily do so, for there are few privet hedges without their inhabitants, who may keep out of sight, but can be brought tumbling to the ground by some sharp taps administered to the stems of the bushes.

In the winter the chrysalis may be obtained by digging under privet bushes. There the caterpillar resorts, and works a kind of cell in the ground for its reception. It is better not to choose a frosty day for the disinterment, or the sudden cold may kill the insect, and the seeker's labour be lost.

Should it be desirable to capture the larva and to keep it alive, the object can be easily attained; for the creature is hardy enough, and privet bushes grow everywhere. In default of privet leaves, it will eat those of the syringa and the ash. When it reaches its full growth, it should be provided with a vessel containing earth some inches in depth. Into this earth it will burrow, and remain there until the moth issues forth.

Care should be taken to keep the earth rather moist, as otherwise the chrysalis skin becomes so hard that the moth cannot break out of its prison, and perishes miserably.

On the same plate, fig. 4, may be seen a moth of a

curious shape, very feathery about the thorax, the head being all but concealed by the dense down, and as difficult to find as the head of a Skye-terrier, were not its position marked by the antennæ. This is the Buff-tip moth, so called on account of the upper wing-tips being marked with buff-coloured scales.

The caterpillar, which is represented immediately above, and marked 4 a, is a very singular creature, its habits being indicated by the marks on its skin. As soon as the young caterpillars are hatched, they arrange themselves in regular order, much after the fashion of the dark stripes, and so march over leaf and branch, devastating their course with the same ease and regularity as an invading army in an enemy's land.

When they increase to a tolerably large size, they disband their forces, and each individual proceeds on its own course of destruction. Were it not for the colours which they assume, these creatures would do great damage; but the ground being yellow and the stripes black, the caterpillars are so conspicuous that sharp-sighted birds soon find them out, and having discovered a colony, hold revelry thereon, and exterminate the band.

Comparatively few escape their foes and attain maturity. When they have reached their full age, they let themselves drop from the branches, and when they come to soft ground, bury themselves therein **to** await their last change. Individuals may often be seen crossing gravel paths, which they are unable to penetrate, and getting over the ground with such speed and in so evident a hurry that they seem to be aware that birds are on the watch and ichneumons awaiting their opportunity.

There is a very pretty moth covered with a downy white plumage even to the very toes, and carrying at the extremity of its tail a tuft of golden silky hair. From this coloured tuft, the creature bears the name of Gold-tailed Moth. It may often be found sticking tightly to the bark of tree stems, its glossy white wings folded roof-like over its back, and the golden tuft just showing itself from the white wings.

This golden tuft is only found fully developed in the female moth, and comes into use when she deposits her eggs. The moth is shown on plate E, fig. 4.

As the eggs are laid in the summer time, they need no guard from cold; but they do require to be sheltered from too high a degree of temperature, and for this purpose the silken tuft is used.

At the very end of the tail the moth carries a pair of pincers, which she can twist about in all directions; and this tool is used for the proper settlement of the eggs. The moth, after fixing on a proper spot, pinches off a tiny tuft of down, spreads it smoothly, lays an egg upon it, covers it over, and finally combs the hair so as

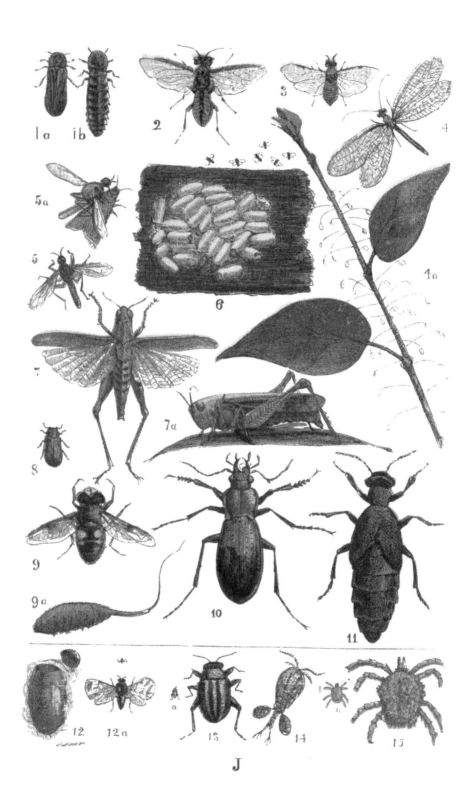

to lie evenly. And when she has laid the full complement, she gives the whole mass some finishing touches, like a mother tucking-in her little baby in the bed-clothes, and smoothing them neatly over it.

The egg masses are common enough, and are readily discovered by means of their bright yellow covering.

The caterpillar of this moth is a very brilliant scarlet and black creature, commonly known by the name of the "palmer-worm," and to be found plentifully of all sizes.

People possessed of delicate skins must beware of touching the palmer-worm, or they may suffer for their temerity. I was a victim to the creature for some time before I discovered the reason of my sufferings. And the case was as follows.

Being much struck with the vivid colours of the caterpillar, I was anxious to preserve some specimens if possible, in a manner that would retain the scarlet and black tints. One mode that seemed feasible was, to make a very small snuff-box, as ladies call a rectangular rent, in the creature's skin, to remove the entire vital organs, to fill the space with dry sand, and then, when the skin was quite dry, to pour out all the sand, leaving the empty skin.

After treating six or seven caterpillars in this fashion, I perceived a violent irritation about my face, lips, and eyes, which only became worse when rubbed. In an hour or so, my face was swollen into a very horrid and withal a very absurd mass of hard knobs, as if a number of young kidney potatoes had been inserted under the skin.

Of course, I was invisible for some days, and after returning to my work, was attacked in precisely the same manner again. This second mischance set me thinking; and on consultation with the medical department, the fault was attributed to the hot sand which I had been using.

So, when I went again to the work, I discarded sand, and stuffed the caterpillars with cotton wool cut very short, like chopped straw. My horror may be conjectured, but not imagined, when I found, for the third time, that my face was beginning to assume its tubercular aspect.

Then I did what I ought to have done before, went to my entomological books, and found that various caterpillars possessed this "urticating" property, as they learnedly called it, or as I should say, that they stung worse than nettles. Since that time, I have never touched a palmer-worm with my fingers.

It was perhaps a proper punishment for neglecting the knowledge that others had recorded. But I always had rather an aversion to book entomology, and used to work out an insect as far as possible, and *then* see what books said about it. Certainly, although not a very rapid mode of work, yet it was a very sure one, and fixed the knowledge in the mind.

On the same plate, fig. 4 *a*, is shown the caterpillar of this moth, a creature conspicuous from the tufts of beautifully-coloured hair which are set on its body like camel-hair brushes.

The caterpillar spins for itself a silken nest wherein to pass its pupa state, and in general there is nothing remarkable about the nest. But I have one in my collection of insect habitations, that is very curious.

I had caught, killed, and pinned out a large dragon-fly, and placed it in a cardboard box for a time. Some days afterwards, a palmer-worm had been captured, and was imprisoned in the same box. I was not aware that such a circumstance had happened, and so did not open the box for a week or two, when I expected to find the dragon-fly quite dry and ready for the cabinet.

When, however, the box was opened, a curious state of matters was disclosed. The caterpillar had not only spun its cocoon, but had shredded up the dragon-fly's wings, and woven them into the substance of its cell. The glittering particles of the wing have a curious effect as they sparkle among the silver fibres.

On plate D, fig. 3 *a*, is represented a creature whose sole claim to admiration is its domestic virtue, for elegance or beauty it has none. It hardly seems possible, but it is the fact, that this clumsy creature is the female Vapourer Moth, the male being represented immediately below fig. 3.

Why the two sexes should be so entirely different in aspect, it is not easy to understand. The female has only the smallest imaginable apologies for wings, and during her whole lifetime never leaves her home, seeming to despise earth as she cannot attain air.

This moth is not obliged to form laboriously a warm habitation for her eggs, for she places them in the silken web which she occupied in her pupal state, and from which she never travels.

Curiously enough, her eggs are not placed within the hollow of the cocoon as might be supposed, but are scattered irregularly and apparently at random over its surface. Even there, though, they are warm enough; for the cocoon itself is generally placed in a sheltered spot, so that the eggs are guarded from the undue influence of the elements, and at the same time protected from too rapid changes of temperature.

In the hot summer months, the leaves of trees are crowded with insects of various kinds, which fly out in alarm when the branches are sharply struck. Oak trees are especially insect-haunted, and mostly by one species of moth, a figure of which is given on plate B, fig. 1.

This little moth is a pretty object to the eyes, but a terribly destructive creature when in its caterpillar state, compensating for its diminutive size by its collective numbers. The caterpillar is one of those called "Leaf Rollers," because they roll up the leaves on which they feed, and take up their habitation within.

There are many kinds of leaf rollers, each employing a different mode of rolling the leaf, but in all cases the leaf is held in position by the silken threads spun by the caterpillar.

Some use three or four leaves to make one habitation, by binding them together by their edges. Some take a single leaf, and, fastening silken cords to its edges, gradually contract them, until the edges are brought together and there held. Some, not so ambitious in their tastes, content themselves with a portion of a leaf, snipping out the parts that they require and rolling it round.

The insect before us, however, requires an entire leaf for its habitation, and there lies in tolerable security from enemies. There are plenty of birds about the trees, and they know well enough that within the circled leaves little caterpillars reside. But they do not find that they can always make a meal on the caterpillars, and for the following reason.

The curled leaf is like a tube open at both ends, the caterpillar lying snugly in the interior. So, when a bird puts its beak into one end of the tube, the caterpillar tumbles out at the other, and lets itself drop to the distance of some feet, supporting itself by a silken thread that it spins.

The bird finds that its prey has escaped, and not having sufficient inductive reason to trace the silken thread and so find the caterpillar, goes off to try its fortune elsewhere. The danger being over, the caterpillar ascends its silken ladder, and quietly regains possession of its home.

Myriads of these rolled leaves may be found on the oak trees, and the caterpillars may be driven out in numbers by a sharp jar given to a branch. It is quite amusing to see the simultaneous descent of some hundred caterpillars, each swaying in the breeze at the end of the line, and occasionally dropping another foot or so, as if dissatisfied with its position.

Each caterpillar consumes about three or four leaves in the whole of its existence, and literally eats itself out of house and home. But when it has eaten one house, it only has to walk a few steps to find the materials of another, and in a very short time it is newly lodged and boarded.

The perfect insect is called the "Green Oak Moth." The colour of its two upper wings is a bright apple green; and as the creature generally sits with its wings closed over its back, it harmonizes so perfectly with the green oak leaves, that even an accustomed eye fails to perceive it. So numerous are these little moths, that

THE EMPIS.

their progeny would shortly devastate a forest, were they not subject to the attacks of another insect. This insect is a little fly of a shape something resembling that of a large gnat; and which has, as far as I know, no English name. Its scientific title is Empis. There are several species of this useful fly, one attaining some size; but the one that claims our notice just at present is the little empis, scientifically Empis tessellata.

I well remember how much I was struck with the discovery that the empis preyed on the little oak moths, and the manner in which they did so.

One summer's day, I was entomologizing in a wood, when a curious kind of insect caught my attention. I could make nothing of it, for it was partly green, like a butterfly or moth, and partly glittering like a fly, and had passed out of reach before it could be approached. On walking to the spot whence it had come, I found many of the same creatures flying about, and apparently enjoying themselves very much.

A sweep of the net captured four or five; and then was disclosed the secret. The compound creature was, in fact, a living empis, clasping in its arms the body of an oak moth which it had killed, and into whose body its long beak was driven. I might have caught hundreds if it had been desirable. The grasp of the fly was wonderful, and if the creature had been magnified to the human size, it would have afforded the very type of a remorseless, deadly, unyielding gripe. Never did miser tighter grasp a golden coin, than the empis fasten its hold on its green prey. Never did usurer suck his client more thoroughly than the empis drains the life juices from the victim moth.

He is a terrible fellow, this empis, quiet and insignificant in aspect, with a sober brown coat, slim and genteel legs, and just a modest little tuft on the top of his head. But, woe is me for the gay and very green insect that flies within reach of this estimable individual.

The great hornet that comes rushing by is not half so dangerous, for all his sharp teeth and his terrible sting. The stag-beetle may frighten our green young friend out of his senses by his truculent aspect and gigantic stature. But better a thousand stag-beetles than one little empis. For when once the slim and genteel legs have come on the track of the little moth, it is all over with him. Claw after claw is hooked on him, gradually and surely the clasp tightens, and when once he is hopelessly captured, out comes a horrid long bill, and drains him dry. Poor green little moth!

Still continuing our research among the oak leaves, we shall find many of them marked in a very peculiar manner. A white wavy line meanders about the leaf like the course of a river, and, even as the river, increases in width as it proceeds on its course. This effect is produced by the caterpillar of one of the leaf-mining insects, tiny creatures, which live between the layers of the leaf, and eat their way about it.

Of course, the larger the creature becomes, the more food it eats, the more space it occupies, and the wider is its road; so that, although at its commencement the path is no wider than a needle-scratch, it becomes nearly the fifth of an inch wide at its termination. It is easy to trace the insect, and to find it at the widest extremity of its path, either as caterpillar or chrysalis. Often, though, the creature has escaped, and the empty case is the only relic of its being.

There are many insects which are leaf-miners in their larval state. Very many of them belong to the minutest known examples of the moth tribes, the very humming-bird of the moths, and, like the humming-birds, resplendent in colours beyond description. These Micro-Lepidoptera, as they are called, are so numerous, that the study of them and their habits has become quite a distinct branch of insect lore.

Some, again, are the larvæ of certain flies, while others are the larvæ of small beetles. Their tastes, too, are very comprehensive, for there are few indigenous plants whose leaves show no sign of the miner's track, and even in the leaves of many imported plants the meandering path may be seen.

There are some plants, such as the eglantine, the dewberry, and others, that are especially the haunts of these insects, and on whose branches nearly every other leaf is marked with the winding path. I have now before me a little branch containing seven leaves, and six of them have been tunnelled, while one leaf has been occupied by two insects, each keeping to his own side.

The course which these creatures pursue is very curious. Sometimes, as in the figure on plate A, fig. 1, the caterpillar makes a decided and bold track, keeping mostly to the central portion of the leaf.

Sometimes it makes a confused tortuous jumble of paths, so that it is not easy to discover any definite course.

Sometimes it prefers the edges of the leaves, and skirts them with strange exactness, adapting its course to every notch, and following the outline as if it were tracing a plan.

This propensity seems to exhibit itself most strongly in the deeply cut leaves. And the shape or direction of the path seems to be as property belonging to this species of the insect which makes it; for there may be tracks of totally distinct forms, and yet the insects producing them are found to belong to the same species.

If the twigs of an ordinary thorn-bush be examined during the winter months, many of them will be seen surrounded with curious little objects, called "fairy bracelets" by the vulgar, and by the learned "ova of Clisiocampa Neustria." These are the eggs of the Lackey Moth, and are fastened round the twigs by the mother insect, a brown-coloured moth, that may be found in any number at the right time.

It is wonderful how the shape of the egg is adapted to the peculiar form into which they have to be moulded, and how perfectly they all fit together. Each egg is much wider at the top than at the bottom; and this increase of width is so accurately proportioned, that when the eggs are fitted together round a branch, the circle described by their upper surfaces corresponds precisely with that of the branch.

These eggs are left exposed to every change of the elements, and are frequently actually enveloped in a coat of ice when a frost suddenly succeeds a thaw. But they are guarded from actual contact with ice and snow by a coating of varnish which is laid over them, and which performs the double office of acting as a waterproof garment and of gluing the eggs firmly together. So tightly do they adhere to each other, that if the twig be cut off close to the bracelet, the little egg circlet can be slipped off entire.

CHAPTER VIII.

LAPPET MOTH—BRIMSTONE MOTH—ITS CATERPILLAR—CURRANT MOTH — CLEAR-WINGS — WHITE-PLUME MOTH — TWENTY-PLUME MOTH—ADELA—AN INSECT CINDERELLA—NAMING INSECTS—THE ATALANTA—AN INSECT CRIPPLE—PEACOCK BUTTERFLY—BLUE AND OTHER BUTTERFLIES.

The cut on the next page is a good representation of a very singular creature called the "Lappet Moth." As may be seen by the engraving, when it is settled quietly upon a leaf with folded wings, it bears a closer resemblance to a bundle of withered leaves than to any living creature. In this strange form lies its chief safety, for there are few eyes sufficiently sharp to detect an insect while hiding its character under so strange a mask.

There are several other examples of this curious resemblance between the animal and vegetable kingdoms, one or two of which will be mentioned in succeeding pages.

The name of "Lappet Moth" is hardly applicable, as it ought rather to be called the moth of the Lappet caterpillar. This title is given to the creature because it is furnished with a series of fleshy protuberances along

LAPPET MOTH.

the sides, to which objects the name of Lappets has been fancifully given.

LAPPET MOTH.

It is generally supposed to be a rare moth; but I have not found much difficulty in procuring specimens either in the larval state or as moths. Both moth and caterpillar are of a large size, the caterpillar being about the length and thickness of a man's finger. Its colour is a tolerably dark grey, but subject to some variation in tint. There is no difficulty in ascertaining this species of the creature, as it is clearly distinguished from caterpillars of a similar shape or line by two blue marks on the back of its neck, as if a fine brush filled with blue paint had been twice drawn smartly across it. The curious "lappets" too are so conspicuous that they alone would be sufficient for identification.

One of the examples of animal life simulating vegetation now comes before us in the person of the Brimstone Moth, or rather its caterpillar.

This is a very common insect, and may be recognised at once by its portrait on plate C, fig. 3.

The caterpillar is represented immediately above, fig. 3 a. This is one of the caterpillars called "Loopers," on account of their peculiar mode of walking.

They have no legs on the middle portion of their bodies, but only the usual six little legs at the three rings nearest the head, and a few false legs by the tail; so when they want to walk, they attain their object by holding fast with their false or pro-legs as they are called, and stretching themselves forward to their fullest extent. The real legs then take their hold, and the pro-legs are drawn up to them, thus making the creature put up its back like an angry cat.

The grasp of the pro-legs is wonderfully powerful, and in them lies the chief peculiarity of the creature. The surface of the body is of a brownish tint, just resembling that of the little twigs on which it sits; there are rings and lines on its surface that simulate the cracks and irregularities of the bark, and in one or two places it is furnished with sham thorns.

Trusting in its mask, the caterpillar grasps the twig firmly, stretches out its body to its full length, and so remains, rigid and immovable as the twigs themselves. People have been known to frighten themselves very much by taking hold of a caterpillar, thinking it to be a dead branch.

The only precaution taken by the creature is to have a thread ready spun from its mouth to the branch, so that if it should be discovered, it might drop down suddenly, and when the danger was over, climb up its rope and regain its home.

The commonest of the loopers is the well-known caterpillar of the Currant or Magpie Moth, plate E, fig. 3. This creature is remarkable from the circumstance that its colours are of the same character throughout its entire existence; the caterpillar, chrysalis, and perfect moth showing a similar rich colour and variety of tint, as seen on figs. 3 *a* and 3 *b*.

It is a curious fact that almost every stratagem of animals is used by man; whether intuitively, or whether on account of taking a hint, I cannot say.

For example, Parkyns, the Abyssinian traveller, tells an amusing tale of a party of Barea robbers, who when pursued got up a *tableau vivant* at a moment's notice.

One man personated a charred tree-stump, and the others converted themselves into blackened logs and stones lying about its base.

It seemed so impossible for human beings to remain so still, that a rifle-ball was sent towards the stump, and caused it to take to its heels, followed by the logs and stones.

I have heard of a similar stratagem that was put in force by a robber who was interrupted on his way into the tent by the appearance of its inmate, an officer. He was so completely deceived, that he actually hung his helmet on one of the branches, which branch was in fact the robber's leg. The joke was almost too good, but the stump stood fast, until the officer leaned his back against it. Officer and stump came to the ground together, and the stump escaped, carrying off the helmet as a trophy. I think that he deserved it.

I conclude this chapter with a short notice of five beautiful and curious little moths.

The first of these, the "Currant Clear-wing," is frequently mistaken for a gnat or a fly, and it is sometimes a difficult task to persuade those who are unaccustomed to insects, that it can be a moth. As a general rule, the wings of moths are covered with feathers, and many are even as downy in their texture as the plumage of the owl. But there is a family of moth, called the clear-wings, whose wings are as trans-

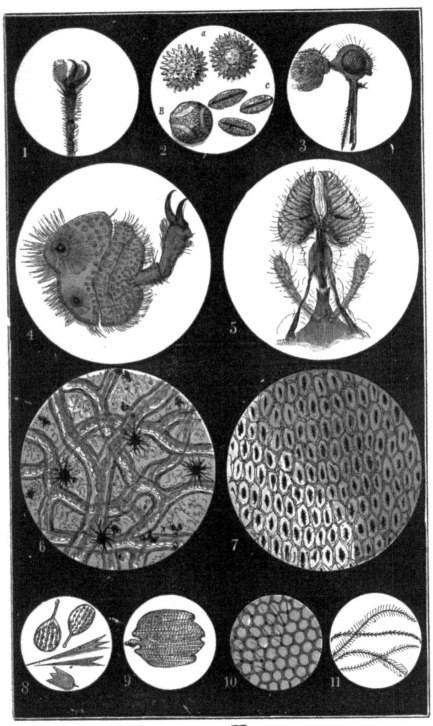

K

parent as those of bees or flies. Some of these are as large as hornets, and resemble these insects closely in general aspect.

Some fourteen or fifteen species of these curious creatures are found in England; and each of them bears so close a likeness to some other insect, that it is named accordingly. For example, the species which we are now examining is called the "gnat-like Egeria," another is the "bee-like," another the "hornet-like," another the "ant-like," and so on. Plate A, fig. 3.

The currant clear-wing may be found on the leaves of currant bushes, where it loves to rest. In 1856 I took a great number of them in one small garden, often finding two or more specimens on one currant bush.

Next come two beautiful examples of the Plume Moth, the White Plume and the Twenty Plume.

The first of these insects is very common on hedges or the skirts of copses, and comes out just about dusk, when it may be easily captured, its white wings making it very conspicuous. See Plate H, fig. 9.

The chief distinguishing point in the plume moth is that the wings are deeply cut from the point almost to the very base, and thus more resemble the wings of birds than those of insects.

In the white plume there are five of these rays or plumes, three belonging to the upper pair of wings and two to the lower.

From the peculiarly long and delicate down with which the body and wings are covered, it is no easy matter to secure the moth without damaging its aspect. The scissors-net is, perhaps, the best that can be used for their capture; for, as they always sit on leaves and grass with their wings extended, they are inclosed at once in a proper position, and cannot struggle. A sharp pinch in the thorax from the forceps, which a collector ought always to have with him, kills the creature instantly; for it holds life on very slender tenure. The slender entomological pin can then be passed through the thorax, while the net is still closed, and thus the head of the pin can be drawn through the meshes of the net when it is opened.

In this way the moth may be preserved without the least injury to its appearance, or without ruffling the vanes of one of its beautiful plumes.

Of all the plume moths this is the largest, as a fine specimen will sometimes measure more than an inch across the wings. There is a brown species, nearly as large, and quite as common; but which is often overlooked on account of its sober colouring; and as often mistaken for a common "daddy-long-legs," to which fly it bears a close resemblance.

The Twenty-plume Moth (plate C, fig. 9) is hardly named as it deserves; for as the wings on each side are divided into twelve plumes. it ought to be named the twenty-four plume. A better title is that of the "Many-plume Moth."

It is very much smaller than either of the preceding "plumes;" and its radiating feathers are so small and so numerous, that at a hasty glance it scarcely seems to present any remarkable structure. It must be examined with the aid of a magnifying glass before its real beauty can be distinguished.

The moth is common enough, and may be easily caught, as it has a strange liking for civilised society, and constantly enters houses. As insects generally do, it flies to the window, and scuds unceasingly up and down the panes of glass, just as if it wished to make itself as conspicuous as possible.

The last of our moths is the beautiful Long-horn, for a figure of which see plate H, fig. 4. Another Long-horn Moth, the Green Adela, is shown on plate C, fig. 10. It is nearly as common as the last-mentioned insect.

It is a horrid name, for its agricultural associations are so potent, that the idea conveyed to the mind by the term "Long-horn," is that of a huge bovine quadruped, with sleek solid sides telling of oil-cake, with horns that are long enough to spike four men at once, two on each horn, and with a ponderous tread that rivals that of the hippopotamus.

Whereas, our little moth is the epitome of every

fragile, fairy-like beauty, and seems fitter for fairy tale, "once upon a time," than for this nineteenth century. Its "horns," as the antennæ are called, are wondrously long and slender. I have just taken measurement of one of these moths, and find that the body and head together are barely a quarter of an inch in length, while the antennæ are an inch and a quarter long. It is hardly possible to conceive any living structure more delicately slender than their antennæ. The moth delights in sunny glades, as so sunny a creature ought to do; it sits on a leaf, basking in the glaring sunbeams, while its antennæ, waving about in graceful curves, are only to be traced by the light that sparkles along them. They are as slender as the gossamer threads floating in the air, and like them only seen as lines of light. They are too delicate even for Mab's chariot traces. The grey-coated gnat might use one of them as his whip; but it would only be for show, as beseemeth the whip of a state-coach; for it could not hurt the tiniest atomy ever harnessed.

And yet the little Adela, for such is her scientific title, flies undauntedly among the trees, threading her way with perfect ease through the thickest foliage, her wondrous antennæ escaping all injury, and gleaming now and then as a stray sunbeam touches them.

There is nothing very striking in the Adela's external appearance; she is just a pretty, unobtrusive, bronze-coloured little thing, from whom many an eye would turn with indifference, if not with contempt. Truly, in vain are there pearls, while the swinish nature prefers dry husks.

Place this quiet, bronze-coloured little creature under a microscope, and Cinderella herself never exhibited such a transformation. The mind of man has never conceived a robe so gorgeous as that which enwraps a small brown moth. Refulgent golden feathers cover its body and wings, sparkling gem-like points scatter light in all directions, while on the edges of each feather rainbow tints dance and quiver. It seems as if the creature wore two robes—a loose golden-feather vesture above, and the rainbow itself beneath. Each fibre of the fringe that edges the wings is a prism, and even the slender antennæ are covered with golden feathers. Words cannot describe the wondrous beauty of this creature.

Methinks a view of these earthly creatures can the better enable one to appreciate the ineffable glories of the heavenly beings. Even the earth-insect is beautiful beyond the power of words to describe—how much more so the heavenly angel!

When the study of entomology first rose to the dignity of a science, it was found necessary that each insect should be distinguished by a definite title. Formerly, it was necessary to describe the insect when speaking of it; and in consequence both cabinets and memories were overloaded with words.

For example, the Meadow-brown Butterfly was named "Papilio media alis superioribus superne media parte rufis." In English, "The middle-sized butterfly, the centre of whose upper wings are reddish on the upper surface." Cromwell's Puritan soldier might have taken a lesson in nomenclature from an entomologist cabinet; and it is not easy to say which would occupy the greater time in reading, the list of butterflies or the regimental roll-call. These difficulties being patent, the nomenclators leaped at once, as is the habit of human nature, into the opposite extreme; and so, instead of making an insect name an elaborate description of its appearance, gave it a title which did not describe it at all, and would have been just as applicable to any other insect. Old Homer's pages afforded a valuable treasury of names; and, accordingly, Greek and Trojan may reasonably be astonished to find their names again revived on earth.

Even our British butterflies have appropriated Homeric titles. For example, the two first on the list are named Machaon and Podalirius, known to students of Homer as the two medical officers that accompanied the Greek army.

Numerous, however, as are the Homeric heroes and heroines, the insects far outnumbered them. So, after exhausting Homer, the dramatists were called into requisition, and plundered of their "personæ." Fiction failing, history, or that which is dignified by the name of history, was next sought; and kings, queens, generals, and statesmen lent their names to swell the insect catalogue.

The Latin authors now are required to make up the deficiency, Terence being especially useful. We have in our English list Davus, Pamphilus, and Chrysis, all out of one play, the "Andria."

At last, when Greek and Latin, prose and verse, history and mythology, had been quite exhausted, some enterprising and imaginative men boldly invented new names for new insects. The import of the name was of no consequence to them, and any harmonious combination of syllables was all that they required. Many a valuable hour have they wasted, or rather caused others to waste, in seeking through lexicons and dictionaries for the purpose of discovering the derivation of those unmeaning and underived names.

At last, men of science began to see that the name ought to be descriptive of the creature, or its habits, and yet as short as possible; and when this idea was matured, true nomenclature began. In the reformed system, insects are gathered together in societies, through which some general characteristic runs, and each individual bears the name of its genus, as the society is called; and also a second name that distinguishes its species.

The first butterfly which will be mentioned in these pages, is seen figured on plate D, fig. 4; and very appropriately bears the name of Atalanta. Those skilled in mythology, or Mangnall's skimmings thereof, will remember that Atalanta was a young lady, so swift of foot that she could run over the sea without splashing her ankles, or on the corn-fields without bending an ear of corn under her weight. The flight of this butterfly is so easy and graceful, that poetical entomologists invested it with the name of the swift-footed Atalanta.

Also it is called the Scarlet-Admiral, in which two names is to be seen the confusion respecting sexes which is found in nautical matters generally. Perhaps the discrepancy might have been avoided by calling the butterfly Cleopatra, that lady being her own Admiral.

Few insects are so conspicuous, or have so magnificent an effect on the wing, as the Atalanta; its velvety-black wings, with their scarlet bands, white spots, and azure edges, presenting a bold contrast of colour that is seldom seen, and in its way cannot be surpassed. It is certainly a grand insect; and it seems to be quite aware of its own beauty as it comes sailing through the sunny glades, gracefully inclining from side to side, as if to show its colours to the best advantage. Perhaps its best aspect is when it sits upon a teazle-head, quietly fanning its wings in the sun; for the quiet purple and brown tints of the teazle set off the magnificent pure colours of the insect.

These brilliant colours are only found on the upper surface of the wings, the under surface being covered with elaborate tracery of blacks, browns, ambers, sober blues, and dusky reds, so that when the wings are closed over the creature's back, it is hardly to be distinguished from a dried leaf, unless examined closely.

This distinction of tint often proves to be the insect's best refuge; for, if it can only slip round a tree or a bush, it suddenly settles on some dark spot, shuts up its wings, and there remains motionless until the danger is past. The rough, brown elm bark is a favourite refuge under these circumstances; and it takes a sharp eye to discover the butterfly when settled.

Sometimes the creature is not quite so magnificent, and even appears shorn of its fair proportions. I have now such a specimen before me, which I found on a sandy bank, unable to fly.

My attention was drawn to it by observing a curious fluttering movement of the grasses that covered the bank; and on going up to the spot to see what was the cause, I discovered an Atalanta butterfly that had apparently lost both wings of the left side, and was endeavouring to fly with the remaining pair. Of course it could only make short leaps into the air, turn over, and again fall to the ground. Wishing to put it out of pain, I killed it, and on examination found that it had never been endowed with wings on its left side, and that those organs had still remained in the undeveloped state in which they had lain under the chrysalis case. Even the right pair had not attained their full development; but in every other respect the insect was perfect.

I suppose that the caterpillar must have selected too dry a spot for its habitation when it became a pupa; and that in consequence the pupa shell was so dry and hard that the butterfly could not make its escape in proper time. I have often seen similar examples in my own caterpillar-breeding experiences.

There are also in one of my insect cases two specimens of the little white butterfly, who have met with even a worse fate; for they have not been able to escape at all out of the chrysalis, and so present the curious appearance of a chrysalis, furnished with head, antennæ, wings, and legs. The cause of the disaster was probably the same in both cases.

The caterpillar of the Atalanta is shown on plate D, fig. 4 *a*, and is a creature worthy of notice.

It is a well-known saying, that " what is one man's meat, is another's poison ;" and the proverb holds good in the case of the Atalanta caterpillar. For its meat is the common stinging-nettle, which is, undoubtedly, poisonous enough to qualify any such proverb.

The colour of the caterpillar is green-black, and along each side runs a spotty yellowish band. Its general shape and appearance can be seen by referring to the figure.

After passing through the usual coat-changing common to all caterpillars, it begins, just before its last change, to prepare a spot where it may pass its pupal state. Its mode of so doing is very curious, and is briefly as follows :—

The chrysalis is intended to remain in an attitude which we should think singularly uncomfortable, but which seems to suit the constitution of certain creatures, such as bats and chrysalides; namely, with its head downward. Why some insects should be thus suspended, while others lie horizontally, is not known as yet. But there can be no doubt but that some purpose is served by the various positions and localities assumed by insects in their pupal state.

Any one of a reflective mind, on hearing that a chrysalis was to be suspended by its tail, would feel some perplexity as to the means by which such a position could be attained. For the old caterpillar's skin has to be shed, and thus the legless, limbless chrysalis is left without any apparent power to suspend itself. The attitude which it assumes may be seen on plate D, fig. 4 b. On examining the chrysalis itself, and the leaf or twig to which it is suspended, it will be seen that a little silken mound is fastened to the leaf, and the chrysalis is furnished with some hooked processes on its tail, which are hitched upon the silken threads, and thus hold the creature in the proper position.

The Peacock butterfly, plate H, fig. 8, is an insect of very similar habits and manners. The under-side of the wings is very dark, and when they are closed over the back, the butterfly looks more like a flat piece of brown paper than an insect. The spots on the upper surface of the wing are especially beautiful; and the mode in which those spots are coloured by their feathers is shown in plate L, fig. 4, where a portion of the wing-spot is slightly magnified. This figure shows also the manner in which the feather-dust of the butterfly's wing is arranged. The larva of this beautiful insect is shown on fig. 8 b. Like that of the Atalanta, it feeds on the stinging-nettle.

On plate D, fig. 1, is drawn a very lovely insect, one of the numerous blue butterflies that may be seen flitting about the flowers in a garden, themselves of so flower-like an azure, that they may often be mistaken for a blue blossom. The caterpillar, fig. 1 b, is, as may be seen, rather curious in shape, and the pupa, fig. 1 c, is hardly less so.

CABBAGE BUTTERFLY.

Among the scales of this insect occur certain specimens called from their shape "battledore" scales, some of which may be seen on plate K, fig. 8, contrasted with the ordinary scales.

On the same plate as the blue butterfly, fig. 2, is seen a very pretty and common insect, called the "Orange-tip," on account of the colour of the wings. Only the male butterfly possesses these decorations, the female having wings merely white above, although she retains the beautiful green speckling of the under-wings.

Two more butterflies, and those the commonest of all, will complete this chapter. One will be at once recognised from the drawing, plate I, fig. 4, as the White Cabbage butterfly. The specimen here represented is the female; the male is smaller and has darker spots.

This is the parent of those green and black caterpillars which devastate our cabbage-beds, make sieves of the leaves, and are so disagreeably tenacious of their rights of possession. Pest as it is to the gardeners, to cooks, and sometimes, alas! to consumers, it would be a hundredfold worse but for the exertions of a fly so small as hardly to be noticed, but by its effects. This insect belongs to the same order as the bees, and is shown upon plate J, fig. 6. Small though it be, one such insect can compass the destruction of many a caterpillar, though not one-thousandth part of the size of a single victim. While the caterpillar is feeding, the ichneumon fly, as it is called, settles upon its back, pierces its skin with a little drill, wherewith it is furnished, and in the wound deposits an egg. This process is repeated until the ichneumon's work is done.

As each wound is made, the caterpillar seems to wince, but shows no farther sense of uneasiness, and proceeds with its eating as usual. But its food serves very little for its own nourishment, because the ichneumon's eggs are speedily hatched into ichneumon grubs, and consume the fatty portions of the caterpillar as fast as it is formed.

In process of time the caterpillar ought to take the chrysalis shape, and for that purpose leaves its food and seeks a convenient spot for its change.

That change never comes, for the ichneumons have been growing as fast as the caterpillar, with whose development they keep pace. And no sooner has their victim ceased to feed, than they simultaneously eat their way out of the doomed creature, and immediately spin for themselves a number of bright yellow cocoons, among which the dying caterpillar is often hopelessly fixed. Sometimes it has sufficient strength to escape, but it never survives.

In the later summer months, these cocoon masses may be seen abundantly on walls, palings, and similar spots.

Plate I, fig. 3, shows the Brimstone Butterfly, one of the first to appear as the herald of spring.

CHAPTER IX.

STAG-BEETLE—MUSK-BEETLE—TIGER-BEETLE—COCK-TAIL—VARIOUS BURYING-BEETLES—ROSE-BEETLE—GLOW-WORM—GROUND AND SUN-BEETLES, ETC.—HUMBLE-BEES, HORNETS AND THEIR ALLIES—DRAGON-FLIES—CADDIS-FLY—WATER BOATMAN—CUCKOO-SPIT—HOPPERS, EARWIG, AND LACE-FLY.

OF the remaining objects, only a very brief description can be given. Enough, however, will be said to assist the observer in identifying the object, and to serve as a guide to its locality and manners. We will first take the beetles; and as the largest is the most conspicuous, the great Stag-beetle shall have the precedence.

This insect (plate E, fig. 5) is quite unmistakeable; and, from its very ferocious aspect, would deter many from touching it. But it is very lamb-like in disposition, and sometimes as playful as a lamb. Its numerous jaws can certainly pinch with much violence; but are not used for the purpose of killing other creatures, as might be supposed.

The food of the stag-beetle is simply the juices of plants, which it sweeps up with that little brush-like organ that may be seen in the very centre of the jaws.

In winter it buries itself in the ground, and then, making a smooth vault, abides the winter's cold unharmed.

Only the male beetle possesses these tremendous jaws; those of the female being hardly one-tenth of their size, but so sharp at their points that their bite is just as severe.

The insect that next comes under notice is the Musk-beetle (plate I, fig. 7), a beautiful and conspicuous insect, of a rich green colour above, and a purplish blue below. Its name of musk-beetle is derived from the fragrant scent which it emits; a scent, however, not the least like musk, but more resembling that of roses. It is so powerful that the presence of the insect may often be detected by the nostrils, though it is hidden from the eyes. It may be found chiefly on willow-trees.

There is another beetle that gives out a sweet scent, much resembling that of the verbena leaf. This is the Tiger-beetle (plate D, fig. 8). With the exception of the white spots on the wing covers, the colours of this insect are much the same as those of the musk-beetle.

Its name seems hardly commensurate with its aspect; but never was a title better deserved. And, space allowing, I could here draw a terrible character; but as brevity is enforced, I can but say that this sparkling

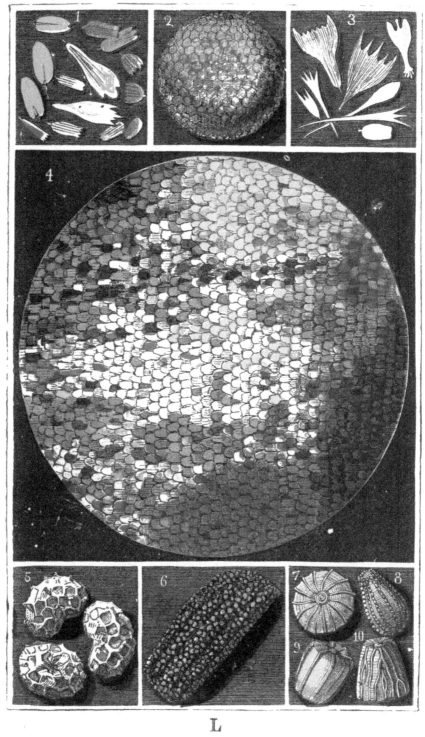

L

and beautiful insect seems to have the spirit of twenty tigers compressed into its little body.

All things have their opposites; and opposed to these perfume-bearing beetles are some who are just insect skunks. Chief among these is the common black Cocktail, a creature of truly diabolical aspect. It is a carrion eater, and intensifies the carrion odour. Still, repulsive as it is, it has its beauties. Its wings are very beautiful, and the mode in which these organs are packed away under their small cases is most wonderful. It is to aid in this process that the cock-tail possesses the faculty of turning its tail over its back. Plate H, fig. 12.

Another beetle of an abominable odour is the Buryingbeetle, one of which is shown on plate C, fig. 8. There are many burying-beetles, but this species is the most common.

Their name is derived from their habit of burying any piece of meat or dead animal that may be lying on the surface of the earth, not so much for the sake of themselves as for their progeny. In the buried animal their eggs are laid, and its putrefying substance affords them nourishment. The rapidity with which these and similar insects will consume even a large animal is marvellous. I have seen a large sheep stripped to the very bones in three days, nothing but bones and wool being left to mark the spot where it had lain.

Another kind of burying-beetle is seen on plate B,

fig. 7 ; but instead of dead meat, it buries the droppings of living animals, those of the cow being preferred. For this purpose it drives a perpendicular shaft into the ground, makes up a round ball of the droppings, puts an egg into the middle of the ball, rolls it into the hole, and after pushing some earth after it, sets to work at another shaft.

It is evident how beneficial the labours of these insects must be ; for by their means the earth is pierced with passages for air,—part is thrown out on the surface, where it becomes regenerated by the atmosphere,—noxious substances are removed from the surface, where they would do harm, and placed deep in the ground, where they do good.

The popular name for this beetle is the Watchman, because in the dusk of the evening it "wheels its drowsy flight," much as watchmen made their sleepy rounds. It belongs to the same family of insects as the sacred Scarabæus of the Egyptians.

On plate C, fig. 11, is depicted the common Rose-beetle, so called because it is an insect of refined habits, and chiefly dwells in the bosom of white roses. Yet it loves earth too, and in pursuance of its mission falls from its rose to earth, and there digs a receptacle for its future progeny. But though in earth, it is not of earth ; and, burrow as it may, it returns to its rose without a stain upon its burnished wings.

The curious Glow-worm, as it is called falsely, it being a beetle, and not a worm, is shown on plate J, fig. 1. Both the male and female insect give out this light, as I have often seen, though that of the female is the more powerful. The two sexes are very different in appearance, as may be seen by reference to the plate, fig. 1 being the male, and 1 a the female. The object of the light is by no means certain, nor the mode in which it is produced.

On the same plate, fig. 11, is seen the Oil-beetle, an eccentric kind of insect, who, when frightened, pours a drop of oil out of every joint, just as if it were a walking oil-barrel with self-acting taps.

One of the commonest beetles, the Ground-beetle, is seen on the same plate, fig. 10. There are very many ground-beetles, but this is one of the handsomest and most conspicuous. The embossment of its upper surface is worth a close examination, and its colouring is peculiarly rich and deep.

Hot sunny days always seem to bring out a host of insects, among which the Sun-beetles are notable examples. One of these insects is shown in plate D, fig. 6. They are beautifully brilliant as they run among the gravel-stones or over paths, their smooth surface glittering in the sun resplendently.

As an aquatic balance to the terrene Sun-beetles, the Whirligigs (plate F, fig. 4) make their appearance on the surface of the water on any light sunny day. What rule they observe in their mazy dance is more difficult to comprehend than the "Lancers," or a cotillon; but that there must be a rule, is clear from the wonderful way in which they avoid striking against each other in their passage.

Every one knows the Lady Bird, with its pretty red wings and black spots. Its larva (plate B, fig. 8) is a very singular creature, and destructive withal, spearing and eating Aphides as ruthlessly as Polyphemus impaled and devoured the captured sailors. It has a curious history, but there is no room for it here.

On plate H, fig. 7, is represented one of the many Skipjack-beetles, who afford such amusement to juveniles by their sudden leaps into the air, when laid on their backs. This feat is performed by means of a sudden blow of the head and thorax. Farmers, however, are not all amused by it; for it is the parent of the terrible "wire-worm," so deadly a foe to corn and potatoes.

Some insects prefer corn when placed in granaries; and these are the Weevils, whose grubs populate sea-biscuit, and run races across plates for wagers. Nuts also fall victims to the weevil represented on plate I, fig. 9, or rather to its grub, "Time out of mind the fairies' coachmaker."

There is a very common little green weevil shown on plate C, fig. 7, which, although ordinary enough to the unassisted eye, yet under the microscope glows with jewels and gold. It is, in truth, the British Diamond-beetle. An idea of its appearance may be obtained from plate L, fig. 6, but to give the real glory of the colouring is impossible.

One of the little insects called Death-watches is shown on plate J, fig. 8. There are many insects that go by this name, because they make a slight tapping sound with their heads, probably to call their mates; and which sound has been thought to prognosticate death rather than marriage.

The curious Tortoise-beetle is depicted on plate C, fig. 6. Its chief peculiarity is in its larval state, when it carries a kind of parasol, formed from the remains of the leaves on which it has been feeding.

Last and least of the beetles comes one as destructive as it is small, the Turnip-hopper. This little animal, no larger than a small pin's head, does great damage to the turnip crops, and is therefore hated by farmers. It is shown, much magnified, on plate J, fig. 13.

From the beetles we proceed to the Bee tribe; and first take the common Humble-bee, several of which are shown on plate H, fig. 10, representing the " Red-hipped Humble-bee," which mostly makes its nest among stone-heaps. Fig. 11 is the common Humble-bee, that burrows in the ground, and there builds its thimble-like cells. These cells are very irregular in shape, and are affixed to each other without any definite order. Of these two insects, the latter is harmless enough; but the former becomes very fierce if its nest is approached too closely.

A magnified view of some hairs of the Humble-bee is given on plate K, fig. 11.

There are some bees which make their nests in old walls, where they either dig for themselves a hole, or oftener take advantage of a nail-hole, and so save themselves much trouble. One of these bees is shown on plate H, fig. 2, and is chiefly remarkable for the beautifully tufted extremities of its middle pair of legs.

On plate D, fig. 7, is seen the common Hornet, one of the really terrible of our insects. It mostly makes its nest in hollow trees, and it behoves one to keep very clear of the neighbourhood. The nest is made of wood-fibre, nibbled, and made into a primitive papier-maché.

Two of the Saw-flies may be seen on plate J. Fig. 2 is the common green Saw-fly, and fig. 3 the dreaded Turnip-fly. These are called Saw-flies because they are furnished with saw-like implements, by means of which they cut grooves in certain plants, and in those grooves lay their eggs.

Mention has already been made of the little Ichneumon fly. One of these insects is shown magnified on plate J, fig. 12 a, and one of the large species is depicted on plate H, fig. 3. The threefold appendage to the tail is the ovipositor, or instrument by means of which they pierce their victims and deposit their eggs.

There are some allied insects that pierce vegetables instead of insects; and one of their works may be seen figured on plate A, where a bramble-branch has been perforated by them. The well-known oak-apples, plate B, fig. 6, are caused by a Cynips, as the little creature is called; and so is the common Bedeguar of the rose, seen on plate C, fig. 2.

The last of these insects that will be named is the beautiful Fire-tail, plate D, fig. 5, one of the most brilliant insects that our island can boast. There are many British species of this insect, but they all much resemble each other, and are insect cuckoos, laying their eggs in the nests of other insects.

From the bees, we pass to the Flies; and first take a most singular insect, shown on plate H, fig. 5. This insect is found on the blackberry blossoms, and the upper part of its body is so transparent that the leaf on which it sits can be seen through it. It is swift of wing and wary, requiring a quick eye and hand for its capture.

On the same plate, fig. 6, is shown one of the traveller's pests, a fly that bites, or rather bores, the skin, and that with such virulence that it can even strike its poisoned dart through a cloth coat, and make its victim to lament for many an hour after.

One of the various hoverer-flies is shown on **plate J**, fig. 9. The larva of this insect is very remarkable, on account of its curious breathing apparatus. The larva is properly called the Rat-tailed Maggot, and is shown on the same plate, fig. 8 *a*. The body of the creature is found buried in the mud at the bottom of stagnant pools or cisterns, and the respiration is carried on through the telescopic tail, which is long enough to protrude through the mud, and to convey the necessary oxygen to the system through two flexible air-tubes that pass through the " tail."

It will be remembered that in mentioning the Green Oak Moth, the Destroying Empis was also noticed. One of these flies is shown on plate J, fig. 5, with the poor Tortrix in its grasp. Plate K, fig. 1, shows its foot, and fig. 3 its head, together with its long beak.

The beak of this fly somewhat reminds one of the corresponding portion of the Gnat, which insect is not itself depicted, though on plate F, fig. 10, is shown the wonderful little egg-boat which it makes. This insect glues together its eggs in such a manner that they are formed into a true life-boat, which cannot be upset, or sunk, or filled with water, but floats securely on the surface until the young are hatched. That object accomplished, the gnat-larvæ tumble into the water, and there undergo their transformation.

The last of the two-winged flies that will be mentioned is the common Daddy Long Legs, or Crane-fly, which seems to set such little value on its limbs. It is a very injurious insect in its larval state, feeding on roots, and doing great damage. Plate H, fig. 1, shows a very pretty species, covered with yellow rings.

Every one must have noticed the beautiful and active insects that are with great truth called Dragon-flies. Their habits and peculiarities would demand a volume; and here they can but be mentioned. Plate F, fig. 6, shows the common Flat Dragon-fly, that may be seen chasing and following flies of all sizes, and even butterflies. Fig. 8 is the elegant Demoiselle, the male of which is shown here, with its dark purple spots on the wings, and dark blue body. The female is of a uniform green. Its larva is shown at fig. 8 a, where the singular leafy gills may be seen at the end of its tail. Fig. 7 shows another very common Dragon-fly, very thin and ringed with blue circlets.

On the same plate, fig. 12, may be seen several varieties of the objects known to fishermen as "Caddis" cases. These are residences built by the larva of the common Caddis, or Stone-fly, which is represented on the same plate, fig. 9.

Still keeping to plate F, and referring to fig. 1, is seen the horrid-looking Water-scorpion, a creature which, though it does not sting, has much of the scorpion nature, and so bites. Fig. 1 *a* shows the same insect as it appears when flying.

At fig. 3 is seen the Water Boatman, so called because it lies on its back, which is ridged like the keel of a boat, and then rows itself about by means of its middle pair of legs, which closely resemble oars.

Fig. 5 shows a very curious insect which is common enough on the margin of pools, and runs on the surface of the water as if it were dry land. When alarmed, it shuts up all its legs, and looks just like a piece of dry grass or thin stick.

Another insect much resembling it, is the common Gerris, seen on plate I, fig. 6. It may be seen on every pond or still water, running over its surface, and is furnished with wings wherewith it can fly to great distances. I have found specimens on the tops of hills, far from any water, and hiding under stones out of the sun's heat. Fig. 1 shows the common May-fly.

All gardeners have been annoyed with the curious production called the Cuckoo-spit. This proceeds from the larva of one of the hoppers, and on removing the frothy substance, the little soft, greenish insect may be found within. The perfect insect is shown on plate C, fig. 1 *a*, and the exudation itself at fig. 1.

There is another hopper seen on plate B, fig. 2, called from its colour the Scarlet Hopper. It is common enough on ferns, and may be found chiefly in the open spots of forests where ferns abound.

On plate J, fig. 7 a, is the common Green Grasshopper, as it appears when standing ; and on fig. 7, the same insect as it appears when using its wings.

The common Earwig, plate I, fig. 8, is introduced for the purpose of showing the very beautiful wing which this insect possesses, and which is seen expanded at fig. 8 a.

The very lovely, though ill-odoured Lace-wing Fly, is shown on plate J, fig. 4, and its very remarkable eggs at 4 a. Each egg is placed at the end of a footstalk, whereby it is kept out of the reach of certain predacious insects.

Various shells are drawn on one or two of the plates, but there is not space for any description. Their names may be found on the Index to Plates. Plate G contains certain fungi and mosses. Fig. 1 is that peculiar plant which rein-deer scrape from under the snow in the winter time. Fig. 2 was once dreaded by rustics as "Witch's butter." Fig. 6 shows the curious Earth-star, chiefly remarkable for its resemblance to the marine Star-fish.

INDEX.

Adela, 154.
Admiral Butterfly, 158.
Armadillo, 89.

Bat, 3.
Bedeguar, 173.
Bird-nest moss, 177.
Blindworm, 45.
Blue Butterfly, 162.
Brimstone Butterfly, 164.
Brimstone Moth, 148.
Buff-tip Moth, 133.
Burnet Moth, 115.
Burying Beetle, 167.

Cabbage Butterfly, 163.
Caddis, 175.
Clear-wing Moth, 150.
Cock-tail Beetle, 167.
Crane-fly, 175.
Cray-fish, 83.
Cuckoo-spit, 176.
Cup Moss, 177.
Cynips, 173.

Death-watch, 171.
Drinker Moth, 124.
Dragon-flies, 175.

Earwig, 177.
Eft, 67.
Elephant Hawk Moth, 131.
Emperor Moth, 127.
Empis, 141.

Field Mouse, 17.
Firetail, 173.
Frog, 55.

Galls, 172.
Gerris, 176.
Glowworm, 169.
Gnat Eggs, 174.
Gold-tailed Moth, 134.
Grasshopper, 177.
Ground Beetle, 169.

Harvest Mouse, 21.
Hopper, scarlet, 176.
——— Cuckoo, 176.
Hornet, 172.
Hoverer Fly, 174.
Humble-Bee, 171.
Hydrometra, 176.

Ichneumon, Microgaster, 164.

Lacewing Fly, 177.
Lackey Moth, 145.
Lady-bird, 170.
Lampern, 80.
Lappet Moth, 146.
Leaf-Miners, 144.
Lizard, 43.
Longhorn Moths, 153.

Magpie Moth, 149.
Mason Bee, 172.
May-fly, 176.

Mole, 33.
Mouse, Field, 17.
——— Harvest, 21.
——— Shrew, 26.
Musk Beetle, 166.

Newt, 67.

Oak-apples, 173.
Oak Egger-moth, 116.
Oak Moth, 140.
Oil Beetle, 169.
Orange-tip, 163.

Peacock Butterfly, 162.
Pill Millepede, 89.
Plume Moths, 151.
Privet Hawk-moth, 131.
Puss Moth, 114.

Rat, Water, 22.
Rat-tailed Maggot, 174.
Reindeer Moss, 177.
Rose Beetle, 168.

Shrew, 26.
——— Water, 32.
Shrimp, Fresh-water, 87

Skip-jack Beetle, 170.
Snake, 50.
Stag Beetle, 165.
Stickleback, 75.
Sting-fly, 173.
Stone-fly, 175.
Sun Beetle, 169.

Tiger Beetle, 166.
——— Moth, 97.
Toad, 59.
Tortoise Beetle, 171.
Turnip Fly, 172.
——— Hopper, 171.

Vapourer Moth, 138.
Viper, 48.
Volucella, 173.

Watchman Beetle, 168.
Water Scorpion, 175.
——— Boatman, 176.
Weasel, 40.
Weevils, 170.
Whirligig Beetle, 169.
Witch Butter, 177.
Woodlouse, 89

INDEX TO PLATES.

A. (Front.)

1. Tubercled Gall on Bramble-stem.
2. Track of Leaf-Miner on Bramble-leaf.
3. Gnat-Clearwing Moth.
4. Buff-tip Moth.
— a. Caterpillar of do.
5. Privet Moth.
—a. Caterpillar of do.
6. Snail (*Helix nemoralis*).
7. Do. (*Helix nemoralis*) var.
8. Do (*Helix cantiana*).
9. Do. (*Helix ericetorum*).
10. Do. (*Helix lapicida*).
11. Shell (*Cyclostoma*).
12. Do. (*Zonites*).
13. Do. (*Helix caperata*).
14. Do (*Pupa*).
15. Do. (*Clausilia*).

B.

1. Green Oak-Moth (*Tortrix*).
2. Scarlet Hopper (*Cercopis*).
3. Burnet Moth.
—a. Cocoon of do.
4. Puss Moth.
—a. Caterpillar of do.
5. Tiger-Moth (*Arctia*).
—a. Caterpillar of do.
—b. Cocoon of do.
6. Oak-galls.
7. Watchman-Beetle (*Geotrupes*).
8. Lady-bird (*Coccinella*).
—a. Larva of do.

C.

1. Cuckoo-spit.
  a. Cuckoo Hopper (*Tettigonia*).
2. Bedeguar of Rose.
3. Brimstone Moth.
—a. Caterpillar of do.
4. Emperor Moth.
—a. Caterpillar of do.
—b. Cocoon of do.
5. Elephant Hawk-Moth.
—a. Caterpillar of do.
6. Tortoise Beetle (*Cassida*).
7. Green Weevil.
8. Burying-Beetle (*Necrophorus*).
9. Twenty-Plume Moth.
10. Green Adela.
11. **Rose-Beetle.**

D.

1. Blue Butterfly (*Alexis*).
—a. Do. Wings closed.
—b. Caterpillar of do.
—c. Pupa of do.
2. Orange-tip Butterfly.
3. Vapourer Moth, Male.
—a. Do. Female.
4. Red Admiral.
—a. Caterpillar of do.
—b. Pupa of do.
5. Firetail (*Chrysis*).
6. Sun-Beetle.
7. Hornet.
8. Tiger Beetle.
—a. Do. Flying.

E.

1. Drinker Moth, Male.
—a. Do. Female.
—b. Do. Caterpillar.
—c. Do. Cocoon.
—d. Do. Chrysalis.
—e. Do. Eggs.
2. Humble-bee Fly (*Bombylius*).
3. Magpie Moth.
—a. Do. Chrysalis.
—b. Do. Caterpillar.
4. Gold-tailed Moth (*Porthesia*).
—a. Do. Caterpillar.
5. Stag-Beetle.

F.

1. Water Scorpion.
—a. Do. flying.
2. Amber Shell (*Succinea*).
3. Water Boatman.
4. Whirligig Beetle.
5. Hydrometra.
6. Dragon-Fly (*Libellula*).
7. Do. (*Agrion*).
8. Do. Demoiselle (*Calepteryx*).
—a. Do. Larva.
9. Stone-Fly (*Phryganea*).
10. Eggs of Gnat.
12. Caddis-cases, composed—
  a. Of flat stones.
  b. Of bark.
  c. Of sand.
  d. Of grass.
  e. Of grass-stems.
  f. Of shells.

INDEX TO PLATES.

13. Water shells.
    g. Planorbis.
    h. Ancylus.
    i. Lymnæus.
    k. Paludina.

G.

1. Rein-deer Moss (*Cladonia*).
2. Witch-butter (*Tremella*).
3. Polytrichum.
4. Bird-nest Moss (*Nidularia*).
5. Xylaria.
6. Earth-star (*Geastrum*).
—a. Do. closed.
7. Arscyria.
8. Cup-moss (*Cenomyce*).
9. Scarlet Cup-moss (*Peziza*).
10. Marchantia.

H.

1. Crane-fly.
2. Mason Bee (*Megachile*).
3. Ichneumon (*Pimpla*).
4. Adela Long-horn.
5. Volucella.
6. Sting-fly (*Chrysops*).
7. Skip-jack Beetle (*Elater*).
8. Peacock Butterfly.
—a. Do. wings closed.
—b. Do. Caterpillar.
9. White Plume-Moth.
10. Red-tailed Humble-Bee.
11. Common do.
12. Cock-tail Beetle (*Goërius*).

I.

1. May-fly (*Ephemera*).
2. Scorpion-fly.
3. Brimstone Butterfly.
4. Cabbage White Butterfly.
—a. Do. Caterpillar.
5. Oak-Egger Moth, female.
—a. Do. Cocoon.
6. Gerris.
7. Musk Beetle.
8. Earwig.
—a. Do. flying.
9. Nut Weevil.

J.

1a. Glowworm, male.
—b. Do. female.
2. Green Sawfly (*Tenthredo*).
3. Turnip-fly.
4. Lacewing Fly.
—a. Eggs of do. on lilac branch.

5. Empis.
—a. Do. killing Oak-moth.
6. Ichneumon (*Microgaster*) and cocoons.
7. Grasshopper, flying.
—a. Do. walking.
8. Death-watch (*Anobium*).
9. Hoverer-fly.
—a. Rat-tailed Maggot.
10. Ground Beetle (*Carabus*).
11. Oil Beetle.
12. Cocoon of Microgaster, magnified.
—a. Microgaster, magnified.
13. Turnip-hopper (*Haltica*), magnified.
—a. Do. natural size.
14. Cyclops, magnified, showing egg-sacs.
15. Scarlet Spider (*Trombidium*), magnified.
—a. Do. natural size.

K.
MICROSCOPICAL.

1. Foot of Empis.
2. Pollen—a. Sunflower.
         b. Passion Flower.
         c. Lily.
3. Head of Empis.
4. Foot of Male Water-Beetle (*Dyticus*).
5. Trunk of Blue-bottle Fly.
6. Foot of Frog, showing circulation.
7. Petal of Geranium, showing stomata.
8. Battledore Scales of Blue Butterfly.
9. Scale of Fritillary Butterfly.
10. Eye of Butterfly.
11. Hairs of Humble-Bee.

L.
MICROSCOPICAL.

1 and 3. Scales of various Butterflies.
2. Eye of Hemerobius.
4. Wing of Peacock Butterfly.
5. Poppy seeds.
6. Wing-case of Green Weevil.
7. Egg of Red Underwing Moth.
8. —— of Small White Butterfly
9. —— of Tortoiseshell Butterfly.
10. —— of Lathonia Butterfly.

R. CLAY, SON, AND TAYLOR, PRINTERS.

BOTANY AND GARDENING.

REEVE'S
POPULAR NATURAL HISTORIES.
(BOTANICAL DIVISION.)

Price **7s. 6d.** each, cloth gilt, with Coloured Illustrations,

"A popular series of scientific treatises, which from the simplicity of their style, and the artistic excellence and correctness of their numerous illustrations, has acquired a celebrity beyond that of any other series of modern cheap works."—*Standard*.

GARDEN BOTANY: containing a Familiar and Scientific Description of most of the Hardy and Half-hardy Plants introduced into the Flower-Garden. By AGNES CATLOW. With 20 Pages of Coloured Illustrations, embracing 67 Plates.

GREENHOUSE BOTANY: containing a Familiar and Technical Description of the Exotic Plants introduced into the Greenhouse. By AGNES CATLOW. With 20 Pages of Coloured Illustrations.

FIELD BOTANY: containing a Description of the Plants common to the British Isles. By AGNES CATLOW. With 20 Pages of Coloured Illustrations, embracing 80 Plates.

ECONOMIC BOTANY: a Description of the Botanical and Commercial Characters of the principal articles of Vegetable origin, used for Food, Clothing, Tanning, Dyeing, Building, Medicine, Perfumery, &c. By THOMAS C. ARCHER, Collector for the Department of Applied Botany in the Crystal Palace, Sydenham. With 20 Pages of Coloured Illustrations, embracing 106 Plates.

THE WOODLANDS: a Description of Forest Trees, Ferns, Mosses, and Lichens. By MARY ROBERTS. With 20 Pages of Coloured Plates.

GEOGRAPHY OF PLANTS: or, a Botanical Excursion round the World. By E. M. C. Edited by CHARLES DAUBENY, M.D., F.R.S., &c. With 20 Pages of Coloured Plates of Scenery.

PALMS AND THEIR ALLIES: Containing a Familiar Account of their Structure, Distribution, History, Properties, and Uses; and a complete List of all the species introduced into our Gardens. By BERTHOLD SEEMANN, Ph.D., M.A., F.L.S. With 20 Pages of Coloured Illustrations, embracing many varieties.

REEVE'S NATURAL HISTORIES, 7s. 6d. EACH—*continued.*

BRITISH FERNS AND THE ALLIED PLANTS: comprising the Club-Mosses, Pepperworts, and Horsetails. By THOMAS MOORE, F.L.S. With 20 Pages of Coloured Illustrations, embracing 51 subjects.

BRITISH MOSSES: comprising a General Account of their Structure, Fructification, Arrangement, and Distribution. By ROBERT M. STARK, F.R.S.E. With 20 Pages of Coloured Illustrations, embracing 80 subjects.

BRITISH LICHENS: comprising an Account of their Structure, Reproduction, Uses, Distribution, and Classification. By W. LAUDER LINDSAY, M.D. With 22 Pages of Coloured Plates, embracing 400 subjects.

BRITISH SEAWEEDS: comprising their Structure, Fructification, Specific Characters, Arrangement, and General Distribution, with Notices of some of the Fresh-water ALGÆ. By the Rev. D. LANDSBOROUGH, A.L.S. With 20 Pages of Coloured Illustrations, embracing 80 subjects.

In small post 8vo, price 5s. cloth, or 5s. 6d. gilt edges,

A TOUR ROUND MY GARDEN. By ALPHONSE KARR. Revised and Edited by the Rev. J. G. WOOD. The Third Edition, finely printed. With upwards of 117 Illustrations by W. HARVEY.

"There are few writers who have shown such keen perception of character, such true delicacy of feeling, and such real originality of thought, as are to be found in every page of this charming author."—*Editor's Preface.*

"Have you ever read 'A Tour Round My Garden,' by Alphonse Karr? You should read it, it is a book of deep philosophy."—*Blackwood.*

Cloth limp, price 1s.,

FAVOURITE FLOWERS: How to Grow Them; being a Complete Treatise on the Cultivation of the Principal Flowers, with Descriptive Lists of all the best varieties in cultivation. By ALFRED GILLETT SUTTON, F.H.S., Editor of "The Midland Florist."

Fcap. cloth gilt, price 2s. 6d.,

THE KITCHEN AND FLOWER GARDEN; or, the Culture in the open ground of Roots, Vegetables, Herbs, and Fruits, and of Bulbous, Tuberous, Fibrous, Rooted, and Shrubby Flowers. By EUGENE SEBASTIAN DELAMER.

Also, sold separately, each 1s.,

THE KITCHEN GARDEN. | THE FLOWER GARDEN.

BOTANY AND GARDENING.

In fcap. 8vo, price **3s. 6d.** cloth gilt, or with gilt edges, **4s.**,

WILD FLOWERS: How to See and how to Gather them. With Remarks on the Economical and Medicinal Uses of our Native Plants. By SPENCER THOMSON, M.D. A New Edition, entirely revised, with 171 Woodcuts, and 8 large Coloured Illustrations by NOEL HUMPHREYS.

Also, price **2s.** in boards, a Cheap Edition, with Plain Plates.

Fcap., price **3s. 6d.** cloth gilt, or **4s.** gilt edges,

OUR WOODLANDS, HEATHS, AND HEDGES: a Popular Description of Trees, Shrubs, Wild Fruits, &c., with Notices of their Insect Inhabitants. By W. S. COLEMAN, M.E.S.L. With 41 Illustrations printed in Colours on 8 Plates.

*** A Cheap Edition, with Plain Plates, fancy boards, **1s.**

In Fcap. 8vo, cloth gilt, price **3s. 6d.**, or **4s.** gilt edges,

BRITISH FERNS AND THEIR ALLIES: comprising the Club-Mosses, Pepperwort, and Horsetails. By THOMAS MOORE, With 40 Illustrations by W. S. COLEMAN, beautifully printed in Colours.

A Cheap Edition of the above with Plain Plates, price **1s.**, fancy boards.

In square 16mo, price **2s. 6d.** cloth,

VEGETABLE PRODUCTS OF THE WORLD: a Description of the Botanical and Commercial Characters of the Chief Articles of Vegetable Origin, used for Food, Clothing, Tanning, Dyeing, Building, Medicine, Perfumery, &c. For the Use of Schools. By THOMAS C. ARCHER. With 20 Pages of Plates, embracing 106 Figures.

"An admirable and cheap little volume, abounding in good illustrations of the plants that afford articles of food or applicable to purposes of manufacture. This should be on the table of every family, and its contents familiar with all rising minds."—*Atlas*.

Fcap., cloth limp, price **1s.**,

FLAX AND HEMP: Their Culture and Manipulation. By E. SEBASTIAN DELAMER. With Illustrations.

"We may, if we choose, grow our own hemp to quite an indefinite extent, and hold ourselves independent of foreign supply. The soil of Ireland alone is capable of sending forth an enormous export."—*Preface*.

AGRICULTURE AND FARMING.

MECHI'S SYSTEM OF FARMING.
Price **3s.** boards, or **3s. 6d.** half-bound,

HOW TO FARM PROFITABLY; or, the Sayings and Doings of Mr. Alderman Mechi. With a Portrait and three other Illustrations, from Photographs by Mayall. New Edition, with additions.

The above work contains: Mr. Mechi's account of the Agricultural improvements carried on at the Tiptree Estate—His Lectures, Speeches, Correspondence, and Balance Sheets—and is a faithful history of his Agricultural career during the last fifteen years.

*** In this Edition is incorporated Mr. Mechi's valuable Pamphlets on TOWN SEWAGE and STEAM PLOUGHING.

In 1 vol. price **5s.** half-bound,

RHAM'S DICTIONARY OF THE FARM. A New Edition, entirely Revised and Re-edited, with Supplementary Matter, by W. and HUGH RAYNBIRD. With numerous Illustrations.

This book, which has always been looked up to as a useful and general one for reference on all subjects connected with country life and rural economy, has undergone an entire revision by its present editors, and many new articles on agricultural implements, artificial manures, bones, draining, guano, labour, and a practical paper upon the subject of animal, bird, and insect vermin inserted, which at once render it an invaluable work for all who take pleasure in, or make a business of, rural pursuits.

Limp cloth, price **1s.**,

SMALL FARMS. A Practical Treatise intended for Persons *inexperienced* in Husbandry, but desirous of employing time and capital in the cultivation of the soil. By MARTIN DOYLE.

In fcap. 8vo, price **2s.** cloth boards, or **1s. 6d.** cloth limp,

AGRICULTURAL CHEMISTRY. Comprising Chemistry of the Atmosphere, Chemistry of the Soil, Water, Plants—Means of Restoring the impaired Fertility of Land, exhausted by the growth of Cultivated Crops, and of improving Land naturally Infertile—Vegetable and Animal Produce of the Farm. By ALFRED SIBSON (Royal Agricultural College, Cirencester), with a Preface by Dr. Augustus Voelcker, Consulting Chemist of the Royal Agricultural Society. With Illustrations.

"This is an excellent treatise—comprehensive, full of most useful, and, to the agriculturist, necessary information. It enters fully into the composition and various physical conditions of soil, atmosphere, and other agents which contribute to agricultural results. It should be in the possession of every agriculturist in the kingdom."—*Weekly Times.*

AGRICULTURE AND FARMING.

Second Edition, in demy 8vo, price **10s. 6d.** half-bound,

THE FARMER'S CALENDAR. By ARTHUR YOUNG. Entirely Re-written to Present date, by JOHN CHALMERS MORTON, Author of "The Cyclopædia of Agriculture," "Farmer's Almanac," &c. Illustrated by numerous Wood-engravings.

GENERAL CONTENTS.

The Cultivation of all kinds of Soil.	Rent, Lease, Tenant's Right, Wages,
The Cultivation of all Farm Crops.	Improvements of Land by Build-
The Breeding, Rearing, and Feeding	ings, Roads, Drains, Fences, &c.
of all the Live Stock of the Farm.	The Influence of Weather, month by
The use of all kinds of Agricultural	month, in reference to 30 or 40 sta-
Tools and Machines.	tions in England, Scotland & Ireld.

In post 8vo, price **2s. 6d.** cloth extra,

SCIENTIFIC FARMING MADE EASY. By THOMAS C. FLETCHER, Agricultural and Analytical Chemist.

GENERAL CONTENTS.

Habits and Food of Plants.	Inorganic Constituents of Plants.
Carbonic Acid.	Manures. Artificial ditto.
Constituents of Water.	Chemistry of the Dung-hill.
Ammonia.	Gas Refuse, Lime, Bones, &c.
What Plants derive from Carbon.	Cattle Feeding, Appendix, &c.

Price **1s.** cloth limp,

HINTS FOR FARMERS AND AGRICULTURAL STUDENTS. By ROBERT SCOTT BURN, Editor of the "Year-Book of Agricultural Facts."

GENERAL CONTENTS.

Rotation, Soils, Weeds and Seeds.	Manures.
Cereal and Seed Crops.	Cattle Food.
Root and Forage Crops.	Analysis of Feeding Materials, &c.

In post 8vo, price **1s. 6d.** cloth limp, or **2s. 6d.** half bound,

BRITISH TIMBER TREES. By JOHN BLENKARN, C.E. The work is essentially practical—the result of long experience and observation, and will be found of especial service to Landed Proprietors, Land Agents, Solicitors, Landscape Gardeners, Nurserymen, Timber Merchants, Timber Valuers, Architects, Auctioneers, Civil Engineers, and all persons having the management of, or in any way connected with, the improvement of Landed Estates.

"A *practical, able,* and *instructive book.*"—*The Builder.*

"A valuable book, of the greatest service to all proprietors and managers of wooded estates, as well as those engaged in planting trees, or buying and selling timber."—*The Athenæum.*

In demy 8vo, price **18s.** half-bound, 600 pages,

THE HORSE, IN THE STABLE AND IN THE FIELD. His Varieties: Management in Health and Disease: Anatomy: Physiology, &c. By J. H. WALSH (Stonehenge), Editor of "The Field," Author of "British Rural Sports." Illustrated with 160 engravings, by BARRAUD, WEIR, ZWECKER, &c. Second Edition.

> "The work most effectually meets a universal want among those connected with the horse."—*Bell's Weekly Messenger.*
> "This is a splendid book, and one of immense attraction and value."—*Sunday Times.*
> "An admirable and comprehensive volume. . . . . Certain of public favour and success."—*Era.*
> "A book of permanent interest and usefulness. A first-class book of reference on all equine matters."—*Globe.*
> "It is by far the most elaborate work on the horse we have yet seen; it is also much the best; it is not only good as compared with others, but it is absolutely good in itself."—*Sporting Life.*

Fcap., price **2s. 6d.** half-bound,

THE HORSE: its History, Varieties, Conformation, Management in Health and Disease. With 8 Illustrations. By WILLIAM YOUATT. A New Edition by CECIL, with Observations on Breeding CAVALRY HORSES.

⁎ A Cheap Edition, thin paper and limp cloth, price **1s.**

Post 8vo, price **10s. 6d.** half-bound,

THE GENTLEMAN'S STABLE MANUAL; a Treatise on the Construction of the Stable; also on the Feeding and Grooming of Horses; on the Treatment of the Sick Horse; on Shoeing; on the Management of the Hunter; and on Equine Diseases and Accidents, with the most Scientific Modes of Treatment. By WILLIAM HAYCOCK, V.S. and M.R.C.V.S. 3rd Edition. With 36 wood-engravings.

> "Of the work, as a whole, we can scarcely speak in too high a strain of eulogy. It is a work which ought to be read by the gentleman and the groom; no horse-keeper ought to be without it; and to the author we owe a deep debt of gratitude both for the matter and the manner of the book."—*Daily Telegraph.*
> "We heartily recommend every owner of a horse to purchase and study it." —*Chester Record.*

In fcap. 8vo., boards, price **1s.**,

THE LIFE OF A NAG-HORSE. With Practical Directions for Breaking and Training Horses, and some Hints on Horsemanship. By FREDERICK TAYLOR. Editor of "The Irish Sporting Times," &c., &c.

CONTENTS:—Breaking in—Sale to a London Dealer—Rotten Row—Lameness—Sold to a Gang of Copers—In the Dragoons—Cheshire Hunt—Sold to a Quaker—Sold to a Commercial Traveller—In an Omnibus—Stolen—The Four-in-Hand Club—My present benevolent Owner.